Materials Horizons: From Nature to Nanomaterials

Series Editor

Vijay Kumar Thakur, School of Aerospace, Transport and Manufacturing, Cranfield University, Cranfield, UK

Materials are an indispensable part of human civilization since the inception of life on earth. With the passage of time, innumerable new materials have been explored as well as developed and the search for new innovative materials continues briskly. Keeping in mind the immense perspectives of various classes of materials, this series aims at providing a comprehensive collection of works across the breadth of materials research at cutting-edge interface of materials science with physics, chemistry, biology and engineering.

This series covers a galaxy of materials ranging from natural materials to nanomaterials. Some of the topics include but not limited to: biological materials, biomimetic materials, ceramics, composites, coatings, functional materials, glasses, inorganic materials, inorganic-organic hybrids, metals, membranes, magnetic materials, manufacturing of materials, nanomaterials, organic materials and pigments to name a few. The series provides most timely and comprehensive information on advanced synthesis, processing, characterization, manufacturing and applications in a broad range of interdisciplinary fields in science, engineering and technology.

This series accepts both authored and edited works, including textbooks, monographs, reference works, and professional books. The books in this series will provide a deep insight into the state-of-art of Materials Horizons and serve students, academic, government and industrial scientists involved in all aspects of materials research.

Review Process

The proposal for each volume is reviewed by the following:

1. Responsible (in-house) editor
2. One external subject expert
3. One of the editorial board members.

The chapters in each volume are individually reviewed single blind by expert reviewers and the volume editor.

Dattatray J. Late · Chandra Sekhar Rout
Editors

Science and Technology of 2D MoS_2

Springer

Editors
Dattatray J. Late
Department of Physics
Federal University of Lavras
Lavras, Minas Gerais, Brazil

Chandra Sekhar Rout
Centre for Nano and Material Sciences
Jain University
Bengaluru, Karnataka, India

ISSN 2524-5384 ISSN 2524-5392 (electronic)
Materials Horizons: From Nature to Nanomaterials
ISBN 978-981-97-7366-4 ISBN 978-981-97-7367-1 (eBook)
https://doi.org/10.1007/978-981-97-7367-1

This Springer imprint is published by the registered company Springer Nature Singapore Pte Ltd.
The registered company address is: 152 Beach Road, #21-01/04 Gateway East, Singapore 189721, Singapore

Contents

About the Editors

Prof. Dr. Dattatray J. Late obtained a B.Sc., M.Sc., and Ph.D. degree in Physics, Materials Science and Nanotechnology from Savitribai Phule Pune University, formerly the University of Pune (India), in 2008. He was awarded the DST Post-doctoral Fellowship in Nanoscience and Nanotechnology at JNCASR Bengaluru under Prof. C. N. R. Rao and then IUSSTF postdoctoral fellowship (2010) at Northwestern University (USA) under Prof. V. P. Dravid. He was then awarded the highly prestigious Ramanujan Fellowship (2013) and a Young Research Scientist Career research Award (2014) by the Science and Engineering Research Board (Government of India) at the CSIR-National Chemical Laboratory, Pune (India). He is the recipient of the INUP Best Project Award (2016) from the IITB and the Top Peer Reviewer Award given by the Publons in recognition of his journal review work for 2019–2022. Currently working as Visiting Professor of Physics at Federal University of Lavras, Lavras, Brazil. He was elected as a fellow of the Maharashtra Academy of Sciences in October 2020, a fellow of the Royal Society of Chemistry (FRSC) London, UK, in December 2020 and fellow of International Society for development and sustainability (FISDS) in March 2024. He has filed 20 patent applications and has authored more than 270 research papers with > 21,000 citations in international journals with current h index of 65 and i10 index 200. He has authored four scientific books. He is an Associate editor of AIP Advances (USA) and also serves on the editorial boards of various scientific journals.

Prof. Chandra Sekhar Rout is a full professor at the Centre for Nano and Material Sciences (CNMS), Jain University. Before joining CNMS, He was a DST-Ramanujan fellow at IIT Bhubaneswar, India (2013–2017). He received his B.Sc. (2001) and M.Sc. (2003) degrees from Utkal University and his Ph.D. from JNCASR, Bengaluru (2008), under the supervision of Prof. C. N. R. Rao, Bharat Ratna. His research interests include preparing and characterizing 2D layered materials and their hybrids for chemical sensors and biosensors, supercapacitors and energy storage devices, field emitters, and electronic devices. He has authored more than 200 research papers in international journals and edited six books.

Chapter 1
Introduction

Chandra Sekhar Rout and Dattatray J. Late

Two-dimensional materials (2D) are gaining more popularity in the field of nanomaterials. They are essentially layered materials with one dimension limited to a single atom layer, but they are also referred to as monolayer and few-layer nanomaterials in the literature. Based on the thickness and morphology they can be classified into nanosheets or nanoflakes, nanoribbons, nanoplates, and nanowalls, where the lateral dimensions being one or more orders of magnitude larger than the thickness [1, 2]. They are typically, but not always, prepared by exfoliating their bulk counterparts, which have weak interaction between the neighboring layers. Aside from atomic arrangement and composition, dimensionality plays a crucial role in determining the properties of these stacked nanomaterials. Furthermore, due to quantum confinement, 2D materials exhibit distinct chemical and physical properties that differ from bulk materials. After the mechanical exfoliation of graphene from its bulk analog graphite, the golden era of 2D materials began [3, 4].

Among these 2D materials, Transition metal dichalcogenides (TMDs), particularly MoS_2, have been indispensable throughout the twentieth century and are very promising materials for a wide range of optoelectronic applications and chemical transformations, as evidenced by increased research activity [5]. In 1972, Peter Hjelm for the first time recognized the presence of naturally occurring molybdenite, later J. J. Berzelius in 1826 for the first time synthesized molybdenum by passing hydrogen sulfide through the salt solution containing molybdenum. Later the crystal structure of these molybdenites was determined by Dickinson and Pauling in 1963 based on

C. S. Rout
Centre for Nano and Material Sciences, Jain (Deemed-to-be University), Jain Global Campus, Ramanagaram, Bangalore 562112, India

D. J. Late (✉)
Department of Physics, Federal University of Lavras, Campus Universitário, PO Box 3037, Lavras, Minas Gerais, Brazil
e-mail: datta099@gmail.com

D. J. Late and C. S. Rout (eds.), *Science and Technology of 2D MoS_2*,
Materials Horizons: From Nature to Nanomaterials,
https://doi.org/10.1007/978-981-97-7367-1_1

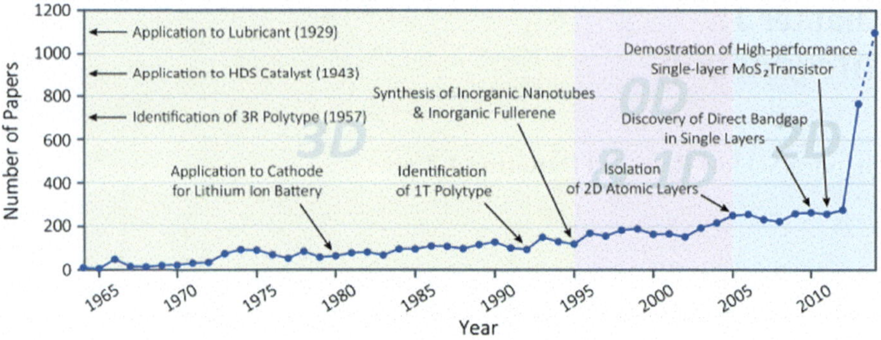

Fig. 1 History of MoS_2 research between 1964 and 2014 [7]

the hexagonal group P6$_3$ mmc. Using carbonate flux, Bell and Herfert synthesized MoS_2 that has a different crystal structure from that of natural molybdenite. Later this structure was resolved by Jellinek, Wildervanck, and Semiletov who described it as rhombohedral crystal structure. Later Pauling's and Guichard's studies demonstrated that triangular crystal of MoS_2 can be synthesized by fusing sulfur, ammonium molybdate, and potassium carbonate [5]. Even though the initial studies of MoS_2 were focused as a lubricant, the burgeoning research on MoS_2 really began after its demonstration as a high-performance transistor in 2011 using single-layered MoS_2 because of its direct bandgap nature. This direct band gaps are essential for the transformation of light onto energy. Utilizing this potential Fontana et al. demonstrated that MoS_2 can be utilized in the field of photovoltaics also. These studies sparked interest in the MoS_2 that results in large number of publications since 1964. Figure 1 shows the history of MoS_2 between 1964 and 2014 [6]. The detailed history of MoS_2 will be discussed in **Chap. 2** [7].

Properties of materials are generally correlated with their electronic structure, especially the shape and size of fermi surfaces and the fermi level position. For example, in the case of MoS_2, its chemical and physical properties are greatly affected with the number of layers and its structures. In summary, they all are correlated with band engineering for modifying the MoS_2 properties [8]. MoS_2 or 2D MoS_2 has three polytypic structures including 1T (octahedral crystal symmetry configuration), 1T' (distorted 1T), 2H (trigonal prismatic coordination), and 3R (Rhomohedral). The letters R, H, and T in the notation stand for rhombohedral, hexagonal, and triangular crystal structures, respectively, while the number in the notation indicates how many layers are necessary to produce a unit cell. While 1T and 1T' are metals, 2H and 3R are all semiconductors. Each monolayer (or single layer) in all three instances is made up of a layer of molybdenum atoms sandwiched between two layers of sulfur atoms. The three atomic layers' relative positions are what set them apart from one another. The Mo layer contains a mirror plane for each of the 2H and 3R layers, and the sulfur atoms in the 1T phase are lateral opposites of the molybdenum atom. Here 2H-MoS_2 has the most thermodynamically stable arrangement and it can be explained by a hexagonal unit cell with trigonal prismatic MoS6 centers, but other metastable polytypes are possible [5]. Of these, the rhombohedral (3R-MoS_2) polytype is of primary interest as it occurs naturally. It is also the second most stable polytype of MoS_2, lending itself to practical importance in the fields of catalysis [6], engineering (e.g., lubricants), [7, 8], and mineral processing [9]. Additionally, 3R-MoS_2 lacks an inversion center, has a tunable band gap, and has a high piezoelectric constant. These features make it a promising material for optoelectronic and photonic devices [10, 11]. The structural properties of these phases will be discussed in **Chap. 3**. There are several ways used to prepare the 2D MoS_2 including exfoliation, CVD, hydrothermal/solvothermal method, electrodeposition, and so on. While liquid phase exfoliation and solution chemical procedure are less expensive, have acceptable material quality, and are suited for large-scale industrial production, micromechanical exfoliation offers the best material quality but is constrained by yield, making it mostly for fundamental research. MoS_2 thin films are made using CVD and are intended for high-end electronic applications. Because these are more advanced technologies with gas-phase precursors and are anticipated to be the predominant ways in the future for manufacturing MoS_2 films targeted toward actual electronic device applications [2], we have placed a special emphasis on this synthesis techniques and discussed in detail in **Chap. 3**.

Similar to other nanomaterials MoS_2 can also synthesized into 0D [9], 1D [10], 2D [4], and 3D dimensionalities. This changes in the dimensionalities help to bring novel functionalities due to the enhancement in the active sites and in surface area [11]. **Chapter 4** discusses in detail how to change the dimensionalities and its effects in properties in detail.

Previous studies prove that the 2H-MoS_2 grain boundaries have a major effect on the electric, magnetic, and transport properties [12]. Along with the grain boundaries, the 1T/2H interfaces also result in many physical properties which were discussed in **Chap. 5**.

Phase engineering is considered to be critical for some 2D materials such as MoS_2. MoS_2 was revealed to exist as a metastable octahedral (1T phase) with enhanced catalytic activity but limited stability, in addition to the normal 2H phase (semiconductor phase) with insufficient active sites. As a result of its superior stability, the 2H phase is favored for use in energy storage and conversion. However, the 2H phase is limited by its low conductivity and semi-insulating nature. In compared to gapless graphene and 2H phase, 1T-MoS_2 has a densely populated Fermi level, making it more capable of storing and transporting charge. Thus, changing the phase composition of MoS_2 is important for performance optimization in order to manage charge transfer across the interface of MoS_2 heterostructures [13]. Phase hybridization which effectively addresses the drawbacks of both the 1T and 2H phases of MoS_2 was explained in **Chap. 6** [14–16]. The summary of the techniques involved is discussed in the book which is given in Fig. 2.

Among the solid lubricants currently in use MoS_2 holds a special position. The usage of MoS_2 as a solid lubricant dated back to previous century has still been used in modern technology. This lamellar material made up of thin atomic planes can easily slide each other and can be used as a dry lubricant and also not require humid condition to perform well unlike graphene [17]. These properties required for MoS_2

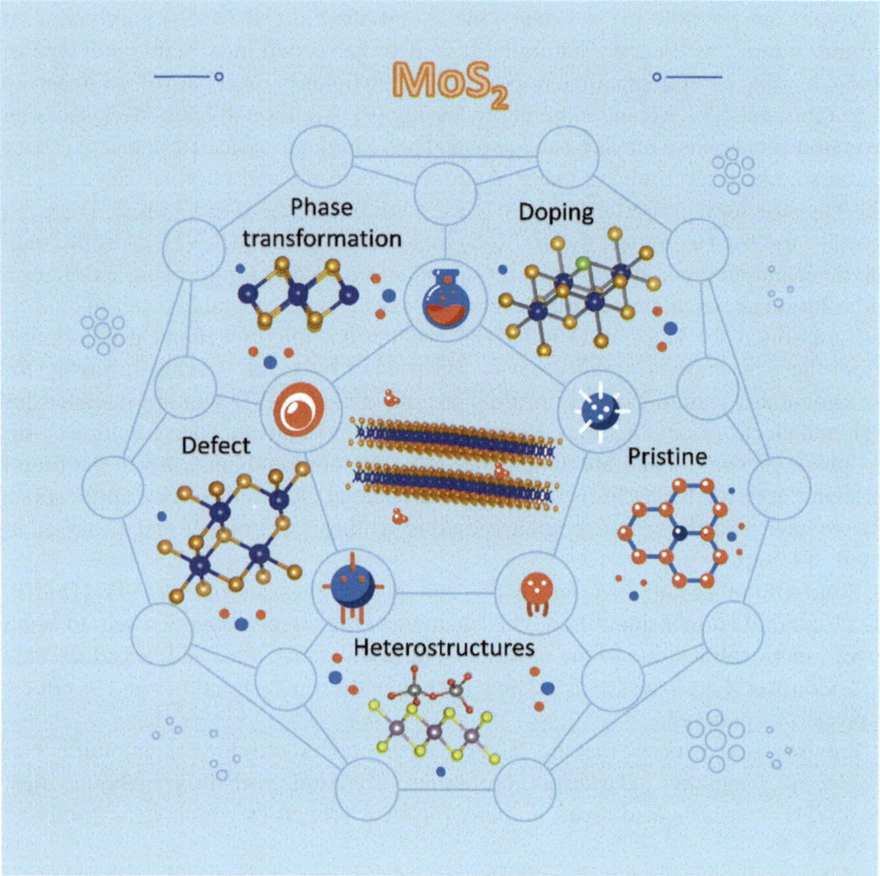

Fig. 2 Summary of the techniques used to change the properties of MoS_2

to act as solid lubricants and its application as a lubricant are summarized in **Chap. 7**.

The assembly of 2D-based van der Waals heterostructures (vdWHs) can be beyond the limitations of processing compatibility and lattice matching due to the lack of direct chemical bonding capacity. Up until now, a lot of work has gone into designing and synthesizing MoS_2/2D, MoS_2/1D, or MoS_2/OD heterostructures that are vertically stacked or grown in-plane is given in **Chap. 8**. Considering all these favorable properties they have been widely explored in different field which will be discussed in other chapters including sensors (**Chap. 9**), photovoltaic devices (**Chap. 10**), field emitters (**Chap. 11**), energy storage and conversion applications (**Chaps. 12 and 13**), and in the field of biomedicals also (**Chap. 14**).

The topics covered in the book are of vital importance in a wide range of modern and emerging layered materials employed or to be employed in most industries,

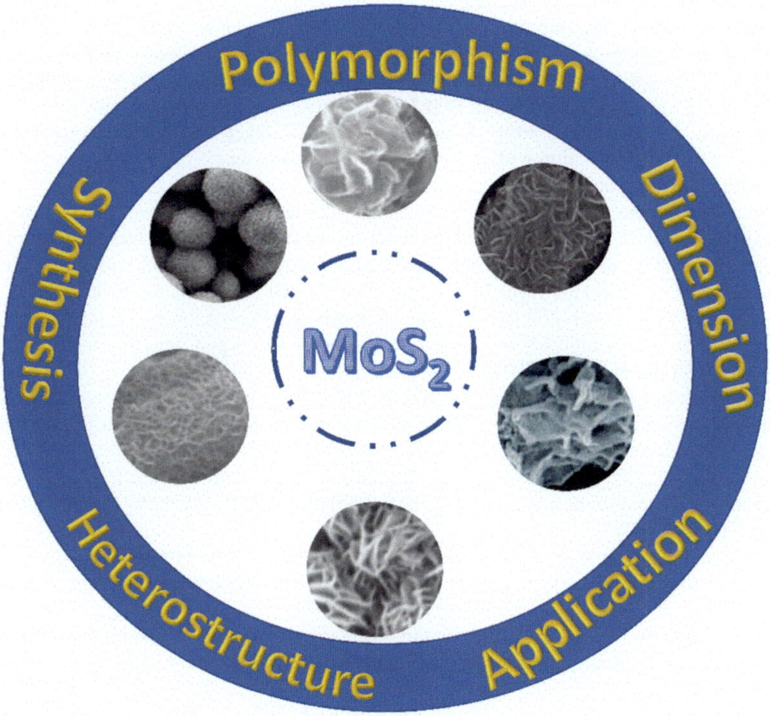

Fig. 3 Summary of the book

academics, health care, food, environment, and research institutes, and consequently have a great impact on our society.

The main objective of the book is to provide the reader with a basic understanding of recent advancements in 2D MoS_2. In this book shown in Fig. 3, we bring together numerous experts in the field of MoS_2 on novel synthesis, usual properties, and applications of this modern material to address fundamental issues prevailing in this thematic domain.

References

1. Gupta A, Sakthivel T, Seal S (2015) Recent development in 2D materials beyond graphene. Prog Mater Sci 73:44–126. https://doi.org/10.1016/j.pmatsci.2015.02.002
2. Sun J, Li X, Guo W et al (2017) Synthesis methods of two-dimensional MoS_2: a brief review. Crystals 7:198. https://doi.org/10.3390/cryst7070198
3. Krishnan U, Kaur M, Singh K et al (2019) A synoptic review of MoS_2: synthesis to applications. Superlattices Microstruct 128:274–297. https://doi.org/10.1016/j.spmi.2019.02.005
4. Singh AK, Kumar P, Late DJ et al (2018) 2D layered transition metal dichalcogenides (MoS_2): synthesis, applications and theoretical aspects. Appl Mater Today 13:242–270. https://doi.org/10.1016/j.apmt.2018.09.003

5. Strachan J, Masters AF, Maschmeyer T (2021) 3R-MoS_2 in review: history, status, and outlook. ACS Appl Energy Mater 4:7405–7418. https://doi.org/10.1021/acsaem.1c00638
6. Fontana M, Deppe T, Boyd AK et al (2013) Electron-hole transport and photovoltaic effect in gated MoS_2 Schottky junctions. Sci Rep 3:1634. https://doi.org/10.1038/srep01634
7. Brischetto M (2018) Slim Moly S makes hydrogen : layer dependent electrocatalysis in hydrogen evolution reaction with individual MoS_2 nanodevices
8. Gan X, Zhao H, Quan X (2017) Two-dimensional MoS_2: A promising building block for biosensors. Biosens Bioelectron 89:56–71. https://doi.org/10.1016/j.bios.2016.03.042
9. Najafi L, Taheri B, Martín-García B et al (2018) MoS_2 quantum dot/graphene hybrids for advanced interface engineering of a $CH_3NH_3PbI_3$ perovskite solar cell with an efficiency of over 20%. ACS Nano 12:10736–10754. https://doi.org/10.1021/acsnano.8b05514
10. Yi Y, Liu Z, Yang P et al (2020) MoS_2 nanorods with inner caves through synchronous encapsulation of sulfur for high performance Li–S cathodes. J Energy Chem 45:18–24. https://doi.org/10.1016/j.jechem.2019.09.032
11. Schneider R, Facure MHM, Chagas PAM et al (2021) Tailoring the surface properties of micro/nanofibers using 0D, 1D, 2D, and 3D nanostructures: a review on post-modification methods. Adv Mater Interfaces 8:2100430. https://doi.org/10.1002/admi.202100430
12. Guo X, Yang G, Zhang J, Xu X (2015) Structural, mechanical and electronic properties of in-plane 1T/2H phase interface of MoS_2 heterostructures. AIP Adv 5:097174. https://doi.org/10.1063/1.4932040
13. Ke Q, Zhang X, Zang W et al (2019) Strong charge transfer at 2H–1T phase boundary of MoS_2 for superb high-performance energy storage. Small 15:e1900131. https://doi.org/10.1002/smll.201900131
14. Yu H, Wang Z, Ni J, Li L (2021) Freestanding nanosheets of 1T–2H hybrid MoS_2 as electrodes for efficient sodium storage. J Mater Sci Technol 67:237–242. https://doi.org/10.1016/j.jmst.2020.08.030
15. Zhang Y, Kuwahara Y, Mori K et al (2020) Hybrid phase 1T/2H-MoS_2 with controllable 1T concentration and its promoted hydrogen evolution reaction. Nanoscale 12:11908–11915. https://doi.org/10.1039/D0NR02525A
16. Wang D, Su B, Jiang Y et al (2017) Polytype 1T/2H MoS_2 heterostructures for efficient photoelectrocatalytic hydrogen evolution. Chem Eng J 330:102–108. https://doi.org/10.1016/j.cej.2017.07.126
17. Vazirisereshk MR, Martini A, Strubbe DA, Baykara MZ (2019) Solid lubrication with MoS_2: a review. Lubricants 7:57. https://doi.org/10.3390/lubricants7070057

Chapter 2
Brief History of MoS_2

Raphael Longuinhos, Jenaina Ribeiro-Soares, and Dattatray J. Late

1 History

Molybdenum disulfide, MoS_2, was first described in its naturally occurring form, molybdenite, in 1782 by Peter Hjelm, one year after he unveiled molybdenum itself [1]. The first synthesis was reported almost 40 years later, in 1826, by J. J. Berzelius, who sinthesized the bulk crystal by reacting a molybdenum salt solution with hydrogen sulfide [1]. In 1923, Pauling and Dickinson determined the crystal structure of the hexagonal molybdenite ($P6_3mmc$) by using X-ray diffraction [2], as illustrated in Fig. 2.1. On the other hand, in 1957, a carbonate flux was used by Bell and Herfert, repeating procedures previously used by Schulten (1889), resulting in a new rhombohedral form (*R3m*) of MoS_2 [3]. This form was then identified as naturally occurring in Canada [4], and since then in several places around the world.

In the following years after the bulk MoS_2 discovery, the community realized that it displays different physical and chemical properties depending on its transition metal coordination and the arrangement of its layers. More recently, the interest in MoS_2 was further boosted by the experimental realization of its low-dimensional forms, as bi-dimensional (few to monolayer MoS_2), one-dimensional (MoS_2 nanotubes), and zero-dimensional (MoS_2 quantum dots) structures, which are represented in Fig. 2.2.

R. Longuinhos · J. Ribeiro-Soares (✉)
Department of Physics, Institute of Natural Science, Federal University of Lavras, Lavras, Minas Gerais 37200-900, Brazil
e-mail: jenaina.soares@ufla.br

R. Longuinhos
e-mail: raphael.lobato@ufla.br

D. J. Late
Department of Physics, Federal University of Lavras, Campus Universitário, PO Box 3037, Lavras, Minas Gerais, Brazil
e-mail: datta099@gmail.com

D. J. Late and C. S. Rout (eds.), *Science and Technology of 2D MoS*$_2$,
Materials Horizons: From Nature to Nanomaterials,
https://doi.org/10.1007/978-981-97-7367-1_2

a)

Reflection Data from Molybdenite

h i k l	λ	Observed angle of reflection	$\frac{1}{n} \times d_{hikl}$	Relative intensities	S u = 0.621
(0001)	0.7078	3° 18′	½ × 12.32	strong	5.05
....	....	. 6° 37′	¼ × 12.28	weak	1.27
....	....	9° 56′	⅙ × 12.32	medium	5.85
....	....	13° 19′	⅛ × 12.28	med. strong	9.17
....	....	16° 43′	$^1/_{10}$ × 12.30	med. weak	4.26
(10$\bar{1}$0)	0.7085	7° 27′	$^1/_1$ × 2.73	....	..
(11$\bar{2}$0)	0.7085	13° 25′	$^1/_1$ × 1.53	....	..

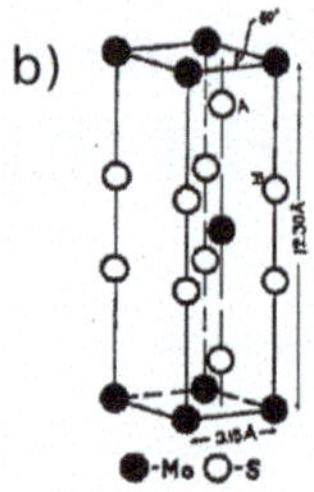

Fig. 2.1 First determination of molybdenite crystal structure. **a** Reflection data acquired by Dickinson and Pauli and **b** the schematic representing the atomic positions. Reprinted (adapted) with permission from [2]

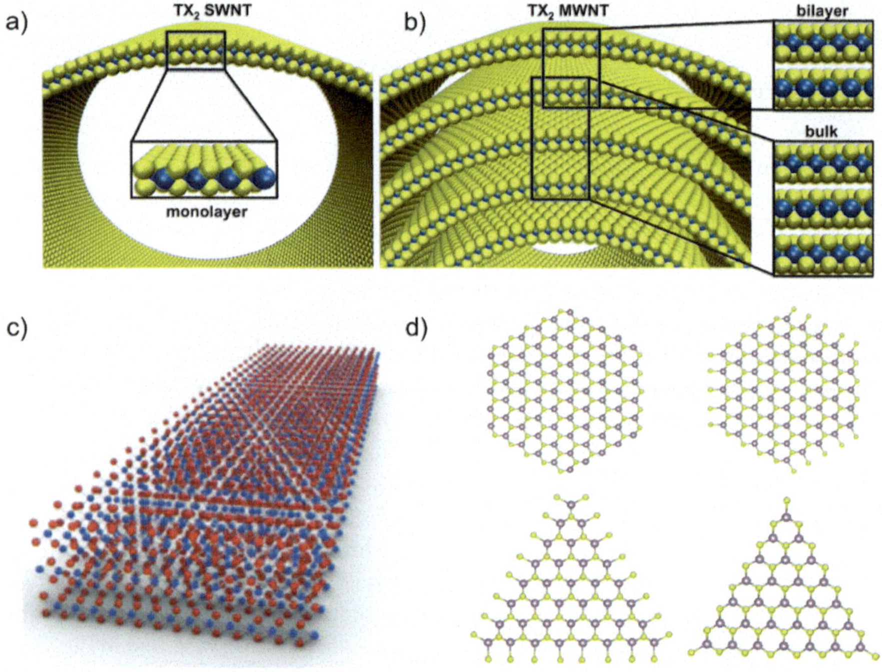

Fig. 2.2 Structural representations of MoS_2 nanotubes, nanowires, and quantum dots In **a** and **b**, the structures of single-wall and multi-wall nanotubes of the MoS_2 family are exemplified, in which "T" represents the transition metal (Mo, W, for example), and "X," the chalcogenide (S, Se, for example). **c** Structure of a MoS_2 nanowire. **d** Possible equilibrium shapes of MoS_2 quantum dots, showing hexagonal dots with mixed Mo and S edges, hexagonal dots with S edges, triangular dots with S edges, and triangular dots with alternating Mo and S edges. Reprinted and adapted with permission from [5–7]

Bulk MoS_2:

The most studied polytypes of MoS_2 are the so-called 1T, 2H, and 3R phases. The 1T phase displays metallic character and has not been found naturally yet, while the man-made samples are metastable, undergoing an irreversible phase transition at 95 °C to disordered 2H-MoS_2 [8]. The 3R polytype is semiconducting, with the advantages of being the second most thermodynamic stable polytype of MoS_2 and naturally available [1]. The 2H phase is also semiconducting and the most thermodynamic stable phase of MoS_2 and abundant in nature [1]. The naming convention of these polytypes comes from the number of layers (1, 2, and 3) and symmetry of their bulk unit cell (T from tetragonal, H from hexagonal, and R from rhombohedral), illustrated in Fig. 2.3. The 1T phase has a single MoS_2 layer (1) in its unit cell, where the transition metal atoms display an octahedral coordination, resulting in the tetragonal (T) symmetries of the D_{3d} point group. The 2H phase has two MoS_2 layers (2) in its unit cell, where the transition metal coordination by sulfur atoms is trigonal prismatic. Considering the single MoS_2 layer as a tri-layer S-M-S, the trigonal prismatic coordination can be seen as the resulting tri-layer geometry after rotating one of the sulfur layers (e.g., the top) by 60 degrees with respect to the other sulfur layer (e.g., the bottom). The stacking of the two MoS_2 layers in the pristine 2H phase is found in two different forms [9–12]. In the 2Ha stacking, the transition metal atom of the MoS_2 layer is vertically aligned with the transition metal atom of the adjacent MoS_2 layers. In the 2Hc, the transition metal atom of the MoS_2 layer is vertically aligned with the sulfur atoms of the adjacent layers. Both 2Ha and 2Hc belong to the hexagonal (H) lattice group and display the symmetries of the D_{6h} point group. The 3R phase has three MoS_2 layers (3) in its unit cell, where the transition metal atoms display the trigonal prismatic coordination, as in the 2H phases. Considering the three layers in the 3R phase unit cell, the transition metal atoms of the bottom MoS_2 layer are vertically aligned with the chalcogen atoms of the middle MoS_2 layer, and the transition metal atoms of this layer are vertically aligned with the chalcogen atoms of the top MoS_2 layer, conferring a rhombohedral (R) lattice for the structure, with the symmetries of the C_{3v} point group.

Two-dimensional (2D) MoS_2:

The two-dimensional basic form of MoS_2 (monolayer) was the subject of theoretical speculation several decades before it was possible to synthesize it in large scale and to study its properties when isolated in only one layer. Seminal works were performed by Yoffe and collaborators in the 60s and 70s related to the interpretation of the bulk counterpart's optical, electrical, and structural properties [9]. They also developed the semiempirical tight-binding modeling of its band structure by using the monolayer approximation, extending it to the three-dimensional case [14]. This development is similar to the observed for the graphite/graphene case, in which the monolayer band structure was studied in 1947 by Wallace [15], for only the monolayer be experimentally realized in 2004 by Novoselov and Gain [16]. Advances in the synthesis of MoS_2 single layers in aqueous suspension by lithiation were presented

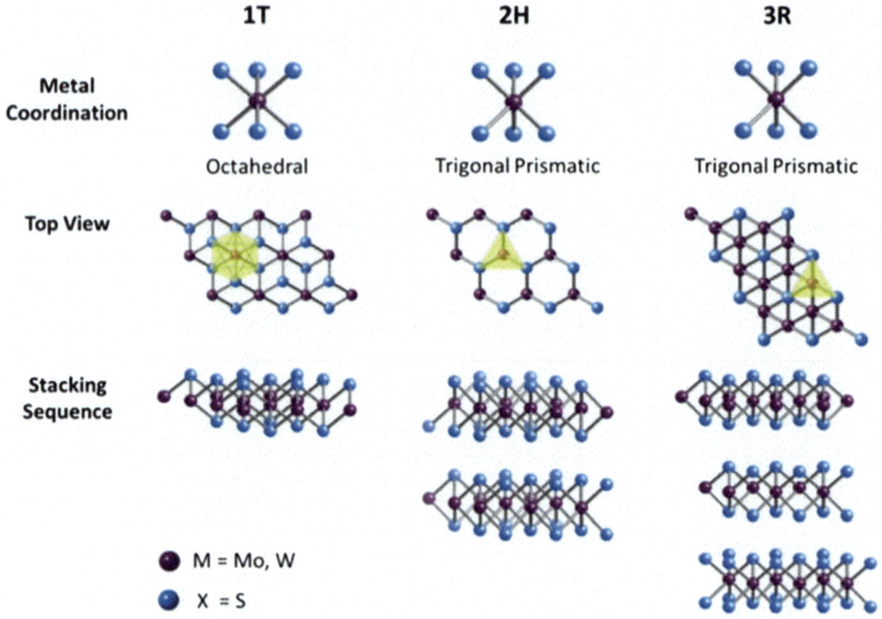

Fig. 2.3 The transition metal coordination and stacking sequences for MoS_2 and WS_2 transition metal dichalcogenides polytypes, for bulk and few layer. The coordination can be octahedral or trigonal prismatic, in which the stacking of the layers can result in the different polytypes, namely, 1T (octahedral monolayer layer stacking), 2H, or 3R (trigonal prismatic monolayer stacking). Reprinted from [13]

by Gordon and collaborators in 2002 [17], for example, but only in 2010, Splendiani showed the emerging photoluminescence in monolayer MoS_2 [18] by using the micromechanical exfoliation developed by Novoselov and collaborators [16] (see Fig. 2.4a–d) for images of the exfoliated flakes images and respective photoluminescence measurements). In this work, the photoluminescence was shown to emerge in mono- and few-layer MoS_2, indicating the existence of new phenomena related to the d-electron system promoting a direct bandgap transition when the indirect bandgap bulk MoS_2 crystal is reduced to the atomic layer thickness, also indicated by Mak et al. [19]. Simultaneously to this discovery, Lee et al. showed that the thickness of few-layer MoS_2 had specific phonon signatures studied by using Raman spectroscopy [20] (see Fig. 2.4e), and that the wavenumber difference among the E_{2g} and A_{1g} as a fast and reliable strategy to identify the number of layers to an atomic-level precision (Fig. 2.4f). In 2011, the single-layer MoS_2 transistor was demonstrated [21], as well as its integrated circuits performing logic operations [22], unveiling the possibility to fabricate the building blocks of low-power nanoelectronic devices. From this moment on, all knowledge about computational simulations, synthesis, characterization, fabrication, and applications developed for graphene studies were quickly applied to MoS_2 and other isostructural materials from its family (WS_2, $MoSe_2$, and so on). The maturity of the graphene research field developed then, motivated additionally by the fact of the direct bandgap of monolayer MoS_2 and its profound application perspectives complementing graphene, resulted in an effervescent research atmosphere resulting in developments that will be explored in the next chapters of this book.

One-dimensional (1D) MoS_2:

The carbon nanotube experimental realization in 1991 [23] opened the door to the research on unidimensional materials, as other layered materials could exist as coaxial cylinders of wrapped hexagonal sheets, such as transition metal dichalcogenides. These speculations were found to be true with the synthesis of cylindrical and polyhedral tungsten diselenide (WS_2) in 1992 [24] and, subsequently, of MoS_2 nanotubes and fullerenes in 1995 [25]. The synthesis of these MoS_2 nanotubes was performed in large quantities by using a turbulent flow regime of MoO_{3-x} and H_2S precursors at high temperatures (800–950 °C), in a quartz tube reactor furnace (see Fig. 2.5 for TEM images of MoS_2 nanotubes from [26]). Since then, the experimental [26] and theoretical [27] understanding of the structure and electronic properties of these tubes evolved significantly [28], especially motivated by their application perspectives ranging from catalysis, hydrogen storage, and lubricants, among others [29]. One-dimensional MoS_2 forms include, beyond nanotubes, there are nanorods, nanowires, and nanoribbons, for example, which were progressively obtained throughout the next years especially as subproducts of nanotube synthesis, and also as intentionally made at different synthesis methodologies.

Zero-dimensional (0D) MoS_2:

After the synthesis and description of quantization effects in small colloidal particles of layered semiconductors like PbI_2 and BiI_3, among others, MoS_2 crystalline particles in the 10–35 Å size range were synthesized by the simple dissolution of its bulk crystal in acetonitrile [30]. Shortly after [31], Gutierrez and Henglein obtained colloid solutions by sonicating water containing powder of MoS_2 under an argon–hydrogen atmosphere, also presenting their absorbance spectra, as shown in Fig. 2.6. At that time, the ultrafine semiconductor colloids attracted considerable attention due to their photocatalytic properties and the perspective of their use in photo- and radiation chemical experiments, where the related materials proved their stability for weeks [31, 32]. The understanding of the size effects and characterization of MoS_2 colloids evolved along the next three decades, especially according to their optical and catalytic properties. They were identified as promising for diverse applications, reaching the limit of the interplay along 0D and 2D structures in the monolayer MoS_2 nanoclusters [33–35]. Although the synthesis and understanding of MoS_2 QD's are in current evolution, their applications are being recently broadly explored in energy (electrocatalysis, solar cells, energy storage), electronic and optoelectronic devices (LEDs, photodetectors, phototransistors), chemical sensors, bioimaging, and cancer treatment, among others [36].

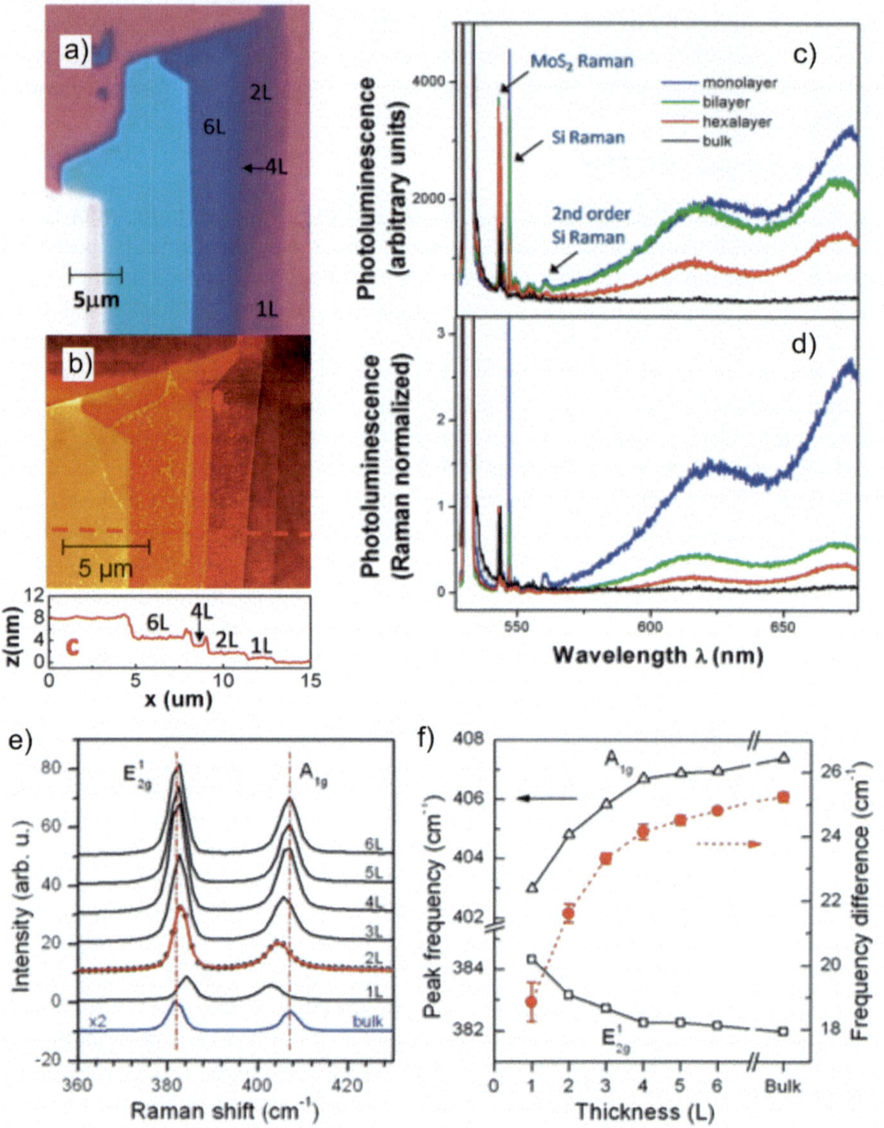

Fig. 2.4 Emerging anomalous photoluminescence and Raman spectra of few-layer MoS_2 samples. In **a**, the optical microscopy of a exfoliated sample on a Si/SiO_2 substrate is shown. In **b**, its respective AFM image is given, with the topographic profile showing the presence of 1, 2, 4, and 6 layers. In **c** and **d**, the layer-number dependent photoluminescence is given for non-normalized and Raman-normalized data. **e** Raman spectra of few-layer MoS_2, from 1 to 6 layers, and bulk. The red solid line for 2L is a double Voigt fit of data. **f** The frequencies of the E_{2g} and A_{1g} Raman modes (right vertical axis), and their difference (left vertical axis). Reprinted (adapted) with permission from [18, 20]

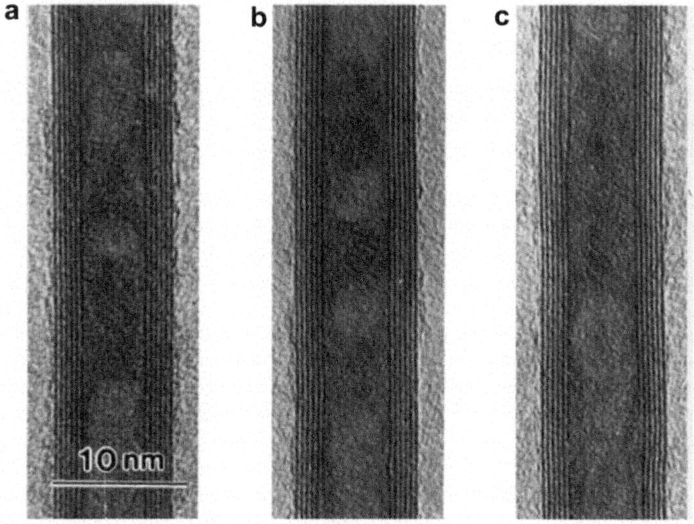

Fig. 2.5 TEM images of MoS_2 non-helical nanotubes consisting of wrapped hexagonal sheets. Reprinted (adapted) with permission from [26]

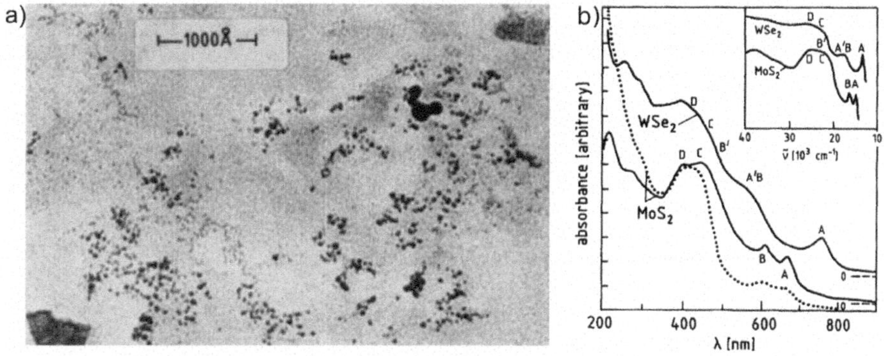

Fig. 2.6 Synthesis and characterization results for MoS_2 nanodots obtained by sonication in water. **a** Electron micrography of a representative sample and **b** absorbance spectra. The inset represents the result for films. Reprinted (adapted) with permission from [31]

2 Conclusion

We overview the history of MoS_2, from three-dimensional to zero-dimensional structures. The reduction of the dimensionality of MoS_2 changed their physical properties, opening opportunities for the development of new nanoscale-based devices, with applications in electronics, optoelectronics, catalysis, clean energy production, and disease treatment [37].

References

1. Strachan J, Masters AF, Maschmeyer T (2021) 3R-MoS_2 in review: history, status, and outlook. ACS Appl Energy Mater 4(8):7405–7418
2. Dickinson RG, Pauling L (1923) The crystal structure of molybdenite. J Am Chem Soc 45(6):1466–1471
3. Bell RE, Herfert RE (1957) Preparation and characterization of a new crystalline form of molybdenum disulfide. J Am Chem Soc 79(13):3351–3354
4. Traill, R. J (1963) A rhombohedral polytype of molybdenite. Can Mineral 7(3):524–526
5. Ghorbani-Asl M et al (2013) Electromechanics in MoS_2 and WS_2: nanotubes vs. monolayers. Sci Reports 3(1):2961
6. Li J et al (2022) Tunable strong coupling in transition metal dichalcogenide nanowires. Adv Mater 34(34):2200656
7. Pei L et al (2015) Structural stability, electronic and magnetic properties of MoS_2 quantum dots based on the first principles. Solid State Commun 218:25–30
8. Wypych F, Schöllhorn R (1992) 1T-MoS_2, a new metallic modification of molybdenum disulfide. J Chem Soc Chem Commun X(19):1386–1388
9. Wilson JA, Yoffe AD (1969) The transition metal dichalcogenides discussion and interpretation of the observed optical, electrical and structural properties. Adv Phys 18(73):193–335
10. Katzke H et al (2004) Phase transitions between polytypes and intralayer superstructures in transition metal dichalcogenides. Phys Rev B 69(8):134111
11. Hromadová L et al (2013) Structure change, layer sliding, and metallization in high-pressure MoS_2. Phys Rev B 87(14):144105
12. Ribeiro-Soares J et al (2014) Group theory analysis of phonons in two-dimensional transition metal dichalcogenides. Phys Rev B 90(11):115438
13. Toh RJ et al (2017) 3R phase of MoS_2 and WS_2 outperforms the corresponding 2H phase for hydrogen evolution. Chem Commun 53(21):3054–3057
14. Bromley RA, Murray RB, Yoffe AD (1972) The band structures of some transition metal dichalcogenides. III. Group VIA: trigonal prism materials. J Phys C: Solid State Phys 5(7):759
15. Wallace, Philip Richard () The band theory of graphite. Physical review 71.9 (1947): 622
16. Novoselov KS et al (2004) Electric field effect in atomically thin carbon films. Science 306(5696):666–669
17. Gordon RA et al (2002) Structures of exfoliated single layers of WS_2, MoS_2, and $MoSe_2$ in aqueous suspension. Phys Rev B 65(12):125407
18. Splendiani A et al (2010) Emerging photoluminescence in monolayer MoS_2. Nano Lett 10(4):1271–1275
19. Mak KF et al (2010) Atomically thin MoS_2: a new direct-gap semiconductor. Phys Rev Lett 105(13):136805
20. Lee C et al (2010) Anomalous lattice vibrations of single-and few-layer MoS_2. ACS Nano 4(5):2695–2700
21. Radisavljevic B et al (2011) Single-layer MoS_2 transistors. Nature Nanotechnol 6(3):147–150
22. Radisavljevic B, Whitwick MB, Kis A (2011) Integrated circuits and logic operations based on single-layer MoS_2. ACS Nano 5(12):9934–9938
23. Ajayan PM, Iijima S (1992) Smallest carbon nanotube. Nature 358(6381):23
24. Tenne R et al (1992) Polyhedral and cylindrical structures of tungsten disulphide. Nature 360(6403):444–446
25. Feldman Y et al (1995) High-rate, gas-phase growth of MoS_2 nested inorganic fullerenes and nanotubes. Science 267(5195):222–225
26. Margulis L et al (1996) TEM study of chirality in MoS_2 nanotubes. J Microsc 181(1):68–71
27. Seifert G et al (2000) Structure and electronic properties of MoS_2 nanotubes. Phys Rev Lett 85(1):146
28. Sinha SS et al (2021) MoS_2 and WS_2 nanotubes: synthesis, structural elucidation, and optical characterization. J Phys Chem C 125(11):6324–6340

29. Krishnan U et al (2019) A synoptic review of MoS_2: synthesis to applications. Superlattices Microstruct 128:274–297
30. Peterson MW et al (1988) Quantized colloids produced by dissolution of layered semiconductors in acetonitrile. J Phys Chem 92(6):1400–1402
31. Gutierrez M, Henglein A (1989) Preparation of colloidal semiconductor solutions of MoS_2 and WSe_2 via sonication. Ultrasonics 27(5):259–261
32. Kamat PV, Dimitrijević NM (1990) Colloidal semiconductors as photocatalysts for solar energy conversion. Solar Energy 44(2):83–98
33. Khairutdinov RF, Rubtsova NA, Costa SMB (1996) Size effect in steady-state and time-resolved luminescence of quantized MoS_2 particle colloidal solutions. J Lumin 68(6):299–311
34. Qiao W et al (2015) Monolayer MoS_2 quantum dots as catalysts for efficient hydrogen evolution. RSC Adv 5(118):97696–97701
35. Wilcoxon JP, Newcomer PP, Samara GA (1997) Synthesis and optical properties of MoS_2 and isomorphous nanoclusters in the quantum confinement regime. Journal of Applied Physics 81(12):7934–7944
36. Kabel J et al (2021) Molybdenum disulfide quantum dots: properties, synthesis, and applications. C 7(2):45
37. Herculano-Houzel S (2009) The human brain in numbers: a linearly scaled-up primate brain. Front Hum Neurosci 3:31

Chapter 3
Synthesis, Structure, and Physical Properties of Bulk MoS_2

Levna Chacko and Dattatray J. Late

1 Introduction

With the discovery of graphene by Novoselov and Geim in 2004, two-dimensional (2D) materials have been extensively researched due to their bizarre promise in the fields of electronics, optics, medical, mechanics, energy conversion, and storage. Especially, 2D-layered materials consisting of atomic sheets stacked together by weak van der Waal forces have received intriguing research interest in recent years. Cutting-edge 2D materials being investigated by researchers include 2D oxides (V_2O_5, MoO_3, $LixCoO_2$), topological insulators (Bi_2Se_3, Bi_2Te_3, HfBr), nitrides (h-BN, MoN, $Ti_4N_3T_x$, W_2N, V_2N), carbides (Ti_3C_2, Ta_4AlC_3), and transition-metal dichalcogenides (MoS_2, WS_2). Research has proved that these materials could counterpart graphene in a variety of fields and applications.

Being the dominant descendants of graphene, layered TMDCs are flexible; transparent; and possess band gap, excellent on/off ratio, and appreciable charge carrier mobility that are comparable to those of ubiquitous silicon. Because of its popularity and alluring potentialities, among TMDCs, MoS_2 is one among the dominant materials and the bulk properties and applications of MoS_2 are reviewed here.

L. Chacko
Department of Physics and Electronics, CHRIST (Deemed to be University), Bengaluru 560029, Karnataka, India

D. J. Late (✉)
Department of Physics, Federal University of Lavras, Campus Universitário, PO Box 3037, Lavras, Minas Gerais, Brazil
e-mail: datta099@gmail.com

D. J. Late and C. S. Rout (eds.), *Science and Technology of 2D MoS_2*,
Materials Horizons: From Nature to Nanomaterials,
https://doi.org/10.1007/978-981-97-7367-1_3

2 Crystalline and Electronic Structure of MoS_2

Being a prototypical transition-metal dichalcogenide, semiconductor MoS_2 has garnered considerable research interest because of its distinctive physical, chemical, electronic, optical, and catalytic properties. Bulk MoS_2 is formed by vertical stacking of 2D S-Mo-S layers having an interlayer separation of ~6.5 Å with a separation between a Mo atom and its nearest S atom is 2.42 Å as shown in Fig. 1 [1, 2]. It is an indirect band gap (~1.23 eV) in its bulk form and is built up of strong covalent bond between the d orbitals of molybdenum and the p orbitals of sulfur atoms. Each individual 2D S-Mo-S layers consists of a hexagonal layer of Mo atoms sandwiched between two hexagonal layers of S atoms wherein the individual layers are coordinated by weak van der Waals forces, facilitating the isolation into single or few layers.

MoS_2 has three distinct polytypes: trigonal prismatic phase (2H), octahedral phase (1T), and rhombohedral phase (3R), depending on the stacking arrangement of each S-Mo-S layers and the coordination between molybdenum and sulfur atoms. Here in the alphanumeric character, the first number specifies the number of layers present

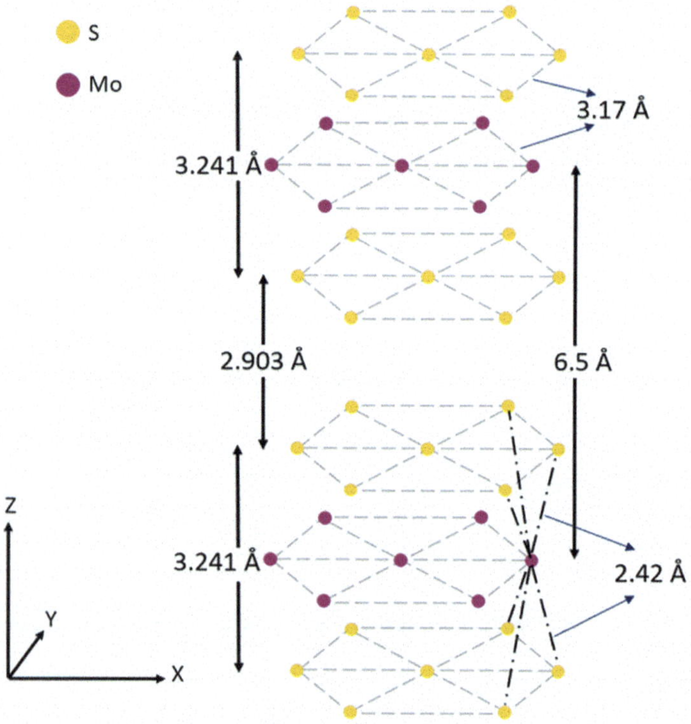

Fig. 1 Crystal structure of MoS_2 specifying distance between different layers [1, 2]

within a single crystallographic unit cell followed by a letter which denotes the crystallographic structure, with T for trigonal, H for hexagonal, and R for rhombohedral form of crystal structure. For instance, the 1T polytype unit cell consists of one S-Mo-S layer, 2H polytype has two S-Mo-S layers, and 3R-MoS2 consists of three S-Mo-S layers in a primitive hexagonal unit cell.

Among these, 2H- and 3R-MoS_2 are thermodynamically stable polytypes in which Mo atoms feature trigonal prismatic coordination (D_{3h}) with six neighboring S atoms with a stacking order of AbABaBAbA with space group $P6_3/mmc$ for 2H-MoS_2 and AbABcBCaCAbA with space group R3m for 3R-MoS_2 polytype. Whereas, in 1T-MoS_2 phase, the Mo atoms are octahedrally coordinated (O_h) to six neighboring S atoms forming a stacking order of AbCAbC and space group P3m1 (Fig. 2a) [3, 4]. 1T-MoS_2 phase is thermodynamically unstable, exhibits metallic and paramagnetic features due to symmetry-induced splitting of Mo-4d orbitals into two groups: dx^2-y^2/dz^2 and dyz/dxz/dxy in which the dyz, dxz, and dxy set of triple degenerate orbitals in the ground state consists of only two electrons (Fig. 2b). The dz^2/dx^2-y^2 degenerate energy level in 1T-MoS_2 is higher than Fermi level and is unoccupied. In 2H- and 3R-MoS_2, the 4d orbital of Mo is split into three groups: dx^2-y^2/dxy, dyz/dxz, and dz^2 within the context of crystal field theory, with fully occupied Mo-4dz^2 in the ground energy level and Mo-4dxy/Mo-4dx^2-y^2 and Mo-4dxz/Mo-4dyz levels which are unoccupied, thereby exhibiting semiconducting behavior and diamagnetic property (Fig. 2b) [5–7].

Splendiani et al. [9] and Mak et al. [10] reported that on varying the thickness of MoS_2 from bulk to monolayer, it undergoes an indirect-to-direct band-gap transition (1.9 eV). Using density functional theory (DFT) calculation Ahmad and Mukherjee [8] studied on the electronic band structure and density of states of bulk MoS_2 and identified that they can be divided into three sets of bands in the electronic band structure and three sets of states in the density of states, separated by gaps as shown in Fig. 3. The first set observed around -14 eV is essentially considered to the presence of 3s orbital of the S atom which is separated from the second set below Fermi energy by a significant gap and exhibits substantial hybridization. The primary contribution of second set comes from 3p orbital of S and 4d orbitals of Mo resulting in a strong hybridization, whereas that of third set above the Fermi energy comes from 4d orbital of Mo atoms. As shown in Fig. 4, bulk MoS_2 is an indirect band-gap semiconductor with the valence band maximum at the high-symmetric Γ-point and the conduction band minima between the Γ- and K-points. As a result of indirect band gap, bulk MoS_2 exhibits negligible quantum yield for photoluminescence (PL). However, monolayer MoS_2 exhibits a direct band gap with the valence band maxima and conduction band minima both at high-symmetric K-points.

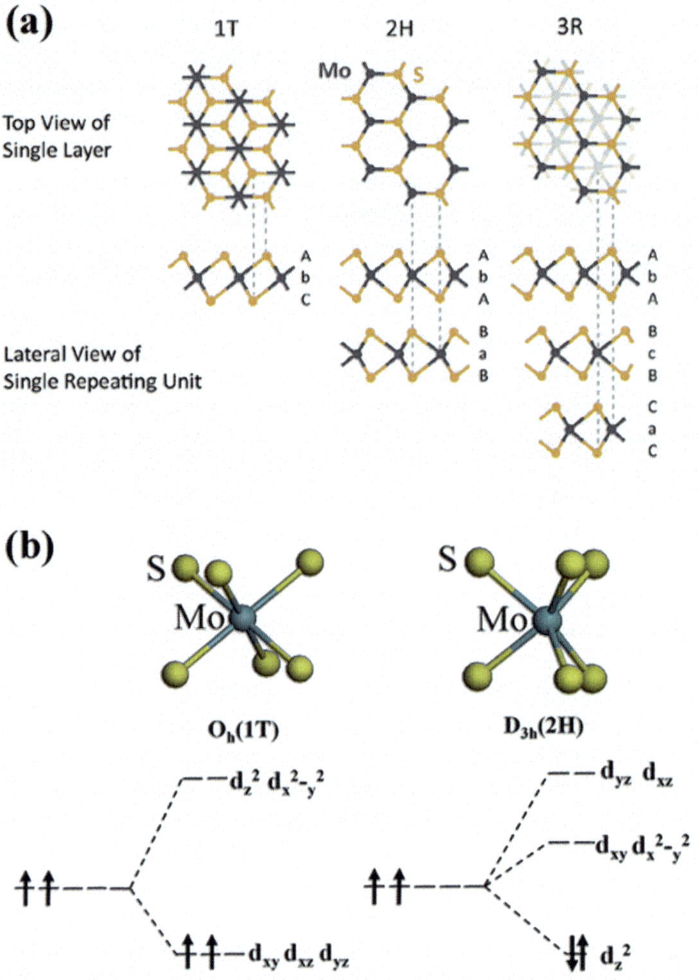

Fig. 2 **a** Different polytypes of MoS_2. Reprinted with permission from Song et al. [3] Copyright 2015, Royal Society of Chemistry, **b** Crystal field splitting of Mo-4d orbits in octahedral and trigonal prismatic symmetry. Reproduced from Ref. [6] with permission from the Royal Society of Chemistry

3 Optical Properties

MoS_2 exhibits excellent optical properties featuring from the strong excitonic effects [10]. The photoluminescence spectrum of bulk MoS_2 consists of two distinct peaks in the spectral region ranging from 1.3 to 2.4 eV cited as A and B excitons, associated with the direct-gap transition between the maxima of split valence bands and minimum of conduction band at Brillouin zone K-point as a result of combined effect of interlayer coupling and spin–orbit coupling especially in multi-layered

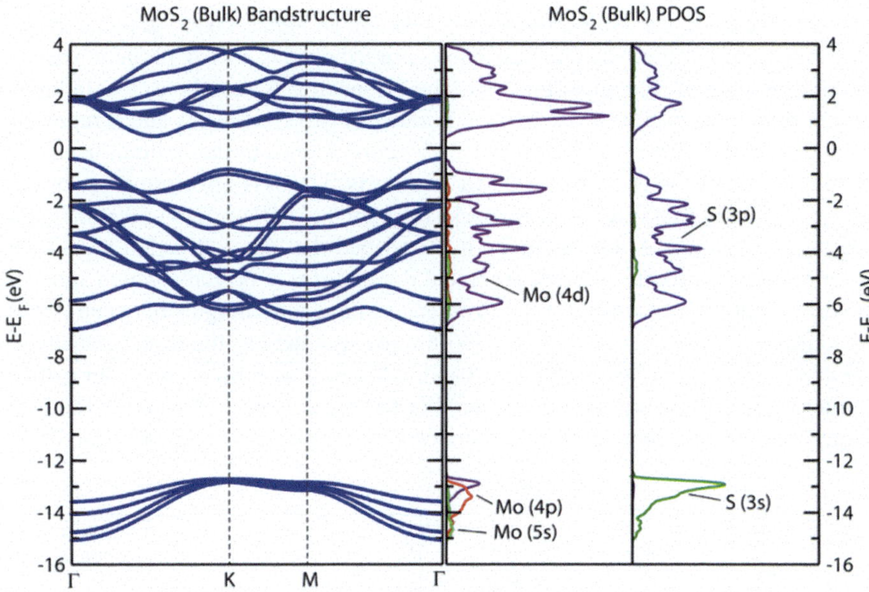

Fig. 3 Electronic band structure and corresponding density of states of bulk MoS_2 [8] Copyright © 2014 by authors and Scientific Research Publishing Inc

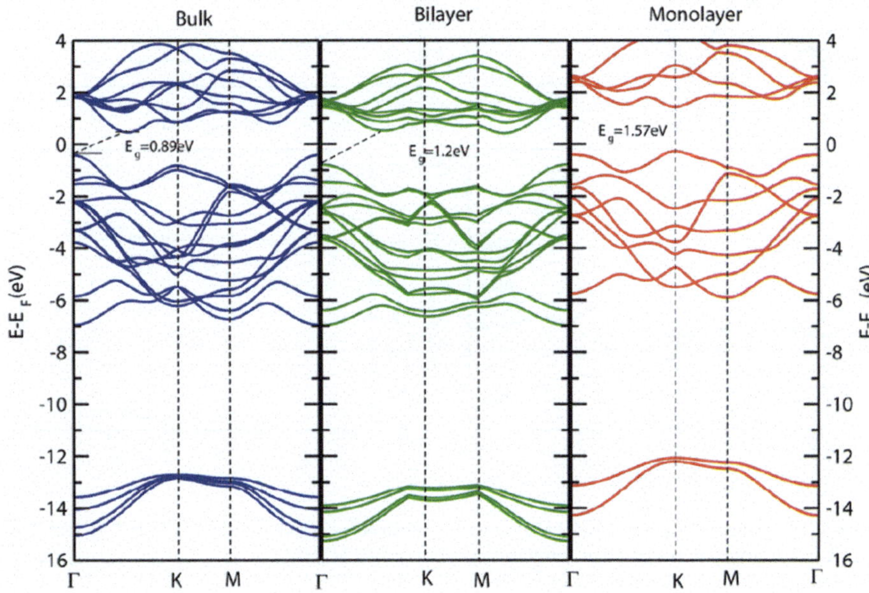

Fig. 4 Band structure of MoS_2 (bulk, bilayer, and monolayer) [8] Copyright © 2014 by authors and Scientific Research Publishing Inc

MoS_2 which exhibit inversion symmetry [3]. Being an indirect band-gap material in bulk form, photoluminescence in bulk MoS_2 is attributed as a weak phonon-assisted process with insignificant quantum yield. The PL spectra also exhibit a clear indirect-to-direct transition with decrease in layer number of MoS_2. Peaks A and B represent the valence band splitting and C (1.2~1.5 eV) corresponds to the indirect band-gap transition in MoS_2 as shown in Fig. 5a. Similarly, UV–Vis absorption spectra of MoS_2 also display four characteristic peaks (Fig. 5b) at ~650 nm and ~590 nm corresponding to excitonic A and B bands. It is ascribed that due to the d–d transitions from the filled d_x orbital to the degenerated d_{xy} and $d_{x^2-y^2}$ orbitals, which split via spin–orbit coupling, resulting in the difference in exciton energies between A and B. The peaks C and D at ~430 and ~300 nm are assigned to the direct transitions from deep valence band to conduction band. Although bulk MoS_2 does not exhibit prominent PL, with decrease in layer number the PL emission of MoS_2 is found to enhance, proving the layer number dependence of PL emission (Fig. 5 (c and d)).

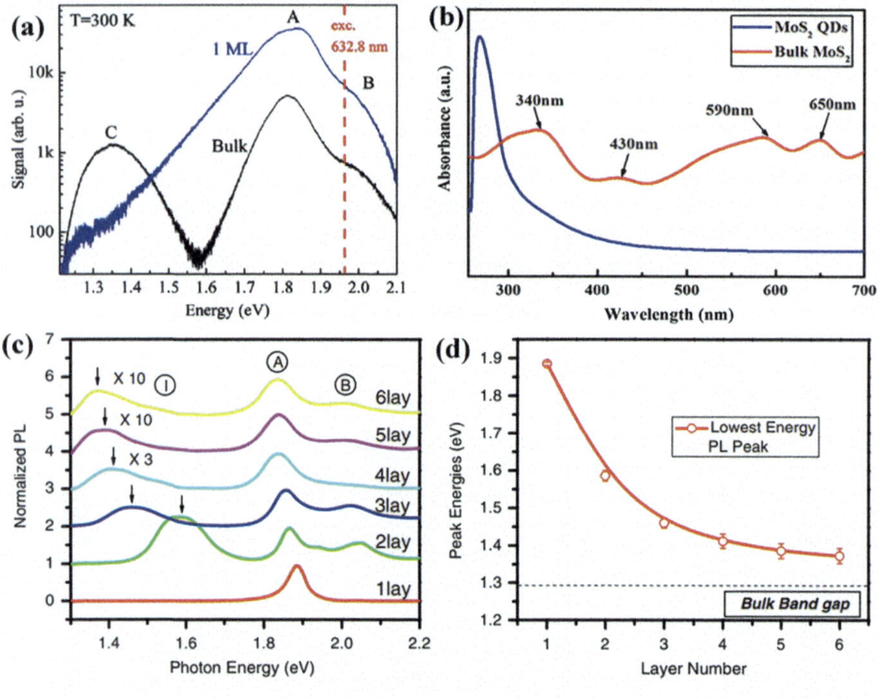

Fig. 5 **a** Photoluminescence (PL) [11] **b** UV–Vis absorption spectra of MoS_2 [12] Copyright © 2017, The Author(s) and **c, d** layer number dependence of PL [10]

4 Raman Spectroscopic Studies

Being a potent non-destructive characterization technique, Raman spectroscopic studies are used to examine the crystalline nature and phonon–phonon interaction of MoS_2. It consists of four main Raman active vibration modes A_{1g} (~408 cm^{-1}), E^1_{2g} (~383 cm^{-1}), E_{1g} (~286 cm^{-1}), and E^2_{2g} (~32 cm^{-1}), infrared active modes A_{2u}, E_{1u}, and four inactive phonon modes B_{1u}, two B_{2g}, E_{2u}. Among these, A_{1g} and E^1_{2g} modes emanate from the interatomic vibrations within the layers and the frequency shift between these modes helps in determining layer number. Being a longitudinal optical (LO) phonon E^1_{2g} mode is assigned to the in-plane vibrations of sulfur atoms and molybdenum atoms in opposite directions, whereas A_{1g} represents ZO phonon with out-of-plane vibration of only sulfur atoms in opposite directions. E^2_{2g} vibration mode corresponds to the interlayer atomic vibrations. However, on the surface perpendicular to the c-axis, E_{1g} mode is forbidden as a result of back-scattering geometry [13, 14].

Lee et al. [14] elucidated the Raman spectroscopic study of single- and few-layered MoS_2 and observed that as thickness of MoS_2 samples increased, the E^1_{2g} vibration undergoes red shift and the A_{1g} vibration undertakes blue shift. In bulk MoS_2, it is reported that with increase in layers the interlayer van der Waals (vdW) interactions are found to suppress the atomic vibrations thereby increasing the effective restoring forces on atom leading to stiffening of the E^1_{2g} and A_{1g} modes [15]. They also observed the variation in the intensities of the Raman peaks, E^1_{2g} and $A_{1g}$with a linear increase in the intensity up to four layers and subsequent decrease with further increase in the layers associated with optical interference effect and variation in the electronic structure of MoS_2 (Fig. 6a and b). Also, the research witnessed a change in the line widths of these modes, especially with A_{1g} mode from 3 to 6 cm^{-1}. As the bulk inner layers of MoS_2 dominate the intensity of Raman modes, line widths in bulk MoS_2 are found to decrease and also attributed to the inherent variation in symmetry (Fig. 7a and b).

In another report, Sun et al. [16] thoroughly examined the second-order and combination Raman modes in MoS_2, including single-layer, bi-layer, quad-layer, and bulk MoS_2, using four excitation laser lines as shown in Fig. 8a–d. It was noticed that the intensities of second-order and combination modes, 2 E^1_{2g} and 2A_{1g} (range from 750 to 840 cm^{-1}) in few-layer MoS_2 are significantly increased under 325 nm laser excitation, attributed to triply resonant Raman scattering (TRRS) based on Fröhlich interaction and deformation potential interaction involved electron–two-phonon coupling and in which the presence of spin–orbit (SO) splitting creates a channel for the resonance of overtone and combination phonon modes. In contrast, in bulk MoS_2, the probability of TRRS which is very sensitive to energy splitting is reduced because of the greater value of SO splitting and it was unable to detect the distinct high-order Raman peaks.

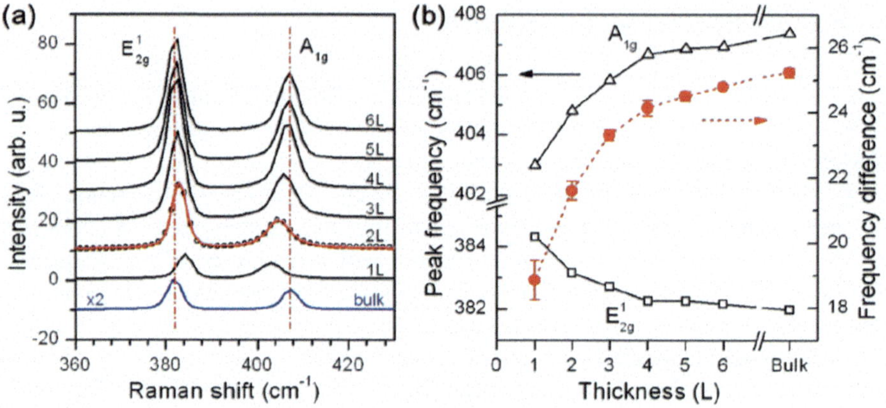

Fig. 6 **a** Raman spectroscopic image of bulk and different layer MoS_2 ranging from one layer to six layers and **b** frequency shift of E^1_{2g} and A_{1g} modes as a function of thickness [14]

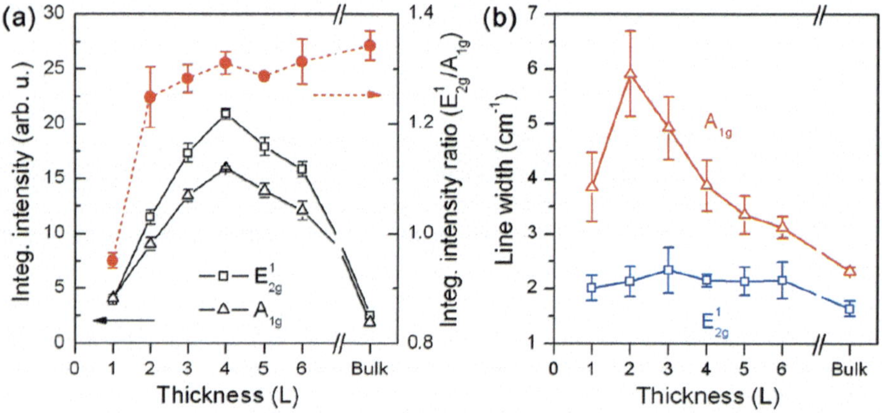

Fig. 7 **a** Variation of Raman intensity and **b** line width of E^1_{2g} and A_{1g} modes of vibrations of MoS_2 as a function of layer thickness [14]

5 Magnetic Properties

In principle semiconducting, bulk MoS_2 is observed to exhibit diamagnetic behavior. Although several research works have reported magnetic behavior in bulk MoS_2 ascribed to the presence of zigzag edges and vacancies. Despite the fact that diamagnetism dominates the magnetic response of bulk MoS_2, experimentally Tongay et al. [17] have shown that apart from the diamagnetic magnetic response, bulk MoS_2 also contains ferromagnetic behavior (Fig. 9), expected due to the presence of zigzag edges and related magnetism at grain boundaries. Compared to single-layered or few-layered MoS_2, the ferromagnetic signal vanishes in the bulk limit because the

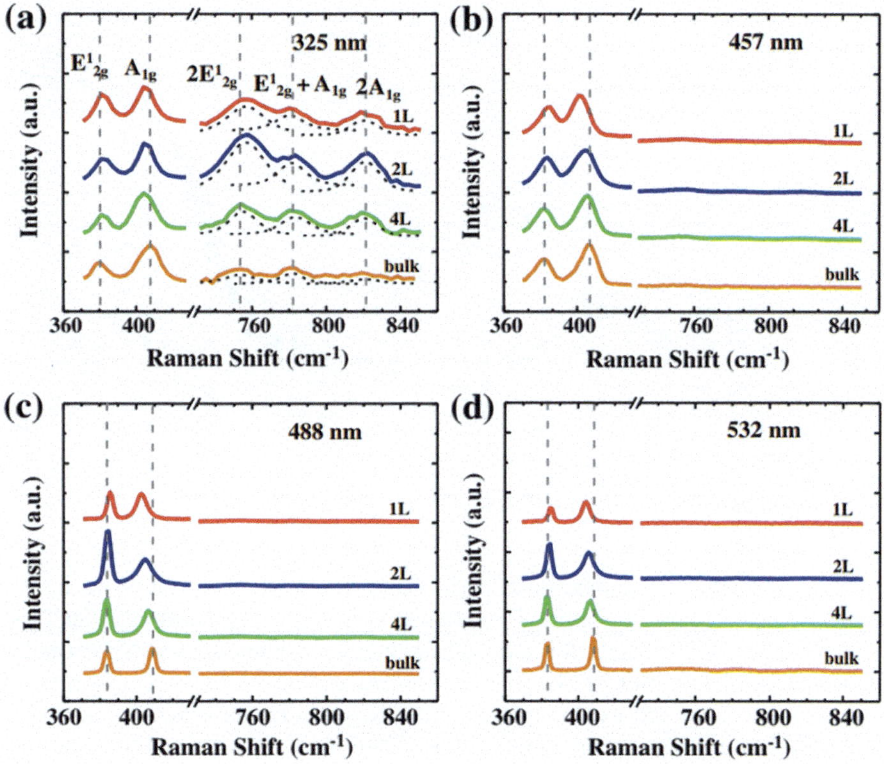

Fig. 8 The Raman spectroscopic images of MoS_2 samples with various thicknesses as excited by different laser lines at **a** 325, **b** 457, **c** 488, and **d** 532 nm [16]

net magnetic moment resulting from these zigzag edges is predicted to decrease with increasing grain size.

6 Synthesis and Application of Bulk MoS_2

In a report, Zhang et al. [18] synthesized bulk MoS_2 single crystals having a dimension of 3 mm × 5 mm employing novel Sn flux method with elemental molybdenum, sulfur, and tin as precursors. The starting materials were loaded into a quartz ampoule evacuated to 2.0×10^{-3} Pa and were subjected to slow heating of 1100–1150 °C for 25 h. Further the sample was cooled down to 1000 ºC at a rate of 0.5–6 °C/h and maintained for 20 h allowing the complete crystallization growth of bulk MoS_2. Decreasing the rate of cooling from 5–6 °C/h to 2–4 °C/h to 0.5–1 °C/h was witnessed with an increase in the MoS_2 crystal size, but were polycrystalline MoS_2. Experimental results identified a cooling rate of 2–4 °C/h as appropriate

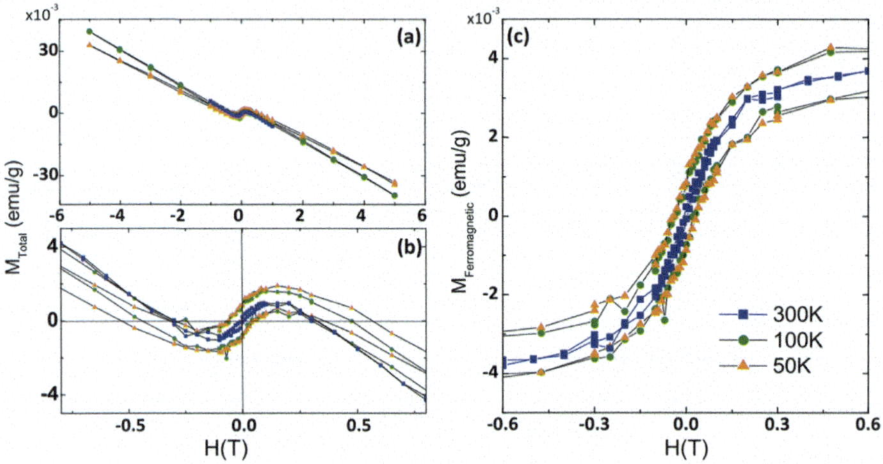

Fig. 9 Magnetization versus applied field (M-H curve) of bulk MoS_2 [17]

rate to form highly crystalline bulk MoS_2 single crystals. Structural characterizations have revealed the crystalline formation of bulk single-crystal MoS_2 with TEM measurements confirming the formation of single-crystal domain for each sheet and is shown in Fig. 10. Ultrathin MoS_2 nanosheets were exfoliated from the as-synthesized MoS_2 single crystal and were employed as MoS_2 saturable absorber in a passively Q-switched laser with a short pulse width of 326 ns at a wavelength of 1.06 μm consisting of Nd: YVO_4 crystal as gain medium.

Despite the fact that numerous methods have been employed to expose or increase the number of active edges, the low electronic conductivity of 2H-MoS_2 still prevents the HER activity from being further enhanced. Among the various strategies involved to enhance the electronic conductivity of MoS_2 is phase engineering of semiconducting 2H-MoS_2 to metallic 1T-MoS_2 having ~10^7-fold higher electronic conductivity than 2H-MoS_2 resulting in appreciable charge transfer efficiency and activation of basal planes. In a report, He et al. synthesized multiphasic 1T/2H-MoS_2 consisting of residual S, N co-doped carbon (SNC) with oxygen functional groups and amorphous characteristics through simple pyrolysis method. Both the percentages of the catalyst 1T-MoS_2 and the amount of SNC residues can be conveniently administered through pyrolysis temperatures and molar ratios of Mo/S. Mo precursor obtained through partial reduction of $(NH_4)_6Mo_7O_{24}.4H_2O$ using DL-tartaric acid and thiourea for different Mo/S feeding molar ratios as ~0.2, 0.26, 0.16 was subjected to facile pyrolysis method at various temperatures, 350, 450, 550, and 650 °C. XRD studies demonstrated the absence of diffraction peak at ~14° attributed to (0 0 2) plane in Mo/S (0.2) at 350 and 450 °C and Mo/S (0.26, 0.16) at 450 °C ascribed to the disordered stacking of S-Mo-S along c-axis. Similarly, the SEM measurement also revealed the "bulk" irregular particle morphology of multiphasic 1T/2H-MoS_2. Analysis of electrocatalytic HER activity of the as-synthesized Mo/S(0.2)-450 °C catalyst unveiled lowest overpotential of η_{10} = ~130 mV to attain a current density

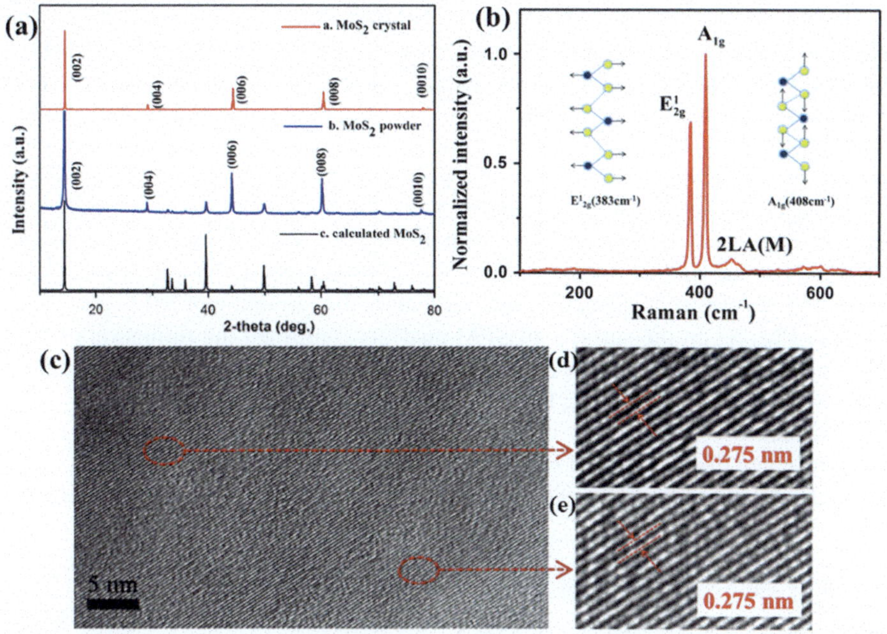

Fig. 10 **a** XRD pattern; **b** Raman spectroscopic images; and **c–e** high-resolution transmission electron microscopic images of bulk MoS_2 crystals synthesized by novel Sn flux method [18]

of 10 mA/cm^2 which is lower than some of the previously reported and synthesized MoS_2 HER catalysts such as Se-doped MoS_2 (140 mV) [19], pristine MoS_2 (727 mV) [20], monolayer MoS_2 with S vacancy (183, 178 mV) [20, 21], Na/iT-MoS_2 (183 mV) [22], heterostructural MoS_2/NiS nanoflowers (158 mV) [23], bare MoS_2 in alkaline (450 mV) and acidic (540 mV) medium, MoS_2/NiO (530 mV) [24], and MoS_2 decorated carbon nanofibers (207 mV) [25]. The Mo/S(0.2)-450 ºC sample also exhibited a Tafel slope of ~77 mV/dec and exchange current density j_0 of ~355.6 μA cm^{-2} characterizing suitable HER kinetics following the Volmer–Heyrovsky reaction mechanism. Of all the resultant catalysts, Mo/S (0.2)-450 °C also showcased highest electrochemical double-layer capacitance, C_{dl} (~3.03 mF), electrochemically active surface area (ECSA) of ~50.50 cm^2, turn over frequency (TOF) of 37.10×10^{-4} s^{-1} at $\eta_{10} = 0$ mV and 0.375 s^{-1} at $\eta_{10} = 200$ mV and appreciable charge transfer efficiency with low charge transfer resistance (R_{ct}) value of ~2.96 Ω confirming the intrinsic HER activity of the "bulk" Mo/S (0.2)-450 ºC catalyst which is also attributed to the presence of residual carbons with S, N dopants and oxygen functional groups. Chronopotentiometry measurement confirmed the durability of Mo/S (0.2)-450 ºC sample with an increase in the overpotential (η_{10}) by ~27 mV whereas an increase of ~36 mV was observed for 20 wt% Pt/C. The observed dominant catalytic activity was assigned to the presence of 1T-MoS2 with active basal and edge sites, disordered stacking of S-Mo-S layers promoting exposure of active sites, and presence of residual carbons with S, N dopants.

Zhang et al. [26] synthesized bulk porous NiMo@MoS_2 electrocatalyst employing microwave sintering process and then subjected to hydrothermal treatment in thiourea solution. Raw materials such as commercially available Ni and Mo powders with atomic ratio 6:4 and 10 wt% NH_4HCO_3 powders were subjected to ball milling to obtain fine powders which were further pressed into green compacts and then sintered at 900 °C in microwave sintering furnace to obtain porous NiMo alloys. After cutting down into NiMo sheets of 1 mm thickness, the acetone, de-ionized water cleaned samples were immersed in 15 mL 1.0 mol/L thiourea solution and kept for hydrothermal treatment at 180 °C for 24 h resulting in the formation of bulk NiMo@MoS_2 electrode. Structural studies feature the formation of NiMo alloy with content of MoS_2 on the surface. FESEM analysis confirmed the formation of bulk catalyst consisting of 3D interconnected pores with stone-like particles distributed on the surface as shown in Fig. 11.

HER measurement using a three-electrode system in alkaline medium (1.0 M KOH) revealed the exceptional HER activity of NiMo@MoS_2 (Fig. 12a and b) catalyst with low overpotential of 89, 153, and 176 mV at current densities of 10, 50,

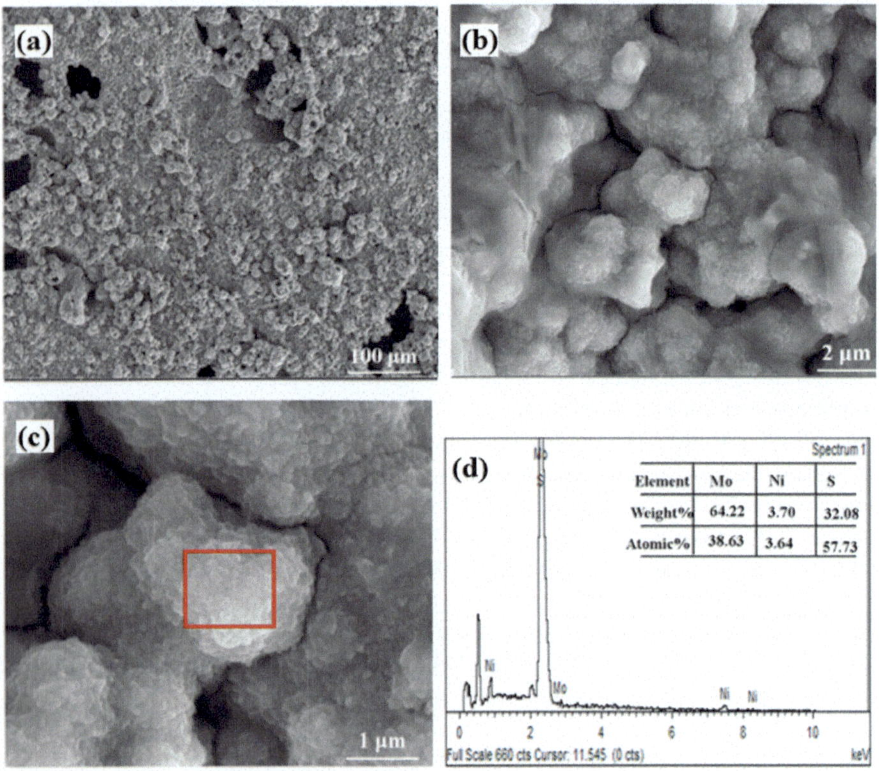

Fig. 11 **a–c** Field emission scanning electron microscopy (FESEM) images and **d** energy-dispersive spectroscopy (EDS) of NiMo@MoS_2 [26]

and 100 mA/cm^2 and low Tafel slope of 84 mV/dec showing faster HER kinetics with Volmer–Heyrovsky mechanism. NiMo alloy presented a low charge transfer resistance (Rct) value of ~5.87 Ω which is much less than that of NiMo@MoS_2 (Rct ~ 36.45 Ω) resulting in poor performance of NiMo@MoS_2 in low current density but exhibited dominant C_{dl} value of 251.8 mF/cm^2 compared to that of the NiMo alloy (12.1 mF/cm^2) as shown in Fig. 12c and d. Bulk porous NiMo@MoS_2 electrode manifested exceptional stability after CV 3000 cycles and long-term durability at constant current density of 10 mA/cm^2 for 25 h.

Aruna et al. [27] synthesized molybdenum disulfide-molybdenum trioxide-rhenium oxide (MoS_2-MoO_3-Re_2O_7) composite employing bulk MoS_2 via a simple thermal treatment method wherein a mixture of bulk molybdenum (IV) sulfide and 10 wt% rhenium (VII) oxide were calcined in air at 300 °C for 4 h. Calcination of the mixture at 300 °C resulted in the partial oxidation of MoS_2 into MoO_3 leading to the formation of MoS_2-MoO_3-Re_2O_7 composite. Bulk molybdenum (IV) sulfide was also subjected to calcination under 300 °C in the absence of rhenium (VII) oxide and was denoted as MoS_2 C. Structural studies (Fig. 13a and b) demonstrated the presence of 2H-MoS_2, orthorhombic Re_2O_7, and orthorhombic MoO_3 with respective phonon vibrations of bulk MoS_2 (375, 403 cm^{-1}), MoO_3 (138, 275, 810 cm^{-1}), and

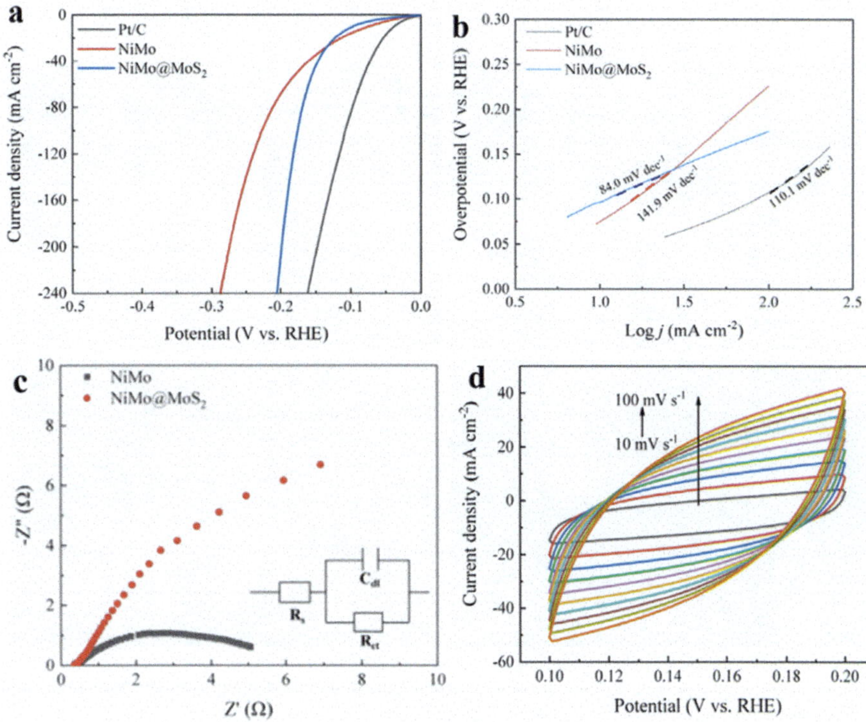

Fig. 12 **a** Linear sweep voltammetry (LSV) curve, **b** Tafel slope, **c** Nyquist plot of NiMo and NiMo@MoS_2, and **d** Cyclic voltammetry curves of NiMo@MoS_2 [26]

Re_2O_7 (984, 330 cm^{-1}). Morphological studies (Fig. 14) revealed the disordered, stacked arrangement featuring the formation of thick layer of bulk MoS_2.

HER electrocatalytic activity examined using linear sweep voltammetry showed an onset potential of −0.35 V for bulk MoS_2 and −0.18 V for MoS_2-MoO_3-Re_2O_7

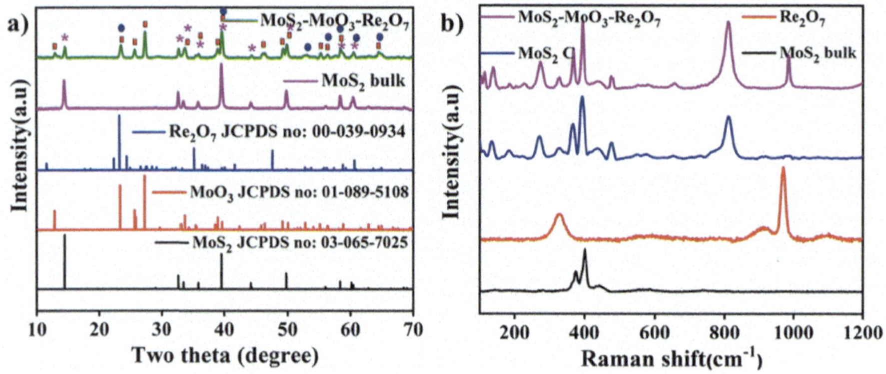

Fig. 13 XRD pattern and Raman spectroscopic images of bulk MoS_2 and bulk MoS_2-MoO_3-Re_2O_7 composite [27]

Fig. 14 Scanning electronic microscopy images of **a** bulk MoS_2, **b** MoS2 C, **c** MoS_2-MoO_3-Re_2O_7 composite, and **d** magnified image of MoS_2-MoO_3-Re_2O_7 composite [27]

composite with overpotential (η_{10}) ~ 0.60 V and ~0.28 V for bulk MoS_2 and MoS_2-MoO_3-Re_2O_7 composite, respectively. The remarkable onset potential and overpotential of MoS_2-MoO_3-Re_2O_7 composite suggested the formation of appropriate heterojunction interface that is advantageous for HER activation and activation of basal planes for hydrogen adsorption with incorporation of Re2O7. Results also validate substantial Tafel slope of 155 mV/dec for MoS_2-MoO_3-Re_2O_7 composite compared to bulk MoS_2(340 mV/dec) with Heyrovsky step as the rate-determining step with excellent stability even after 20,000 s. Photoelectrochemical measurement also revealed excellent hydrogen evolution reaction of MoS_2-MoO_3-Re_2O_7 composite with an increase in the current density of ~126.95 mA/cm^2 with charge transfer resistance of 25.05 Ω in the dark to ~143.30 mA/cm^2 with a charge transfer resistance of 8.20 Ω under light irradiation. It is expected that MoO_3 produced during calcination is anticipated to have a good impact on photoabsorption because of its plasmonic effect and expedites faster transfer of charge contrast to bulk MoS_2. Also, Mott–Schottky results proposed the formation of p–n junction in MoS_2-MoO_3-Re_2O_7 composite between n-MoS_2, n-MoO_3, and p-Re_2O_7 thereby promoting charge carrier separation and electron transfer at the interface.

Late et al. [28] in a report synthesized both single- and multi-layered MoS_2 sheets and fabricated transistor for sensing NO_2, NH_3 and humidity and identified stable current response and superior sensitivity for multi-layered MoS_2 device compared to single-layered attributed to the various electronic structures with change in layer numbers. The sensing response of MoS_2 device was further exploited by applying gate biasing and irradiation of light. Being an electron acceptor accelerating p-type doping in the system, exposure to NO_2 gas led to increase in the resistance, whereas a decrease in the resistance is observed on exposure to NH_3 gas which is an electron donor (n-type doping) (Fig. 15a and b). On application of positive back gate voltage to the MoS_2 transistor-based gas sensor device, sensing response of NO_2 was found to increase due to increased charge carriers (electron) at the MoS_2-SiO_2 interface, while that of NH_3 decreased (Fig. 15c and d). Compared to bilayer, five-layer MoS_2 was found to be more receptive toward sensing response on application of gate bias voltage, further verified by the high on/off ratio of five-layer MoS_2 being 5.5×10^5. Sensitivities around 1372 and 86% were observed for five-layer MoS_2 device on exposure toward 1000 ppm NO_2 and NH_3, respectively, under the influence of gate bias voltage.

Najmzadeh et al. [29] fabricated a 2D MOSFET using bulk MoS_2 by mechanical exfoliation of MoS_2 crystals on SiO_2 substrate and studied various device parameters such as carrier concentration, carrier mobility, flatband voltage, and dielectric constant employing C-V and I-V characteristics measurements. The double-gate bulk MoS_2 MOSFET reported a dielectric constant of 7.8, carrier concentration as 2.1×10^{17} cm^{-3}, electron mobility of 48 cm^2/Vs, and a flatband voltage of −0.45 V.

Interestingly, Krishna et al. [30] reported the phase transition of bulk 2H-MoS2 having hexagonal phase into 1T' phase with distorted octahedral phase through direct injection of high-density electrons via forming heterostructures with a 2D electride material, dicalcium nitride $[Ca_2N]^+.e^-$ crystals consisting of layers of freely mobile electrons. As observed in Figure 16a, the large variation of ~2e V in the work function

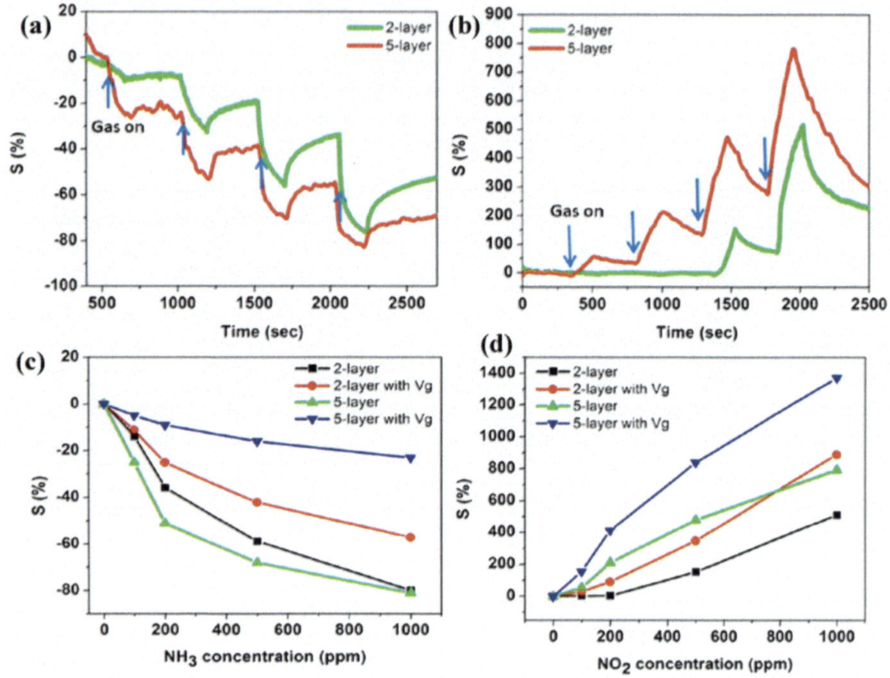

Fig. 15 Gas-sensing performances of bilayer and five-layer MoS_2-based device toward **a** NH_3 gas, **b** NO_2 gas at various gas concentration levels (100, 200, 500, 1000 ppm) and sensitivity-gas concentration dependence of MoS_2 transistor-based gas sensor toward **c** NH_3 and **d** NO_2 gas with and without applying gate bias voltage [28]

between MoS_2 and $[Ca_2N]^+.e^-$ resulted in the fast transfer of electrons in $[Ca_2N]^+.e^-$ into MoS_2 with a high kinetic energy. Figure 16b and c represents the schematic representation of charge transfer mechanism and the optical view of the heterostructure. The penetration of these electrons with high kinetic energy into bulk MoS_2 led to progressive emergence of lattice symmetry and band structure renormalization forming 1T' phase. Synthesis of 2D $[Ca_2N]^+.e^-$ crystals were carried out by pelletizing a mixture of calcium nitride (Ca_3N_2) powders and calcium chips in the molar ratio 1:1, further annealed at 800 °C for 48 h under vacuum (∼10 − 3 Pa). The as-prepared $[Ca_2N]^+.e^-$ crystals were ground into powder in a nitrogen atmosphere and further reannealed at 800 °C for another 48 h to enhance the homogeneity. To form the desired solid-state heterostructure, bulk $[Ca_2N]^+.e^-$ crystals were exfoliated on to a glass slide employing scotch tape method and bulk MoS_2 onto polydimethylsiloxane (PDMS) substrate in an argon gas environment to restrain the oxidation of $[Ca_2N]^+.e^-$. The exfoliated MoS_2 on PDMS substrate was further directly transferred onto the glass slide containing exfoliated $[Ca_2N]^+.e^-$ and was finally heated to 80 °C.

Raman measurement confirmed the significant softening of A_{1g} and E^1_{2g} modes with a maximum red shift of 10 and 8 cm^{-1}, respectively, due to injection of high

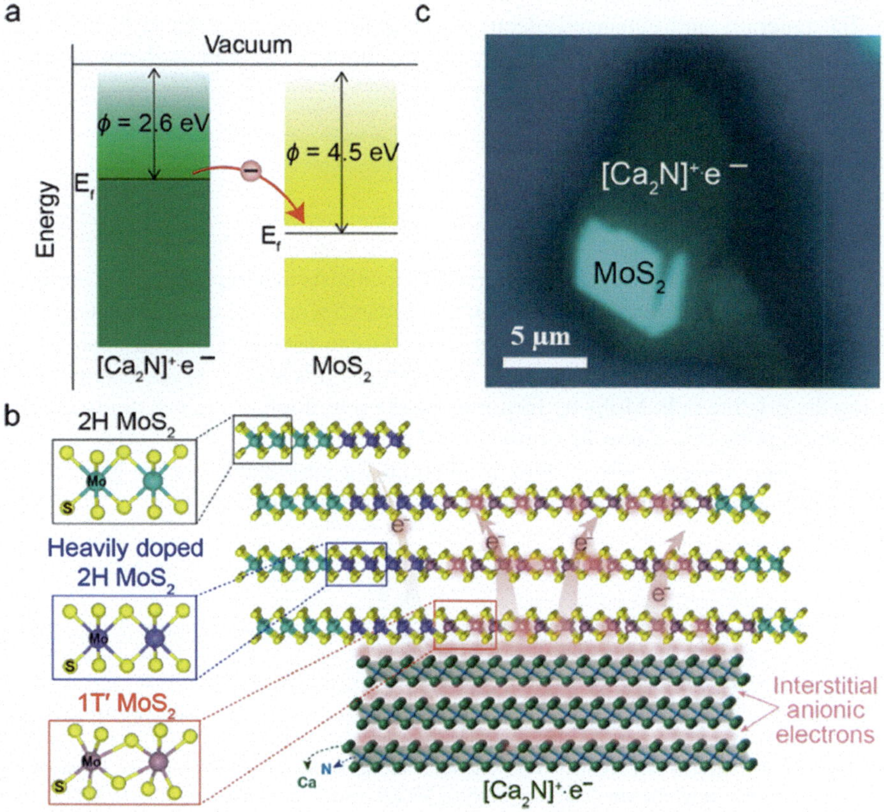

Fig. 16 **a** Band structure of $[Ca_2N]^+.e^-$ and MoS_2 representing the charge injection into bulk MoS_2, **b** schematic representation of electron injection from $[Ca_2N]^+.e^-$ to bulk MoS_2, and **c** Optical view of the solid-state heterostructure [30]

density of electrons from $[Ca_2N]^+.e^-$ to MoS_2 apart from the origin of Raman modes corresponding to 1T' phase. The electron transfer from $[Ca_2N]^+.e^-$ to MoS_2 is observed to be responsible for the red shift in A_{1g} mode and that of E^1_{2g} mode is attributed to presence of strain and electron injection-induced structural changes. Similarly, photoluminescence measurement also confirmed momentous shift (~200 meV lower than original peak position) in the PL peaks with much higher PL intensity in comparison with monolayer MoS_2, especially for heterostructures having substantial phonon softening of A_{1g} and E^1_{2g} Raman modes corroborating the faster electron injection ($\sim 10^{14}$ cm^{-2}) through heterostructure formation resulting in substantiate renormalization in the band gap and phonon modes of bulk MoS_2. The formation of twisted multilayer MoS_2 as a consequence of the decoupling of interlayer interactions may be responsible for the significantly increased PL emission of bulk MoS_2 after heterostructuring with $[Ca_2N]^+.e^-$. The multiphase formation of 2D materials and enhanced electrical, optical properties of these heterostructures can be exploited further for diverse applications in the field of electronics and optoelectronics.

7 Conclusion

In summary, we have presented the properties, synthesis methods, and a few applications of bulk MoS_2. Even though most of the research works are more emphasized on mono- and few-layered MoS_2, from this article we can see that bulk MoS_2 also favors certain properties necessary for several applications. It is also observed that manipulating the pristine bulk MoS_2 by forming composites, doping, etc. can dominate the properties, point into a new direction of band engineering etc., which will contribute to developing a highly active, durable material for sustainable energy generation, sensor fabrication, electronics, and other optoelectronics applications.

References

1. Sun J, Li X, Guo W, Zhao M, Fan X, Dong Y, Xu C, Deng J, Fu Y (2017) Synthesis methods of two-dimensional MoS_2: a brief review. Crystals 7:198 (1–11). https://doi.org/10.3390/cryst7070198
2. Stewart JA, Spearot DE (2013) Atomistic simulations of nanoindentation on the basal plane of crystalline molybdenum disulfide (MoS_2), Model Simul Mater Sci Eng 21:045003 (1–14). https://doi.org/10.1088/0965-0393/21/4/045003
3. Song I, Park C, Choi HC (2015) Synthesis and properties of molybdenum disulphide: from bulk to atomic layers. RSC Adv 5:7495–7514. https://doi.org/10.1039/c4ra11852a
4. Huy VPH, Ahn YN, Hur J (2021) Recent advances in transition metal dichalcogenide cathode materials for aqueous rechargeable multivalent metal-ion batteries, Nanomaterials 11:1517 (1–30)
5. Coogan A, Gun'ko YK (2021) Solution-based "“bottom-up”" synthesis of group VI transition metal dichalcogenides and their applications. Mater Adv 2:146–164. https://doi.org/10.1039/d0ma00697a
6. Xu W, Yan S, Qiao W (2018) Magnetism in monolayer 1T-MoS_2 and 1T-MoS_2H tuned by strain. RSC Adv 8:8435–8441. https://doi.org/10.1039/c7ra10304b
7. Enyashin AN, Yadgarov L, Houben L, Popov I, Weidenbach M, Tenne R, Bar-sadan M, Seifert G (2011) New route for stabilization of 1T-WS_2 and MoS_2 phases. J Phys Chem C 115:24586–24591. https://doi.org/10.1021/jp2076325
8. Ahmad S, Mukherjee S (2014) A comparative study of electronic properties of bulk MoS_2 and its monolayer using DFT technique: application of mechanical strain on MoS_2 monolayer. Graphene 3:52–59
9. Splendiani A, Sun L, Zhang Y, Li T, Kim J, Chim CY, Galli G, Wang F (2010) Emerging photoluminescence in monolayer MoS_2. Nano Lett 10:1271–1275. https://doi.org/10.1021/nl903868w
10. Mak KF, Lee C, Hone J, Shan J, Heinz TF (2010) Atomically thin MoS_2: a new direct-gap semiconductor. Phys Rev Lett 105:136805 (1–4). https://doi.org/10.1103/PhysRevLett.105.136805

11. Golasa K, Grzeszczyk M, Korona KP, Bozek R, Binder J, Szczytko J, Wysmolek A, Babinski A (2013) Optical properties of molybdenum disulfide (MoS_2). Acta Phys Pol A 124:849–851. https://doi.org/10.12693/APhysPolA.124.849
12. Li B, Jiang L, Li X, Ran P, Zuo P, Wang A, Qu L, Zhao Y, Cheng Z, Lu Y (2017) Preparation of monolayer MoS2 quantum dots using temporally shaped femtosecond laser ablation of bulk MoS_2 targets in water. Sci Rep 7:11182 (1–12). https://doi.org/10.1038/s41598-017-10632-3
13. Chakraborty B, Matte HSSR, Sood AK, Rao CNR (2012) Layer-dependent resonant Raman scattering of a few layer MoS_2. J Raman Spectrosc 44:92–96. https://doi.org/10.1002/jrs.4147
14. Lee C, Yan H, Brus LE, Heinz TF, Hone ḰJ, Ryu S (2010) Anomalous lattice vibrations of single- and few-layer MoS_2. ACS Nano 4:2695–2700
15. Li H, Zhang Q, Yap CCR, Tay BK, Edwin THT, Olivier A, Baillargeat D (2012) From bulk to monolayer MoS_2: evolution of Raman scattering. Adv Funct Mater 22:1385–1390. https://doi.org/10.1002/adfm.201102111
16. Sun L, Yan J, Zhan D, Liu L, Hu H, Li H, Tay BK, Kuo J, Huang C, Hewak DW, Lee PS, Shen ZX (2013) Spin-orbit splitting in single-layer MoS_2 Revealed by triply resonant Raman scattering. Phys Rev Lett 111:126801 (1–5). https://doi.org/10.1103/PhysRevLett.111.126801
17. Tongay S, Varnoosfaderani SS, Appleton BR, Wu J, Hebard AF (2012) Magnetic properties of MoS_2: existence of ferromagnetism. Appl Phys Lett 101:123105 (1–4). https://doi.org/10.1063/1.4753797
18. Zhang X, Lou F, Li C, Zhang X, Jia N, Yu T, He J, Zhang B, Xia H, Wang S, Tao X (2015) Flux method growth of bulk MoS_2 single crystals and application as a saturable absorber. CrystEngComm 17:4026–4032. https://doi.org/10.1039/C5CE00484E
19. Ren X, Ma Q, Fan H, Pang L, Zhang Y, Yao Y, Ren X, (Frank)u Li S (2015) A Se-doped MoS_2 nanosheet for improved hydrogen evolution reaction. Chem Commun 51:15997–16000. https://doi.org/10.1039/C5CC06847A
20. Cheng C, Lu A, Tseng C, Yang X, Hedhili MN, Chen M-C, Wei K, Li L (2016) Activating basal-plane catalytic activity of two dimensional MoS_2 monolayer with remote hydrogen plasma. Nano Energy 30:846–852. https://doi.org/10.1016/j.nanoen.2016.09.010
21. Meng C, Lin M, Du X, Zhou Y (2019) Molybdenum disulfide modified by laser irradiation for catalyzing hydrogen evolution. ACS Sustain Chem Eng 7:6999–7003. https://doi.org/10.1021/acssuschemeng.8b06717
22. Attanayake NH, Thenuwara AC, Patra A, Aulin YV, Tran T, Chakraborty H, Borguet E, Klein ML, Perdew JP, Strongin DR (2018) Effect of intercalated metals on the electrocatalytic activity of 1T-MoS for the hydrogen evolution reaction. ACS Energy Lett 3:7–13. https://doi.org/10.1021/acsenergylett.7b00865
23. Zhao X, Bao J, Zhou Y, Zhang Y, Sheng X, Wu B, Wang Y, Zuo C, Bu X (2022) Heterostructural MoS_2/NiS nanoflowers via precise interface modification for enhancing electrocatalytic hydrogen evolution. New J Chem 46:5505–5514
24. Lai B, Singh SC, Bindra JK, Saraj CS, Shukla A, Yadav TP, Wu W, Mcgill SA, Dalal NS, Srivastava A, Guo C (2019) Hydrogen evolution reaction from bare and surface-functionalized few-layered MoS_2 nanosheets in acidic and alkaline electrolytes. Mater Today Chem 14:100207 (1–14). https://doi.org/10.1016/j.mtchem.2019.100207
25. Guo X, Cao G, Ding F, Li X, Zhen S, Xue Y, Yan Y, Liu T, Sun K (2015) A bulky and flexible electrocatalyst for efficient hydrogen evolution based on the growth of MoS_2 nanoparticles on carbon nanofibers foam. J Mater Chem A 3:5041–5046. https://doi.org/10.1039/b000000x
26. Zhang LW, Xu JL, Tang J, Li LL, Luo JM (2022) Synthesis self-supporting bulk porous NiMo@MoS_2 electrocatalyst to enhance hydrogen evolution in alkaline conditions. J Mater Res Technol 17:1167–1175. https://doi.org/10.1016/j.jmrt.2022.01.068
27. Vijayan A, Sandhyarani N (2022) Enhancing the catalytic activity of bulk MoS_2 towards hydrogen evolution reaction by the formation of MoS_2-MoO_3-Re_2O_7 heterostructure. J Colloid Interface Sci 623:819–831. https://doi.org/10.1016/j.jcis.2022.05.107
28. Late DJ, Huang Y-K, Liu B, Acharya J, Shirodkar SN, Luo J, Yan A, Charles D, Waghmare UV, Dravid VP, Rao CNR (2013) Sensing behavior of atomically thin-layered MoS_2 transistors. ACS Nano 7:4879–4891. https://doi.org/10.1021/nn400026u

29. Najmzadeh M, Duarte JP, Khandelwal S, Zeng Y, Hu C (2015) 2D MOSFET operation of a fully-depleted bulk MoS2 at quasi-flatband back-gate. In: 2015 73rd annual device research conference, pp 135–136. https://doi.org/10.1109/DRC.2015.7175592
30. Dhakal KP, Ghimire G, Chung K, Duong DL, Kim SW, Kim J (2019) Probing multiphased transition in bulk MoS_2 by direct electron injection. ACS Nano 13:14437–14446. https://doi.org/10.1021/acsnano.9b08037

Chapter 4
Synthesis and Properties of 0D, 1D, 2D MoS_2

Raphael Longuinhos, Jenaina Ribeiro-Soares, and Dattatray J. Late

1 Introduction

The synthesis methodologies for MoS_2 in different dimensionalities are based on two approaches: the top-down and the bottom-up (see Fig. 1) [25]. The top-down methodologies are based on the progressive reduction of bulk MoS_2 to nanostructures with controlled size or shape, being successful to obtain a broad range of different materials like nanodots, nanoribbons, and thin films. The experimental realization of graphene in 2004 occurred by using the mechanical exfoliation (see Fig. 1(1a)), or the Scotch Tape method [19]. Although important when high-quality single crystals are needed, it is not efficient to produce materials with controlled dimensions, and also in the large scale needed to industrial application perspectives. Laser ablation is another possibility, making it possible to progressively reduce the material to the thin-film limit (Fig. 1(1b)). The more commonly used methodology in this sense is the sonication with liquid exfoliation, which can be combined with ion-intercalation (illustrated in Fig. 1(1c)), for example. Also relevant are the electrochemical exfoliation and thermal ablation.

Bottom-up methods involve constructing materials and desired nanostructures by building from nanoparticles or molecules, in contrast to the previously mentioned top-down approach. In this case, the control over composition and properties is greater, making it possible to achieve different dimensionalities (nanodots, thin films,

R. Longuinhos (✉) · J. Ribeiro-Soares · D. J. Late
Department of Physics, Institute of Natural Science, Federal University of Lavras, Lavras, Minas Gerais 37200-900, Brazil
e-mail: raphael.lobato@ufla.br

J. Ribeiro-Soares
e-mail: jenaina.soares@ufla.br

D. J. Late
e-mail: datta099@gmail.com

D. J. Late and C. S. Rout (eds.), *Science and Technology of 2D MoS_2*,
Materials Horizons: From Nature to Nanomaterials,
https://doi.org/10.1007/978-981-97-7367-1_4

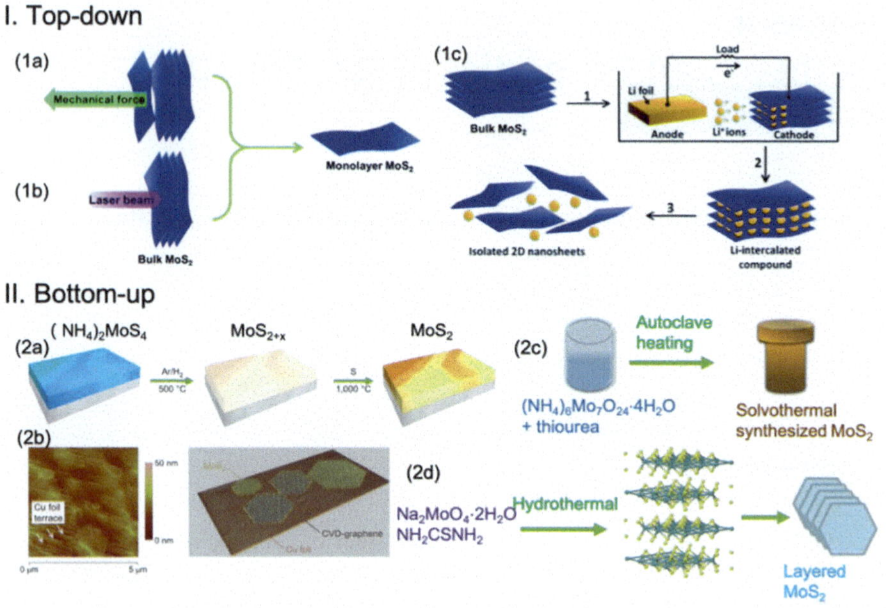

Fig. 1 Fabrication of MoS_2 methods, figure extracted from Ref. [30]. (I) Top-down approaches to prepare the MoS_2 nanosheets via (1a) mechanical exfoliation, (1b) laser exfoliation, and (1c) electrochemical intercalation exfoliation methods. (II) Bottom-up preparation methods including the schematic of (2a) the MoS_2 layer deposited by two-step thermolysis, (2b) MoS_2 nanosheets on CVD graphene/Cu foil, and (2c) and (2d) solvothermal and hydrothermal fabrication of MoS_2 layered structures. Reproduced with permission from [3, 15, 26, 32, 33]

nanowires). In this sense, the thermolysis is used to cover substrates with molecular components coming from the precursor in a heated ambient (Fig. 1(2a)). The principal techniques are Chemical Vapor Deposition (CVD) (Fig. 1(2b)), Physical Vapor Deposition (PVD), and solution-based processes such as hydrothermal and solvothermal synthesis (Fig. 1(2c, d)). These examples will be discussed in the next sections.

2 Synthesis

Top-down synthesis methodology (mechanical exfoliation, intercalation, liquid exfoliation, among others).

Sonication-assisted exfoliation and intercalation (0D, 1D, 2D): In the sonication methodology, the powder of the bulk sample is mixed in a solvent, and intense ultrasonic waves create cavities with high temperature and pressure. When these cavities increase and collapse, the energy is sufficient to break bonds in the layered materials. These bonds are especially those of the weak van der Waals interactions among the layers, exfoliating few or even atomically thin layers, nanodots, etc. The material is then centrifuged to improve the bulk and few layer separation, separating the supernatant for the intended use.

As commented previously in Chap. 2 in the historical overview, MoS_2 nanodots were obtained in 1989 by sonicating bulk MoS_2 powder in water [9], and since then, the methodologies have evolved to control the resulting sample dimensions and properties and testing different solvents. For example, sonication in N,N-dimethylformamide (DMF) for 3 h was used in association with solvothermal treatments with vigorous stirring for 6 h at 140 °C [29], followed by a centrifugation at 2000 rpm for 5 min, separating the supernatant. The process is illustrated in Fig. 2. This material was then characterized according to their UV–Vis spectra and excitation and emission photoluminescence spectra. Extensive work is being continuously performed also evaluating different solvents in this technique as, for example, ethylene glycol (particle size: 4–70 nm, 1 nm thick, emission at 440 nm with 350 nm excitation) [23], sulfuric acid (particle size: 3.5 nm, 1–1.5 nm thick, emission at 425 nm with 300 nm excitation) [5], and isopropyl alcohol (particle size: 1.41 nm, 1.5 nm thick, emission at 440 nm with 360 nm excitation) [18].

One strategy to improve the yield of MoS_2 liquid exfoliation is to preform it with the use of intercalation compounds, adding a solution chemistry step in this formerly only physical sonication process. Lithium (Li) is the most common intercalation compound used to exfoliate MoS_2, entering and enlarging the interlayer spacing, making it easier to separate the layers by sonication. The Li intercalation-assisted sonication of MoS_2 to produce nanosheets was performed with butyllithium in a fast rout, resulting in 2H layers within 1.5 h [6]. Representative Transmission Electron Microscopy (TEM), Scanning Electron Microscopy (SEM), and Atomic Force Microscopy (AFM) of the obtained samples are shown in Fig. 3. In the exfoliation process, the layers lost their semiconductor character (the 2H polytype), converting to the 1T polytype, but the authors were able to recover it by IR laser irradiation.

The intercalated-assisted MoS_2 exfoliation can result also in different materials' dimensionality, going from the 2D nanosheets, producing 2D nanoribbons, porous

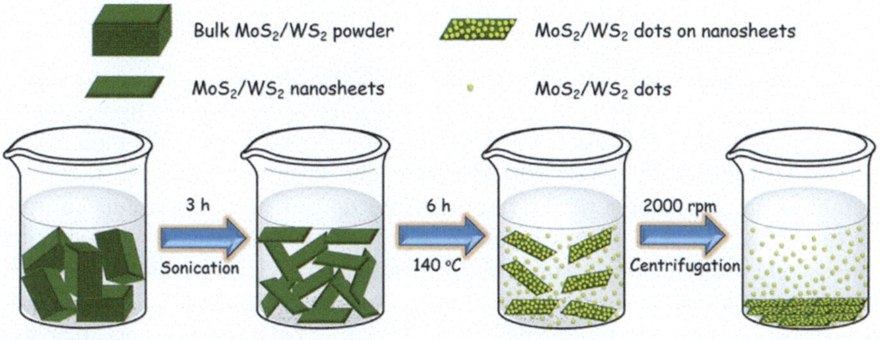

Fig. 2 Sonication process to prepare MoS_2/WS_2 quantum dots, with a solvothermal intermediate treatment. Reprinted from [29]

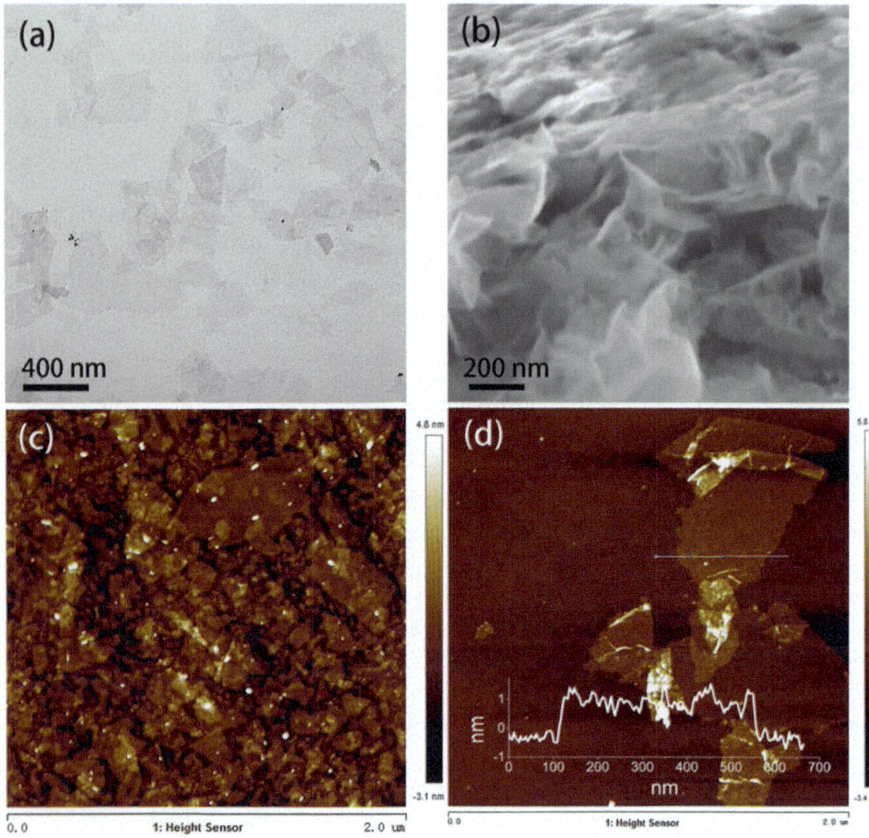

Fig. 3 TEM, SEM, and AFM images obtained from MoS_2 nanosheets synthetized by Li intercalation-assisted sonication of MoS_2 bulk precursors. Reprinted with permission from [6]

nanoribbons, until reaching the MoS_2 nanodot limit. This process is illustrated in Fig. 4, in which the authors used the H_2SO_4 intercalation and sonication to produce the nanodots in the final step [5].

Thermal ablation (0D, 1D, 2D): This methodology, which is based on the vaporization and removal of a material from a surface by diverse means (laser light, temperature, among others), was used to produce MoS_2 nanodots from MoS_2 monolayers previously produced by CVD [20]. The triangular layers were heated to 500 °C in ambient air, and after several days, the MoS_2 domain sizes were reduced from μm to nm [20]. Li et al. were capable of obtaining monolayer MoS_2 nanodots by femtosecond laser ablation of bulk MoS_2 in water, in a clean production process, obtaining nanosheets according to the synthesis parameters used [13]. Figure 5 illustrates the TEM images of the obtained nanodots at different magnifications, their size distribution (inset in b), high-resolution data (d), AFM (e), and topography profile (f).

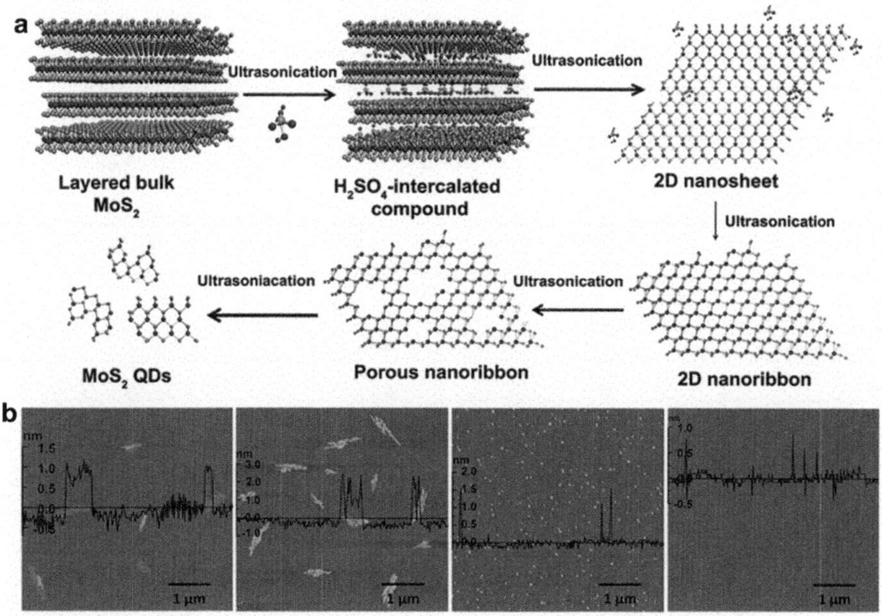

Fig. 4 **a** Representation of the intermediate synthesis stages of the MoS_2 nanodots produced in Ref. [5], exfoliating the precursor to the 2D layer level, reaching the nanoribbon level and, with ultrasonication, producing porous nanoribbons and the aimed nanodots. **b** AFM images and topography profiles of the samples when intercalated, until the nanodot limit. Reprinted with permission from [5]

Electrochemical exfoliation (0D, 1D, 2D): The electrochemical process to obtain MoS_2 in the nanoscale is based on the suspension of the bulk counterpart in an electrolyte, being the system subjected to a potential difference. The electrical current exfoliates the precursor, reaching the few or even monolayer level, being then stabilized by the electrolyte. Figure 6 shows a schematic for the electrochemical process used in Ref. [8] when producing excitation-dependent luminescent few- to single-layer MoS_2 nanodots, with a controlled size distribution ranging from 2.5 to 6 nm. The process consisted of imposing a DC potential of 5 V on MoS_2 bulk pellet samples, in a two-electrode cell, with concentrations of 0.1, 1, and 5 wt% of aqueous LiTFSI or [BMIm]Cl. Figure 6(a, b) displays the TEM results of LiTFSI at 0.1 and 1wt%, with (e, f) showing the HRTEM images with nanodot sizes of 2.6 and 4.6 nm. Figure 6c, d shows the same as the previously described, but for the [BMIm]Cl electrolyte, exhibiting larger particles of 2.8 and 5.8 nm.

Bottom-up synthesis methodology (CVD, PVD, solution chemical processes, etc.)

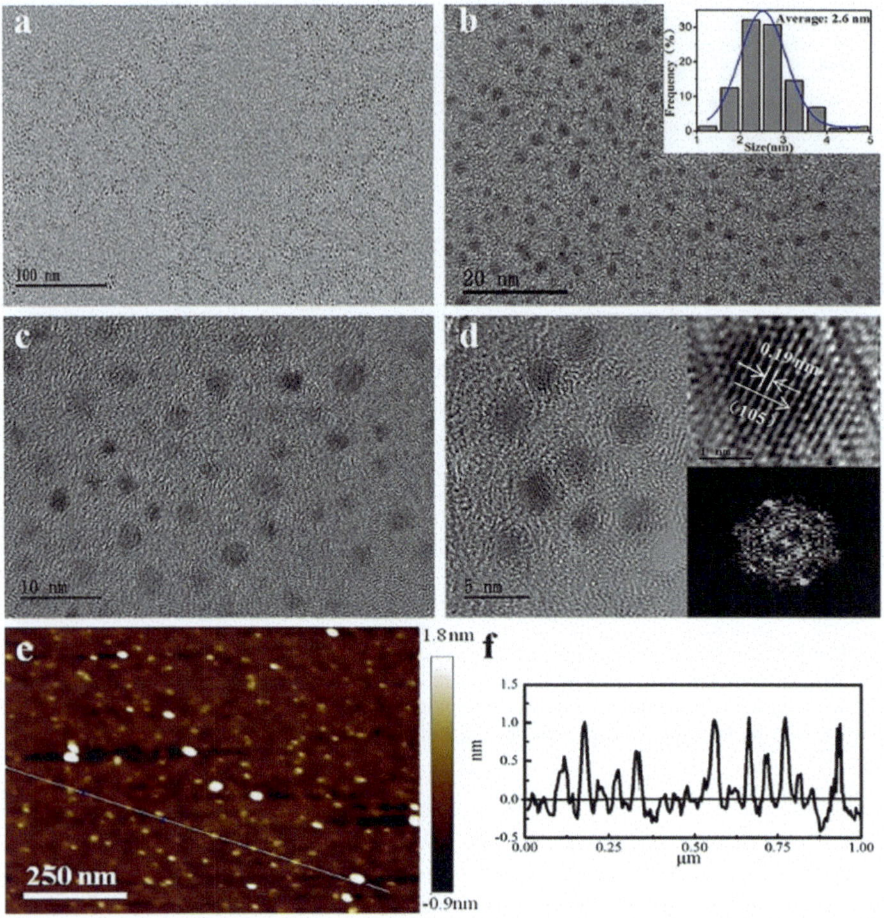

Fig. 5 Representation of the intermediate stages of the synthesis of MoS_2 nanodots produced in Ref. [5], exfoliating the precursor to the 2D layer level, reaching the nanoribbon level and, with ultrasonication, producing porous nanoribbons and the aimed nanodots. Reprinted with permission from [5]

CVD (0D, 1D, 2D): The chemical vapor deposition (CVD) is used to produce crystals, polycrystalline, or amorphous samples from gaseous precursors. In a controlled-environment chamber, gases containing units of the chosen precursors are introduced and heated to usually high temperatures. At these temperatures, the precursors decompose in different phases, reacting and forming solid samples atop of the chosen substrate. The CVD technique also has limitations due to the high temperatures needed in this processing, and also the specialized machinery, making it more expensive. For MoS_2 synthesis, CVD is used when large-area, high-quality MoS_2 thin films are needed, with precise thickness and composition, although defects like domain walls appear in continuous films.

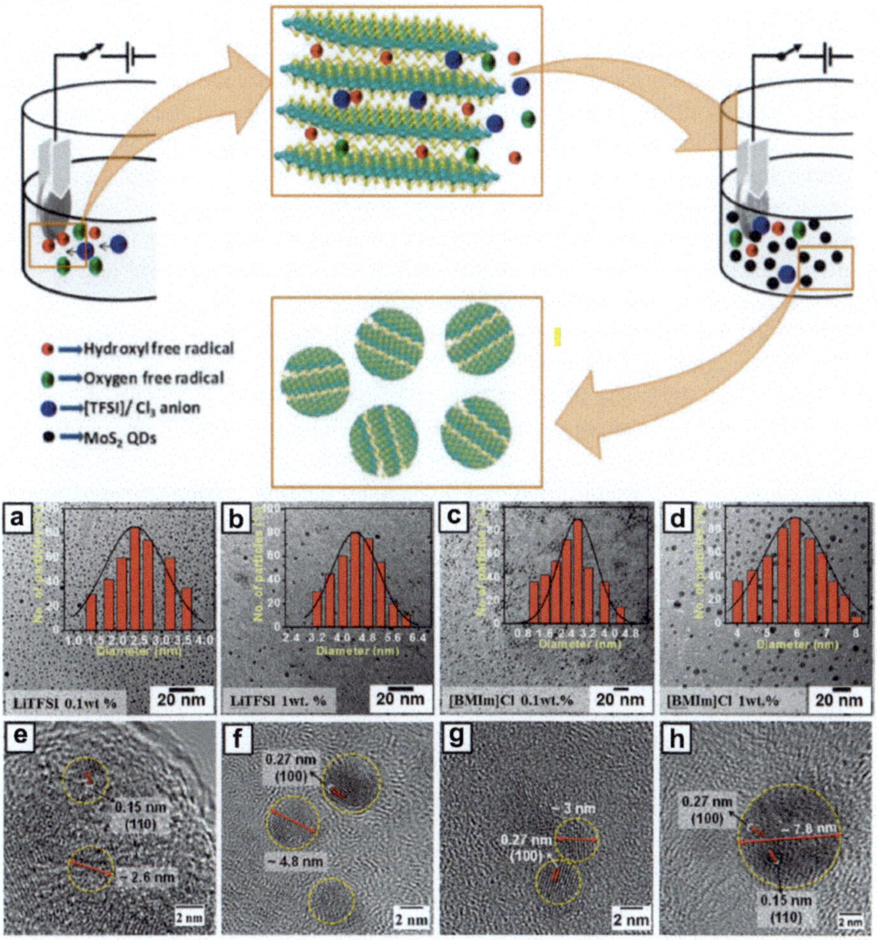

Fig. 6 Schematic of the electrochemical exfoliation process of MoS_2 nanodots. (**a**, **b**) The results for LiTFSI at 0.1 and 1 wt% and their respective HRTEM images (**e**, **f**), with nanodot sizes of 2.6 and 4.6 nm. (**c**, **d**) is the same as the previously explained, but to the [BMIm]Cl electrolyte, showing slighter bigger nanodot sizes (2.8 and 5.8 nm). Reprinted and adapted with permission from [8, 11]

The CVD thin-layer MoS_2 production process is illustrated in Fig. 7. In this case, MoO_3 and S are used as transition metal and chalcogenide sources, respectively, in a process of sulfurization of MoO_3 and MoO_{3-x} vapor phases [4, 12]. The films were synthesized atop of SiO_2/Si substrates in a hot-wall furnace. Graphene oxide (GO) and perylene-3,4,9,10-tetracarboxylic acid tetrapotassium salt (PTAS) were used as growth promoters, with GO produced by a modified Hummers' method [24]. They were spun on the substrate. The MoO_3 and S powder (99 and 99.5% from Aldrich and Alfa, respectively) were placed inside crucibles with the substrate faced down in the MoO_3 region. An annealing at a heating rate of 15 °C/min, 650 °C for 15 min, was sufficient to grow the samples at a N_2 flow of 1 sccm. The experimental setup is illustrated in Fig. 7a. Figure (b) displays the optical images obtained from the grown films. Figure 7c–e represent the Raman and photoluminescence intensity mapping images, and the optical image, respectively, for a region with different number of layers. Figure 7f and g shows the AFM image and respective topography profile of the traced line, showing a monolayer region. Figure 7h and (i) displays the Raman and photoluminescence spectra for one-, two-, and three-layer regions. These results present the success of the CVD technique in creating monolayers or few layers in the micrometric range, with high quality and displaying the structural and optical changes expected by the dimensionality reduction.

Due to its intrinsic nucleation and fast growth nature, it is not considerably successful in producing nanodots.

PVD (0D, 1D, 2D): The Physical Vapor Deposition (PVD) is a methodology to obtain thin films by physically evaporating a material (solid, liquid, or gas) from a source and then condensing it onto different substrates, usually in a vacuum chamber and low pressure. This evaporation can be done by heating at high temperatures to generate the vapor with the material. Other methodologies include sputtering and electron beam evaporation, being the choice of technique related to the specificities needed by the application of the material to be obtained. One disadvantage is the cost-effective production for large-area films, not being suitable for high-volume production.

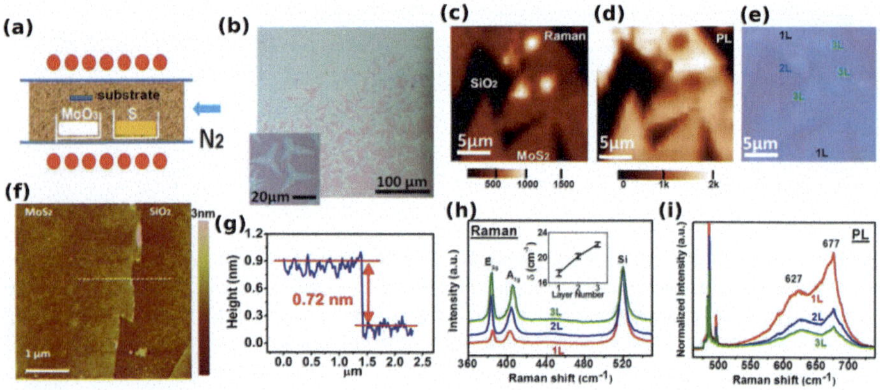

Fig. 7 Production and characterization of few and monolayer MoS_2 films, in the micrometer scale, by the CVD technique. **a** Schematic with the MoO_3 and S powders positioned inside a hot-wall furnace, with the substrate faced down atop of the MoO_3 recipient. A N_2 flow goes through the chamber during the growth process. **b** An optical image of the produced flakes. **c–e** Raman and photoluminescence intensity images, and the optical image for a region with varied number of layers. **f** and **g** The AFM and topography extracted from a layer, showing the monolayer character. **h** and **i** Raman and photoluminescence spectra of one-, two-, and three-layer regions, evidencing the changes according to the layer number. The colored images were adapted from the arXiv version of this paper (1202.5458, https://doi.org/10.48550/arXiv.1202.5458), while the permission for the black and white figures were obtained from [12]

The MoS_2 PVD deposition of nanoclusters was performed by [10] following two stages. Firstly, molybdenum particles were deposited atop of Au(111) substrates, with this substrate specifically chosen due to its noble and chemically inert character, and also by its herringbone reconstruction [14], providing dispersed nucleation sites for metal islands. Figure 8a shows the resulting regular array of one- or two-layer high Mo islands, deposited by e-beam evaporation at 1 × 10–5 mbar H_2S atmosphere at 400 K. The 30 Å Mo clusters were laterally spaced by 73 and 140 Å, reflecting the parameters of the underlaying substrate. The second synthesis stage is the sulfidation of these metal islands, which was performed by annealing them at 673 K for 15 min, maintaining the H_2S. Crystalline one-layer thick triangular MoS_2 nanoclusters can be seen in Fig. 8b, and their orientation was dictated by the hexagonal substrate symmetry. Figure 8c shows the resolved STM image of a triangular cluster of the 2H polytype MoS_2, while Fig. 8d displays the edges of these triangles where sulfur atoms were removed by atomic hydrogen produced by H_2 dissociation, which is especially important to form reactive Mo sites for hydrodesulfurization catalysis.

Recent advances have led to the improvement of PVD strategies to achieve several square centimeter extensions in MoS_2 films, at low temperatures, making realistic the perspective to scale this production for industrial and real-life applications. Muratore et al. produced the continuous high-quality MoS_2 ultra-thin films (3–6 layers) illustrated in Fig. 8e, on top of SiO_2 and highly oriented pyrolytic graphite (HOPG) [17]. These films were produced at 350 °C, in the centimeter scale length, with a magnetron sputtering scalable to the m^2 range, with solid MoS_2 3.3 cm diameter target with 99.95% purity and a 15 mTorr ultra-high purity argon gas at a 25 sccm constant flow rate. Figure 8f shows the Raman spectra for a different positions and number of layers, while Fig. 8g shows the position changes of the main representative Raman out-of-plane A_{1g} (~408 cm^{-1} in bulk) and in-plane E_{2g} (~382 cm^{-1} in bulk) peaks according to the number of layers.

Hydrothermal synthesis (0D, 1D, 2D): The hydrothermal synthesis is based on the use of high temperatures and pressure in a mixture of reactants in an aqueous solution. A sealed container is used to allocate the reactants in the aqueous solvent, and then it is subjected to high temperatures and pressures to stimulate the reaction to originate the expected materials. The result can be solid products presenting specific morphologies and structures difficult to obtain by other methods. The production of materials with small particle sizes and high surface areas, which is challenging with other methodologies, is efficient in hydrothermal synthesis. Its low-cost character, structural and morphology specificity, and scalability to large-scale processes are attractive, but the downside is that it is needed to deal with high-pressure equipment, the necessity to control in a fine level the reaction conditions, and also the necessity of post-synthesis treatments to eliminate side products and impurities to improve the material's quality.

For MoS_2 production with the hydrothermal synthesis, Mo and S sources are exposed to the high temperatures and pressures typical of this technique, resulting in nanowires, nanoparticles, and other structures according to the chosen synthesis conditions. The synthesis of MoS_2 water-soluble quantum dots (QDs) was performed by Wang et al. by using sodium molybdate ($Na_2MoO_4.2H_2O$) and glutathione (GSH)

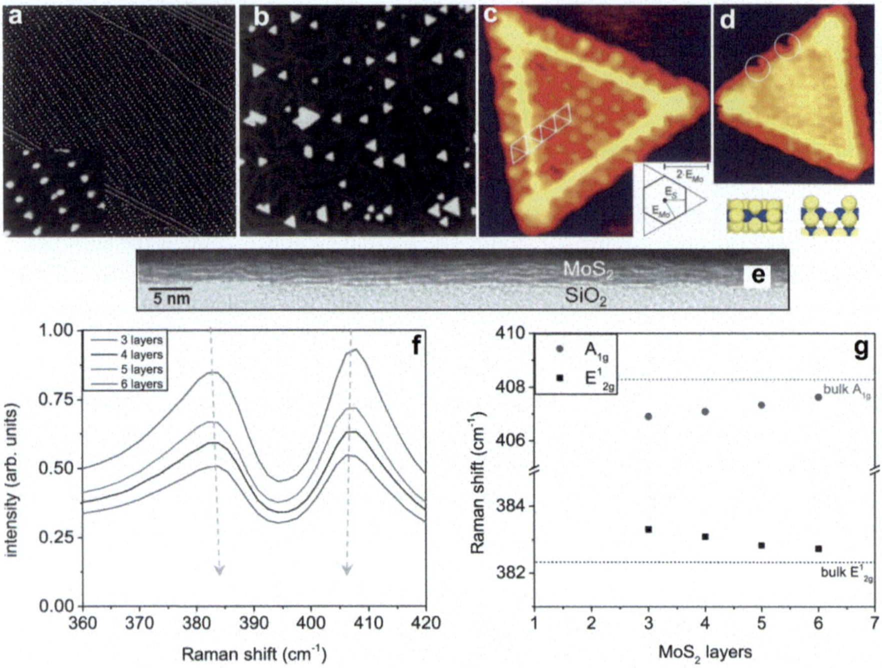

Fig. 8 Synthesis of MoS_2 nanoclusters and few layer MoS_2 films by different PVD techniques. **a** STM image showing regularly spaced Mo islands deposited by e-beam atop of Au(111) substrates, at 307 K and under UHV conditions. **b** STM image of the MoS_2 triangular islands growth by the sulfidation of the Mo islands of **a**. Their array is ordered due to the hexagonal character of the Au substrate. **c** Atom-resolved STM image of the monolayer MoS_2 nanoclusters from **b**. **d** Atom-resolved STM image of an MoS_2 nanocluster with vacant S atoms, intentionally induced by exposing the sample to atomic hydrogen, to study its catalytic sites at the cluster edges. **e** Cross section of a 3–6 layerMoS_2 continuous film growth by magnetron sputtering at 350 °C, atop of a SiO_2 substrate. **f** Raman spectra for films with different thicknesses. The frequency shift of the out-of-plane A_{1g} (~408 cm^{-1} in bulk) and in-plane E_{2g} (~382 cm^{-1} in bulk) representative peaks are related to the number of layers. Reproduced and adapted with permission from [10, 17]

as Mo and S sources to produce glucose and hydrogen peroxide detectors [27]. These sources were dissolved in water and sonicated, adjusting the pH to 6.5 with HCl, and submitted to a Teflon-lined stainless-steel autoclave at 200 °C for 24 h. The supernatant was collected and centrifuged for 5 min at 9000 rpm. The resulting quantum dots present blue emission with high yields (>10%), with a quenching mediated by H_2O_2. H_2O_2 is produced, for example, in the oxidation of glucose by glucose oxidase, being the MoS_2 QDs valuable as glucose sensors.

3 Properties

2D MoS_2 structures:

The electronic properties of MoS_2 are highly dependent of its weakly bonded layered structure, especially when it is reduced to the atomic-layer thickness [16, 22, 31]. Figure 9 shows combined theoretical and experimental studies of the electronic structure of layered MoS_2, which unveil the strikingly thickness dependence of the photoluminescence (PL) spectra and quantum yield (QY) in layered MoS_2, attributed to a thickness-induced indirect-to-direct bandgap transition.

Bulk MoS_2 presents an indirect bandgap semiconductor character (0.88 eV by theory, and 1.2 eV by experiments), with the valence band maximum (VBM) located at the Γ point of the Brillouin Zone (BZ), while the conduction band minimum (CBM) is located along the K-Γ lower symmetry direction [22, 31]. On the other hand, the 2D monolayer bandgap is direct (1.71 eV by theory and 1.9 eV by experiments), with both band edges located at the K point of the BZ [22, 31]. This effect is related to the orbital composition change of the electronic states when reducing the dimensionality, evidencing the role of interlayer coupling. Although the band edge states at the K point of the BZ are mainly composed of the d orbitals from Mo atoms, insensitive to interlayer coupling, those at Γ point of the BZ possess sulfur p_z orbitals, strongly influenced by the interaction with adjacent layers.

Quasi-1D and 1D MoS_2 nanostructures:

MoS_2 nanosheets turn in quasi-1D nanoscrolls by reducing the interaction between the MoS_2 nanosheets and the substrate, which allows the internal van der Waals forces in the nanosheet to scroll onto itself [28].

Molecular dynamic (MD) simulations within the isothermal–isobaric ensemble and using the Stillinger–Weber force field indicate that nanoscrolls with the interlayer distances between 4.7 and 6.5 Å, which is consistent with experiments (see Fig. 10a). Indeed, the MD simulations indicate that MoS_2 nanoscrolls with interlayer distances smaller than 4.5 Å are highly defective, while that with interlayer distances larger than 6.5 Å are collapsed [28].

Band structure calculations of MoS_2 nanoscrolls are absent in literature. However, the electronic structure of MoS_2 nanosheets and nanotubes, as approximations for unrolled and rolled MoS_2 nanoscrolls, respectively, was calculated within density-functional theory with generalized gradient approximation (see Fig. 10b). The first-principles calculations reveal that the bandgap of MoS_2 N-layer and MoS_2 armchair N-walled nanotubes decreases as the number of layers and walls increases [22, 28]. Furthermore, density-functional tight-binding calculations indicate that MoS_2 armchair and zigzag single-walled nanotubes are semiconductors (see Fig. 10c and d), and their direct or indirect bandgaps increase as their diameter increases [21]. In the case of MoS_2 N-layer, all structures are semiconductor materials, while in the case of the MoS_2 armchair N-walled nanotubes, a semiconductor-to-metal transition occurs at the triple-walled nanotube [28].

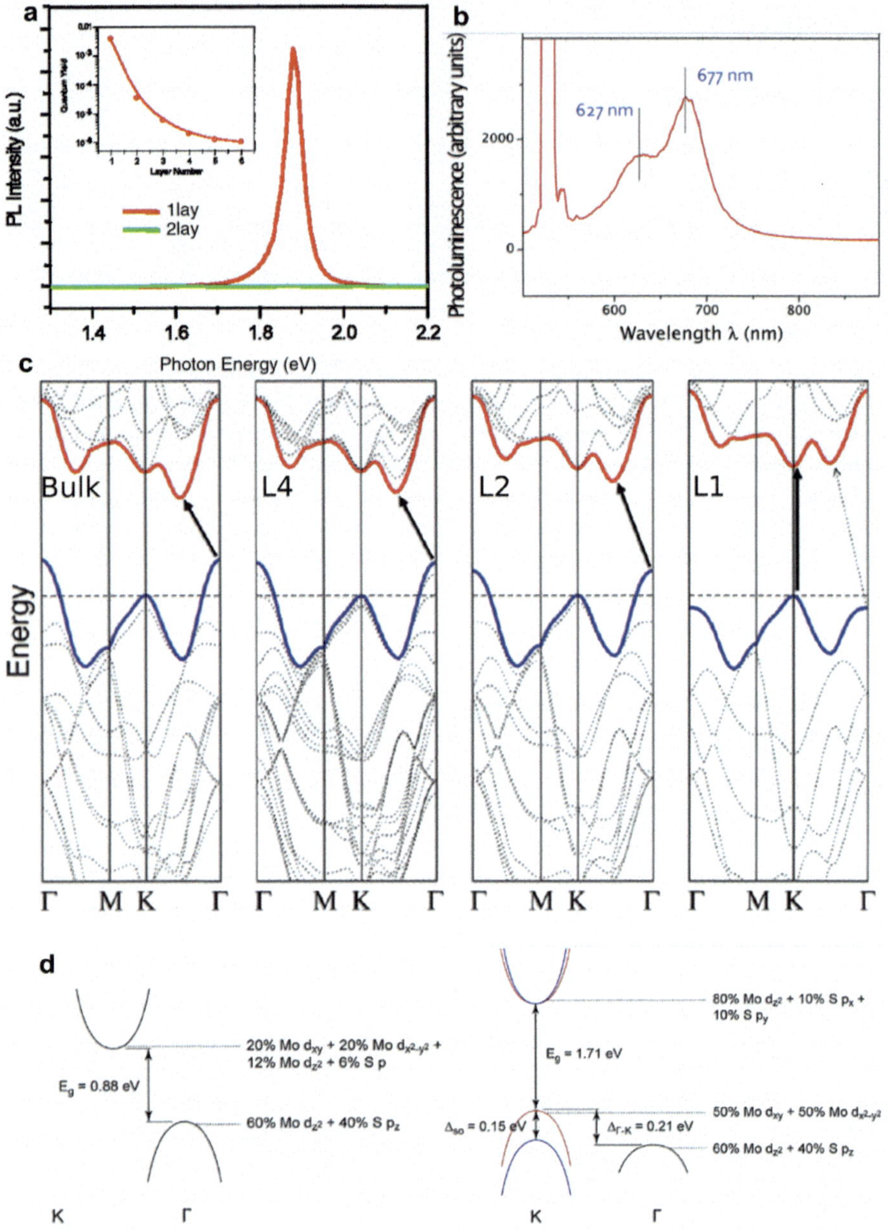
a
PL Intensity (a.u.)
1lay
2lay
1.4
1.6
1.8
2.0
2.2
Photon Energy (eV)
b
Photoluminescence (arbitrary units)
677 nm
627 nm
2000
0
600
700
800
Wavelength λ (nm)
c
Energy
Bulk
L4
L2
L1
Γ M K Γ
Γ M K Γ
Γ M K Γ
Γ M K Γ
d
20% Mo d$_{xy}$ + 20% Mo d$_{x^2-y^2}$ +
12% Mo d$_{z^2}$ + 6% S p
E$_g$ = 0.88 eV
60% Mo d$_{z^2}$ + 40% S p$_z$
K
Γ
80% Mo d$_{z^2}$ + 10% S p$_x$ +
10% S p$_y$
E$_g$ = 1.71 eV
50% Mo d$_{xy}$ + 50% Mo d$_{x^2-y^2}$
Δ$_{so}$ = 0.15 eV
Δ$_{Γ-K}$ = 0.21 eV
60% Mo d$_{z^2}$ + 40% S p$_z$
K
Γ

◂**Fig. 9** Indirect-to-direct bandgap transition due to quantum confinement in layered MoS_2. **a** Photoluminescence spectra of suspended MoS_2 monolayer (red curve) and bilayer (green curve) under excitation at 2.33 eV. Inset: Photoluminescence quantum yield of N-layer MoS_2 where N = 1–6; reproduced with permission from [16]. **b** Photoluminescence spectrum of quartz-supported MoS_2 monolayer under excitation at 532 nm (2.37 eV). The wavelength of the direct excitonic transitions at the Brillouin zone K point is indicated. **c** Calculated band structure of bulk, quadrilayer (L4), bilayer (L2), and monolayer (L1) MoS_2 by using density-functional theory with generalized gradient approximation. Highest valence states and lower conduction states are colored in red and blue, respectively. Arrows with continuous shaft indicate the bandgap of the N-layer structure. In the monolayer band structure, an addition arrow with dotted shaft indicates the indirect bandgap which corresponds to the bulk bandgap; reproduced with permission from [22]. **d** Schematic representation of the valence and conduction band edges in MoS_2 bulk (left drawing) and monolayer (right drawing). Calculated bandgap and orbital composition of the band edge states are indicated. For MoS_2 monolayer, the calculated valence band spin–orbit splitting and the Γ valley band offset are indicated as well. The calculations were performed at the density-functional theory with generalized gradient approximation level of theory; reproduced with permission from [31]

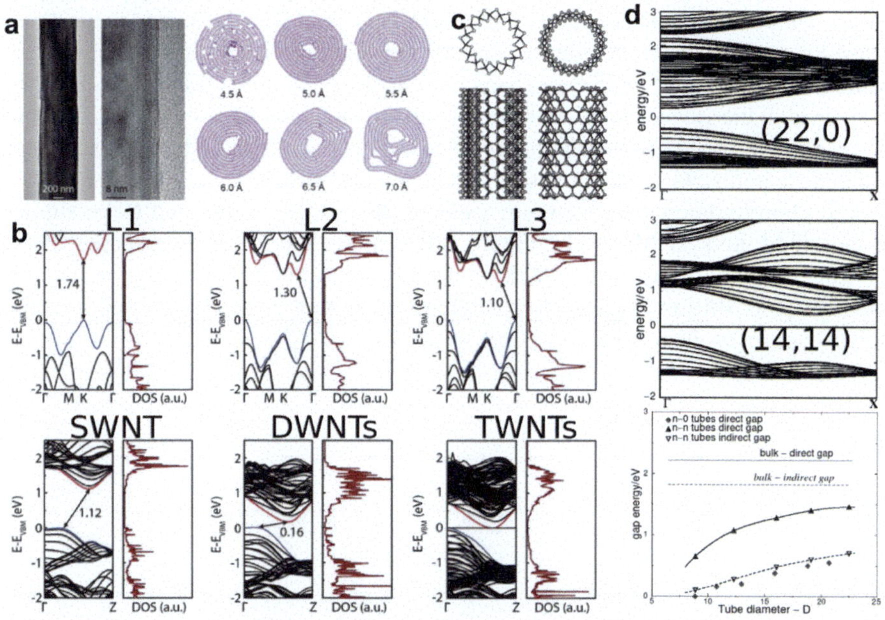

Fig. 10 Structural and electronic properties of MoS_2 nanoscrolls and nanotubes. **a** MoS_2 nanoscroll images obtained from transmission electron microscopy (left image) and high-resolution transmission electron microscopy (right image), and atomistic models of nanoscrolls with several interlayer distances used in molecular dynamic simulations. **b** Band structure and density of states of MoS_2 monolayer (L1), bilayer (L2), and trilayer (L3) nanosheets, and MoS_2 single- (SWNT), double- (DWNTs), and triple-walled (TWNTs) armchair nanotubes; reproduced with permission from [28]. **c** Atomistic models for MoS_2 armchair (left) and zigzag (right) single-walled nanotubes. **d** Density-functional-based tight-binding calculations of the band structure of MoS_2 zigzag (22,0) (upper graphic) and armchair (14,14) single-walled nanotube. Bandgap dependence with the nanotube diameter for MoS_2 armchair and zigzag single-walled nanotubes (lower graphic); reproduced with permission from [21]

The bandgap dependence with the number of layers of MoS_2 nanoscrolls is assumed to resemble that of MoS_2 nanotubes, a hypothesis confirmed by electrical transport experiments on MoS_2 nanoscroll-based field effect transistors, which reveals the channel material to be of metallic character [28].

0D (bidimensional dots):

Layered MoS_2 with a thickness from monolayer to few layers (<10 layers) and with sizes within a few nanometers (<10 nm), i.e., MoS_2 quantum dots (MoS_2 QD), is reported to further blue-shift the bandgap and increase the PL QY of MoS_2. Figure 11a displays a typical size distribution of MoS_2 QD with a few atomic layers. The corresponding bandgap distribution, estimated from the bandgap dependence with size, within quantum confinement theory [1, 2], is shown in Fig. 11b. Figure 11c displays the absorption spectra of MoS_2 microcrystals, nanoflakes (NF), and quantum dots as prepared (QD blue) and with a few-days old, forming larger aggregates (QD no color). The features in the microcrystals marked as A and B are attributed to the direct transition excitons between valence states and conduction states with K wavevectors, while the features at C and D are characteristic direct excitonic transitions from the deep level in the valence band to the conduction band at the M point of the Brillouin Zone. The NF spectra resemble that of microcrystals, while that of QDs reveal very different features, with optical absorption below 400 nm [7].

The photoluminescence quantum yield in MoS_2 QDs is reported to be similar to that of Monolayer MoS_2 [7]. Figure 11d displays the PL of as-prepared MoS_2 QDs in N-methyl pyrrolidone (NMP) under several excitation wavelengths. The PL emission peak shift in response to changing the excitation wavelength confirms the polydispersity of the as-prepared MoS_2 QDs [13].

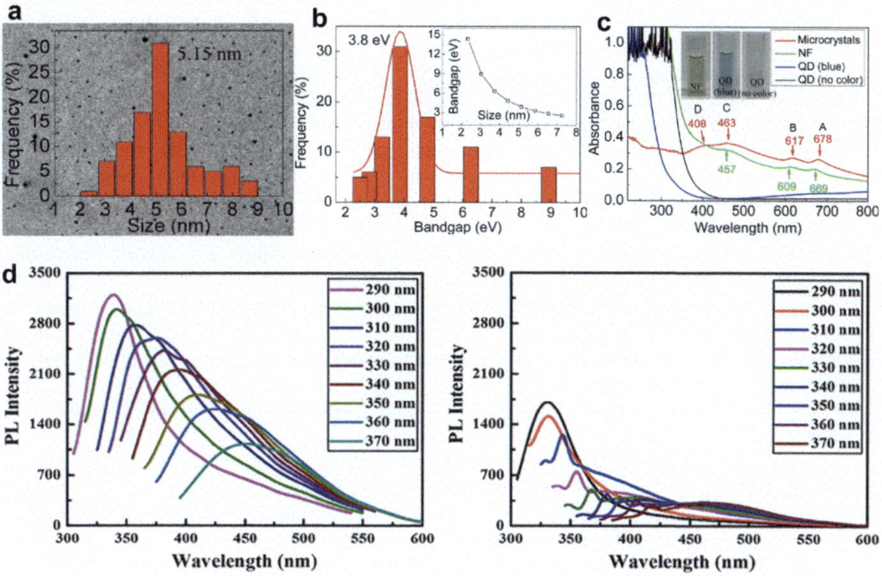

Fig. 11 Electronic properties of MoS_2 quantum dots. **a** Size distribution of MoS_2 quantum dots dispersed in water to about 0.1 mg/ml. The graphic background is the low-magnification transmission electron microscopy image of the MoS_2 quantum dots. **b** Calculated bandgap distribution by considering the previous size distribution and the effective-mass approximation theory. Inset: Bandgap dependence with the size of the MoS_2 quantum dots. **c** Absorption spectra of MoS_2 microcrystals, nanoflakes (NF), quantum dots as prepared (QD blue) and after several days old (QD no color). The exciton features are indicated. Inset: Photograph of the solutions with the NF, QD blue, and QD no color. Reproduced with permission from [7]. **d** Photoluminescence spectra of MoS_2 QD in NMP (left) and of NMP only (right) under several excitation wavelengths. The MoS_2 QDs have lateral size in the range of 1–5 nm. Reproduced under CCA 4.0 from [13]

4 Conclusion

The controlled synthesis of MoS_2 allows to obtain nanomaterials with tunable electronic and optical properties. The exploitation of the dependence of the electronic and optical properties of MoS_2 with the size, due to quantum confinement effects, makes it a versatile material with promising applications in several fields, from the fabrication of next-generation opto-electronic devices, biological probes, and catalysts.

References

1. Brus L (1984) Electron–electron and electron-hole interactions in small semiconductor crystallites: the size dependence of the lowest excited electronic state. J Chem Phys 80(9):4403
2. Brus L (1986) Electronic wave functions in semiconductors clusters: experiment and theory. J Phys Chem 90(12):2555

3. Chhowalla M et al (2013) The chemistry of two-dimensional layered transition metal dichalcogenide nanosheets. Nat Chem 5(4):263–275
4. Chiawchan T et al (2021) CVD Synthesis of intermediate state-free, large-area and continuous MoS_2 via single-step vapor-phase sulfurization of MoO_2 precursor. Nanomaterials 11(10):2642
5. Dai W et al (2015) Tunable fabrication of molybdenum disulfide quantum dots for intracellular microRNA detection and multiphoton bioimaging. Small 11(33):4158–4164
6. Fan X et al (2015) Fast and efficient preparation of exfoliated 2H MoS_2 nanosheets by sonication-assisted lithium intercalation and infrared laser-induced 1T to 2H phase reversion. Nano Lett 15(9):5956–5960
7. Gan ZX, Liu LZ, Wu HY et al (2015) Quantum confinement effects across two-dimensional planes in MoS2 quantum dots. Appl Phys Lett 106(23):233113
8. Gopalakrishnan D et al (2015) Electrochemical synthesis of luminescent MoS_2 quantum dots. Chem Commun 51(29):6293–6296
9. Gutierrez M (1989) Preparation of colloidal semiconductor solutions of MoS_2 and WSe_2 via sonication. Ultrasonics 27:259
10. Helveg S et al (2000) Atomic-scale structure of single-layer MoS_2 nanoclusters. Phys Rev Lett 84(5):951
11. Kabel J et al (2021) Molybdenum disulfide quantum dots: properties, synthesis, and applications. C 7(2):45
12. Lee Y-H et al (2012) Synthesis of large-area MoS_2 atomic layers with chemical vapor deposition. Adv Mater 24(17):2320–2325
13. Li B et al (2017) Preparation of monolayer MoS_2 quantum dots using temporally shaped femtosecond laser ablation of bulk MoS_2 targets in water. Sci Rep 7(1):11182
14. Li P, Ding F (2022) Origin of the herringbone reconstruction of Au (111) surface at the atomic scale. Sci Adv 8(40):eabq2900
15. Liu Y et al (2013) Synthesis of rGO-supported layered MoS_2 for high-performance rechargeable Mg batteries. Nanoscale 5(20):9562–9567
16. Mak KF et al (2010) Atomically thin MoS_2: a new direct-gap semiconductor. Phys Rev Lett 105:136805
17. Muratore C et al (2014) Continuous ultra-thin MoS_2 films grown by low-temperature physical vapor deposition. Appl Phys Lett 104(26):261604
18. Muscuso L et al (2015) Optical, vibrational, and structural properties of MoS_2 nanoparticles obtained by exfoliation and fragmentation via ultrasound cavitation in isopropyl alcohol. J Phys Chem C 119(7):3791–3801
19. Novoselov KS et al (2004) Electric field effect in atomically thin carbon films. Science 306(5696):666–669
20. Park SJ et al (2017) Structural and optical characterization of MoS_2 quantum dots defined by thermal annealing. J Luminescence 183:62–67
21. Seifert G et al (2000) Structure and electronic properties of MoS_2 nanotubes. Phys Rev Lett 85:146
22. Splendiani A et al (2010) Emerging photoluminescence in monolayer MoS_2. Nano Lett 10:1271
23. Štengl V, Henych J (2013) Strongly luminescent monolayered MoS_2 prepared by effective ultrasound exfoliation. Nanoscale 5(8):3387–3394
24. Su C-Y et al (2009) Electrical and spectroscopic characterizations of ultra-large reduced graphene oxide monolayers. Chem Mater 21(23):5674–5680
25. Sun J et al (2017) Synthesis methods of two-dimensional MoS_2: a brief review. Crystals 7(7):198
26. Wang T et al (2017) MoS_2-based nanocomposites for electrochemical energy storage. Adv Sci 4(2):1600289
27. Wang X et al (2017) One-step synthesis of water-soluble and highly fluorescent MoS_2 quantum dots for detection of hydrogen peroxide and glucose. Sens Actuators B: Chem 252:183–190
28. Wang Z et al (2018) Self-scrolling MoS_2 metallic wires. Nanoscale 10:18178–18185
29. Xu S, Li D, Wu P (2015) One-pot, facile, and versatile synthesis of monolayer MoS_2/WS_2 quantum dots as bioimaging probes and efficient electrocatalysts for hydrogen evolution reaction. Adv Funct Mater 25(7):1127–1136

30. Yang F et al (2021) MoS_2 for beyond lithium-ion batteries. APL Mater 9(5):050903
31. Yazyev OV, Kis A (2015) MoS_2 and semiconductors in the flatland. Mater Today 18(1):20–30
32. Zeng Z et al (2011) Single-layer semiconducting nanosheets: high-yield preparation and device fabrication. Angewandte Chemie 123(47):11289–11293
33. Zhang W et al (2015) Synthesis and sensor applications of MoS_2-based nanocomposites. Nanoscale 7(44):18364–18378

Chapter 5
Structural, Mechanical, and Electronic Properties of In-Plane 1T/2H-MoS_2, 2H-MoS_2, and 3R-MoS_2

Siby Thomas, Sreejesh Moolayadukkam, and K. P. Priyanka

1 Introduction

Two-dimensional (2D) materials are generally comprised of a large family of organic and inorganic compounds and a breakthrough to nanomaterials science has happened when Novoselov et al. successfully synthesized different monolayers using mechanical exfoliation [1]. Since then, different classes of planar and buckled monolayers are obtained from their bulk forms including hexagonal boron nitride (h-BN) [2], transition metal dichalcogenides (TMDCs) [3, 4], monochalcogenides, complex oxides, BC_3[5], silicene [6], MXenes [7, 8], etc. Among these 2D materials, the TMDCs possess a vast variety of applications in different areas such as quantum devices [9, 10], nanoelectronics and photodetectors [11, 12], optical communications [13], catalysis [14], and optical filters [15]. TMDCs are layered materials that possess weak out-of-plane and strong in-plane bonding interactions. Similar to graphene and many other 2D materials, the TMDCs can be synthesized from their bulk structure through different techniques such as chemical exfoliation, mechanical exfoliation, solution synthesis, and physical vapor deposition [16, 17]. Among the variety of TMDCs, the MoS_2 is a promising candidate due to its superior properties complementing those of graphene. However, there exist severe challenges in the synthesis routes of MoS_2

S. Thomas (✉)
School of Computation, Information and Technology, Technical University of Munich (TUM), Hans-Piloty-Strasse 1, 85748 Garching, Munich, Germany
e-mail: siby.thomas@tum.de

S. Moolayadukkam
Energy Materials Laboratory, Centre for Nano and Soft Matter Sciences, Jalahalli, Bengaluru 560013, India

K. P. Priyanka
Department of Physics, M S Ramaiah University of Applied Sciences, Bengaluru 560058, Karnataka, India

D. J. Late and C. S. Rout (eds.), *Science and Technology of 2D MoS_2*,
Materials Horizons: From Nature to Nanomaterials,
https://doi.org/10.1007/978-981-97-7367-1_5

and TMDCs in general for their integration in commercial device fabrication. Nevertheless, these challenges can be overcome by controlling their crystalline quality through morphology and defect engineering.

The MoS_2 flakes with different sizes, shapes, and number of layers can be produced on a suitable substrate by following the same micromechanical exfoliation method utilized for the synthesis of graphene. SiO_2 is the common substrate used for the deposition of MoS_2. Besides, for the successful synthesis of TMDCs, single-crystal substrates such as mica or $SrTiO_3$ can be utilized. These substrates can effectively control the orientation and shape of TMDCs through van der Waals epitaxy. Besides, direct sulfurization techniques via a precursor-soluble glass substrate can be utilized for the synthesis of large-area MoS_2 [11]. One of the advantages of the micromechanical exfoliation method is that it produces 2D materials with the highest crystalline quality suitable for improved device performance. In 2013, Lopez-Sanchez et al. synthesized ultrasensitive MoS_2 monolayers using mechanical exfoliation and demonstrated their improved performance for phototransistors [18]. However, the weak van der Waals adhesion of MoS_2 and the SiO_2 leads to the generation of very small MoS_2 flakes with a lateral size of fewer than 10 μm [19]. Even so, by applying a slight modification during the synthesis technique, MoS_2 flakes with several hundreds of microns area can be obtained. However, a typical tape-assisted micromechanical exfoliation method offers a very low yield, which restricts the large-scale production of MoS_2 which is limited to the laboratory-scale analysis of this material. To overcome the issues associated with the mechanical exfoliation of MoS_2, other top-down and bottom-up methods, such as liquid-phase exfoliation, physical vapor deposition, solution chemical process, and chemical vapor deposition, can be employed and the details of the synthesis methods of 2D MoS_2 can be found elsewhere [20].

2 Structural Properties

TMDCs have widely studied materials that possess the general structural formula MA_2, where M represents a group IV-VI transition metal element and A is a chalcogen. The group IV-VI elements (Ti, Hf, Nb, V, W, Mo, etc.) and chalcogens (Se, S, Te, etc.) together form layered structures of the form A-M-A where a chalcogen atom sandwiched between the group IV-VI transition metals. MoS_2 is one of the most celebrated TMDCs and manifests in different polymorphic phases such as metallic 1T, semi-metallic 1T, and semiconducting 1H [21]. Among these phases, the multilayer 1H phase shows different stacking polytypes such as 3R and 2H phases [22]. The structural difference in these polymorphs is mainly aroused due to their interlayer stacking order. Among the different polymorphs of MoS_2, the 2H phase is the thermodynamically stable phase which belongs to the $P6_3/mmc$ space group. In this case, the layers are sandwiched in an A-b-A fashion with the presence of edge-sharing MoS_6 trigonal prisms as shown in Fig. 1a [23]. However, the metastable 1T phase belongs to the *P3m1* space group and was obtained using powder X-ray and electron diffraction and exhibits the A-b-C arrangement of atomic layers with the presence of edge-sharing MoS_6 octahedron as shown in Fig. 1b [24].

Although the 1T phase of MoS_2 can be obtained using different synthesis methods [25], the prepared monolayers always coexist with other phases of MoS_2, such as 2H and 1T, indicating that the synthesis of 1T-MoS_2 with perfect structural geometry is still challenging. However, by employing high-resolution transmission electron microscopy (HR-TEM) analysis, Kim et al. obtained the hexagonal pattern of monolayer 2H or 1T-MoS_2 and found that the Mo-Mo distance of 1T-MoS_2 maintains the same atomic spacing as that of the 2H phase, 3.18 Å [26]. The HR-TEM images of 2H and 1T-MoS_2 are shown in Fig. 1c and d. Besides, X-ray absorption spectroscopy (XAS) indicated that due to the different coordination behavior in 2H-MoS_2, there exist two nearest Mo-Mo distances, such as 2.8 and 3.8 Å, respectively, compared to

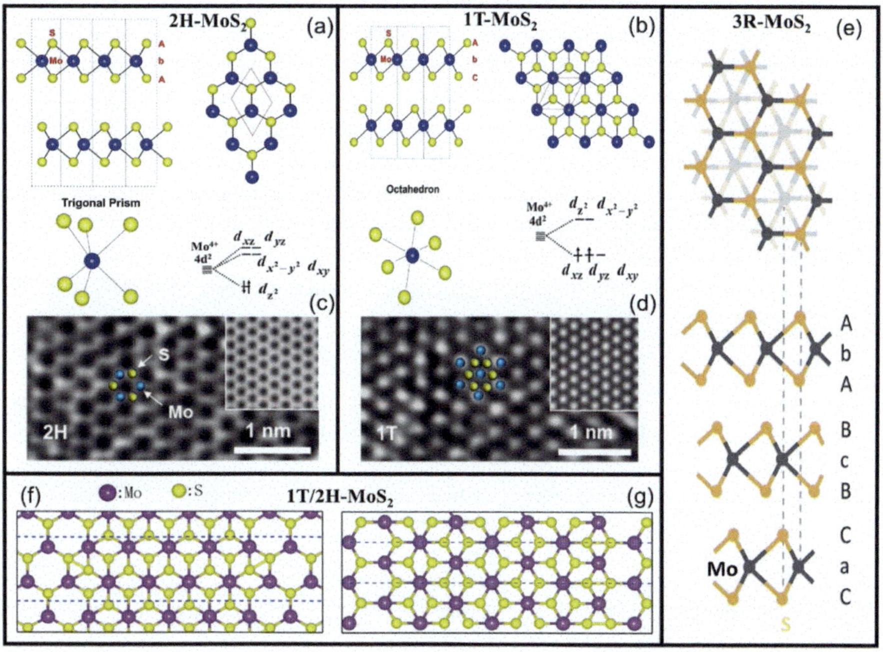

Fig. 1 Crystal structures of **a** 2H-MoS_2 and **b** 1T-MoS_2. The ligand splitting for the 4d orbitals of Mo atoms with trigonal and octahedron prismatic coordination for 2H-MoS_2 and 1T-MoS_2 is also provided (Reproduced with the permission from Ref. [24]; Angew. Chem. 2018, 130 (5), 1246–1249; https://doi.org/10.1002/ange.201710512). The high-resolution TEM images and the multi-slice simulation images (in the inset) of **c** 2H-MoS_2 and **d** 1T-MoS_2. (Reproduced with the permission from Ref. [26]; ACS Nano 2016, 10 (8), 7500–7506; https://doi.org/10.1021/acsnano.6b02267). **e** Schematic crystal structure of 3R-MoS_2 along with the atomic arrangement and stacking order. (Reproduced with the permission from Ref. [28]; RSC Adv., 2015, 5, 7495–7514; https://doi.org/10.1039/C4RA11852A). Schematic models of single-layered MoS_2 and its phase interfaces along the **f** armchair and **g** zigzag edges. The blue dashed line represents the unit cell. (Reproduced with the permission from Ref. [30]; AIP Advances, 2015, 5, 097,174; https://doi.org/10.1063/1.4932040)

the bulk value of 3.16 Å [22]. The structural stability analysis using density functional theory (DFT) has been reported that 2H-MoS_2 possesses more structural stability compared to 1T-MoS_2 [27].

The 3R-MoS_2 is of the other polytypes of MoS_2 based on the stacking order as well as the coordination of S atoms with respect to the central Mo atom (see Fig. 1e). Similar to 2H-MoS_2, the 3R polytypes exhibit trigonal prismatic coordination (D_{3h}) of Mo atoms, but possess different stacking order and belong to the *R3m* space group [28]. The stacking orders of 3R and 2H are AbABcBCa-CAbA… and AbABaBAbA…, respectively. Even so, the crystallographic parameters, the distance between the atoms, thickness, and lattice constant of 2H and 3R, are similar. Besides, the chemical and physical properties of 3R-MoS_2 are almost identical to those of 2H, including the electronic properties. In addition, the great success of graphene/h-BN heterostructures further motivated the possibilities of exploring nanoscale heterostructures composed of MoS_2 and other TMDC-based heterostructures for finding new routes toward photonic and electronic device applications [29]. Since MoS_2 exists in different phases, studies have been reported both lateral and vertical heterostructures of MoS_2 is possible with 2H and 1T phase. Guo et al. reported that the zigzag-type 2H/1T interface is energetically more favorable compared to the armchair-type interface, as shown in Fig. 1f and g [30]. Since the single layers of chemically exfoliated MoS_2 consist of electronically dissimilar but lattice-matched polymorphs including 2H and 1T phases, their heterostructures can be achieved through different experimental techniques described elsewhere [29]. Table 1 summarizes the fundamental structural features of 2H-, 1T-, 3R-, and 2H/1T-MoS_2 polytypes.

Table 1 Summary of the fundamental structural features of 2H-, 1T-, 3R-, and 2H/1T-MoS_2 polytypes. Young's modulus is represented for both zigzag (ZZ)- and armchair (AC)-type orientations

MoS_2 type	Space group	Stacking order	Lattice constants (Å)	Bandgap (eV)	Young's modulus (GPa)	References
2H	P63/mmc	AbABaB	a = b = 3.15, c = 12.30	1.8	170 (ZZ) 194 (AC)	[30–33]
1T	*P3m1*	AbC	a = b = 5.60, c = 5.99	Metallic	183 (ZZ) 188 (AC)	[30, 32]
3R	*R3m*	AbABcBCaC	a = b = 3.17, c = 18.38	1.42	170 ± 10 (ZZ) 194 ± 10 (AC)	[28, 33, 34]
2H/1T	–	–	–	Metallic	189 (ZZ) 163 (AC)	[30]

3 Electronic Properties

The electronic properties of MoS_2 polymorphs are intimately associated with their structural features. Besides experimental analysis, computational approaches are well utilized for the accurate electronic property predictions of different MoS_2 phases. Analysis of the ligand splitting of 4d orbitals in 2H-MoS_2 reveals that the trigonal prismatic coordination completely fills the d_{z^2} orbitals and keeps an empty d_{xy} and $d_{x^2-y^2}$ orbitals, as shown in Fig. 1a. This closed-shell electronic configuration of the 4*d* orbitals of Mo atoms is mainly responsible for the semiconducting electronic properties of 2H-MoS_2. However, in the case of 1T-MoS_2, although the octahedral crystal fields of Mo 4*d* orbitals are split into d_{z^2}, $d_{x^2-y^2}$, d_{xy}, d_{xz}, and d_{xz}, the filled electronic states are mainly due to the d_{xy}, d_{xz}, and d_{xz} orbitals, as shown in Fig. 1b. Unlike 2H-MoS_2, the d_{xy} and $d_{x^2-y^2}$ orbitals of 1T-MoS_2 are empty and the inherent electrons in the t_{2g} orbitals (d_{xy}, d_{xz}, and d_{xz}) induce metallic conductivity. Further, to have an in-depth understanding of the electronic properties of 2H and 1T-MoS_2, band structures computed using the DFT calculations have been discussed. The electronic properties of 2H and 1T-MoS_2 are quite different in which the 2H-MoS_2 is a semiconductor (see Fig. 2a) with a direct bandgap of ~ 1.9 eV [32], whereas the 1T-MoS_2 is metallic as shown in Fig. 2b. However, it is interesting to notice that metallic MoS_2 does not exist in nature. Even so, the intercalation of butyllithium in metallic MoS_2 generates the metastable MoS_2 and after deintercalation, the metastable phase retains the semiconducting behavior.

Since the electronic properties have a direct relation with the optical properties, the distinct electronic bandgap of 2H and 1T-MoS_2 has a direct impact on the optical properties. Consistent with the electronic properties, the ultraviolet–visible (UV) visible spectroscopy studies reported that no visible absorption peaks are observed in metallic 1T-MoS_2. At the same time, the semiconducting 2H-MoS_2 shows two distinct peaks at 604 nm and 667 nm, as shown in Fig. 2c. The dispersion of semiconducting and metallic phases is shown in dark yellow or green and dark gray, respectively, in the inset of Fig. 2c. These peaks are related to the direct bandgap characteristics. Geng et al. also analyzed the optical absorption spectra of 2H and 1T-MoS_2 and found that the semiconducting phase owns two different absorption peaks at 613 and 660 nm, whereas the metallic phase does not show any distinct peaks, as shown in Fig. 2d. The difference in metallic and semiconducting dispersion is shown in gray and green color, respectively. Although this information is consistent with the absorption peaks in Fig. 2c, it should be noted that these two analyses are carried out using different dispersions such as a mixture of water and isopropanol (Fig. 2c) and water (Fig. 2d).

Similar to 2H-MoS_2, the 3R-MoS_2 also possesses semiconducting band structure characteristics, as shown in Fig. 2e. When comparing Fig. 2a and e, it is clear that the material in both 2H and 3R phases behaves as a semiconductor. However, the band orientations and selection of high-symmetric orientations are different in both cases. As shown in Fig. 2e, the valence and conduction bands of 3R-MoS_2 lie in the region of energy values between −6 eV and −16 eV [34]. As detailed before, polymorphism

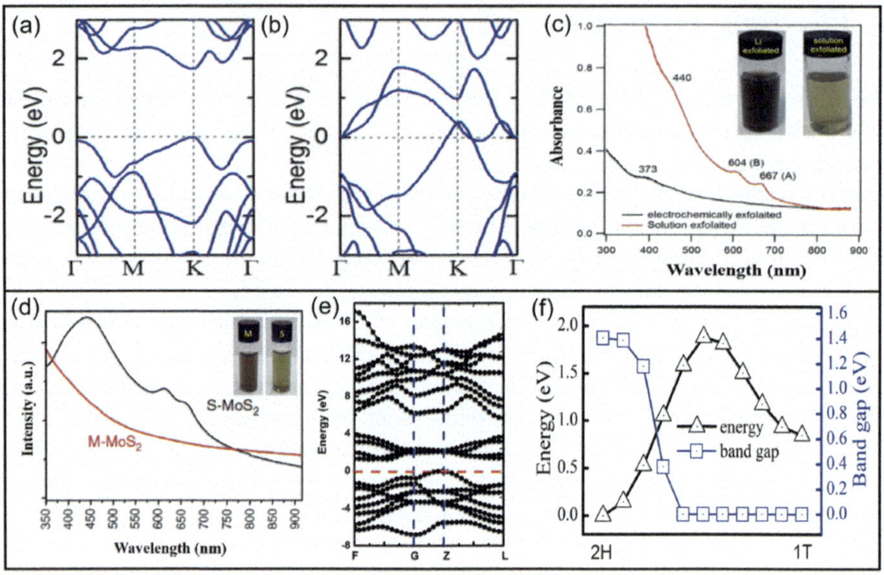

Fig. 2 Electronic band structures of **a** 2H-MoS_2 and **b** 1T-MoS_2 (Reproduced with permission from Ref. [27]; J. Phys. Chem. C 2015, 119, 13,124 – 13,128; https://doi.org/10.1021/acs.jpcc.5b04658). **C** The UV–visible optical absorption spectrum of semiconducting 2H-MoS_2 (red curve) and metallic 1T-MoS_2 (black curve) dispersion in the mixture of water and isopropanol (1:1). **d** UV–visible absorption spectrum of the semiconducting 2H-MoS_2 (black curve) and metallic 1T-MoS_2 (red curve) dispersed in water. (Reproduced with permission from Ref. [36]; Nature Communications, 2016, 7, 10,672. https://www.nature.com/articles/ncomms10672). **e** Band structure of 3R-MoS_2 along the F-G-Z-L high-symmetric directions. (Reproduced with permission from Ref. [34]; Physica E: Low-dimensional Systems and Nanostructures, 2019, 106, 338–345. https://doi.org/10.1016/j.physe.2018.07.003). **f** Evolution of bandgap and energy from 2H-MoS_2 to 2H-MoS_2 phase. (Reproduced with permission from Ref. [30]; AIP Advances, 2015, 5, 097,174.)

in MoS_2 is responsible for distinct electronic properties, especially when it comes to mixed phases. Earlier studies clearly reported that the semiconducting 2H phase is more stable than the 1T phase [35]. However, it is interesting to notice that the gliding of the S plane of the 2H phase to the center of the hexagonal will result in a semiconducting (2H-MoS_2) to metallic transition (1T phase). This transition is also accompanied by a change of energy stability and Fig. 2f depicts the change in energy and bandgap during the transition from 2H to 1T. In addition to this, from Fig. 2f, it is clear that although the 2H/1T mixed phase possesses lower energy stability compared to 2H and 1T phases, they exhibit metallic electronic features.

4 Mechanical Properties

An in-depth analysis of mechanical properties and strain effect of MoS_2 is of paramount importance for potential applications including device fabrication. By employing the atomic force microscopy (AFM) technique, Bertolazzi et al. measured the breaking strength (22 ± 4 GPa) and stiffness of monolayer MoS_2 [31]. They found that the measured Young's modulus of MoS_2 is 270 ± 100 GPa. However, from the theoretical point of view, the mechanical stability of a material is described using the Born–Huang stability criteria [37]. According to this rule, a mechanically stable 2D material, including MoS_2 polymers, should satisfy the relations: $C_{66} = \left(\frac{C_{11}-C_{12}}{2}\right) > 0$ and $C_{11}C_{22} - C_{12}^2 > 0$. A detailed comparative study by Yengejeh et al. reported the mechanical properties of different polymorphs of MoS_2 [33] and found that they all satisfy the mechanical stability conditions. Due to the structural similarities, including the similar prismatic units, of 2H and 3R phases, they possess almost comparable elastic constants. However, the elastic constants of 1T and 2H/1T mixed phases are different from those of 2H and 3R phases of MoS_2 due to their different structural features. As said, due to the similarities of the base structure, both 2H and 3R phases exhibit a similar bulk modulus of 55 GPa, whereas the bulk modulus of 1T is 43 GPa. This indicates that the 2H and 3R hold a higher resistance upon compressive loads and external strains and stresses.

By employing DFT computations, Guo et al. reported that the ideal strengths of monolayer 2H-MoS_2 along the armchair (AC) and zigzag (ZZ) directions are 26.01 GPa and 22.24, respectively [30]. They also reported that Young's modulus of 2H-MoS_2 is 194 GPa and170 GPa, respectively, along the AC and ZZ directions. Besides, they reported that the ideal strength of 1T-MoS_2 is found to be 14.7 GPa and 11.5 GPa, respectively, along the AC and ZZ directions. At the same time, the DFT computed Young's modulus along the AC and ZZ directions which are 188 GPa and 183 GPa, respectively. The ideal strengths of the 2H/1T phase are 12.6 GPa and 10 GPa, respectively, along the AC and ZZ directions with Young's modulus of 163 GPa along the AC direction and 189 GPa along ZZ direction. It is interesting to notice that Young's modulus along the ZZ direction of 2H/1T mixed phase is higher than 2H, 3R, and 1T phases (see Table 1). However, this value is much lower for the case of the 2H/1T along the AC direction and is lower than the other phases of MoS_2. From the mechanical property analysis, it is clear that 2H-MoS_2 possesses superior mechanical stability compared to 1T-MoS_2. Even so, it is clear from the theoretical and experimental analysis that both 2H and 1T-MoS_2 exhibit anisotropic mechanical properties consistent with other 2D materials including graphene and h-BN [38, 39].

5 Conclusions and Outlook

MoS_2 is an exceptional candidate due to its superior physical, chemical, and mechanical properties among transition metal dichalcogenides. Even though there are many reports available on the micromechanical exfoliation synthesis of MoS_2, the restriction on large-scale production demands other synthesis methods of both top-down and bottom-up approaches. This chapter describes many aspects that have been actively studied regarding structural, electronic, and mechanical properties of MoS_2 and its derivatives.

Many studies have reported that the difference in interlayer stacking order in MoS_2 and its polymorphs cause the major structural difference. Among the different polymorphs of MoS_2, the 2H phase is the thermodynamically stable phase than 1T phase of MoS_2. Also, it has been reported that the prepared monolayers of 1T phase always coexist with other phases of MoS_2 such as 2H and 1T, indicating that the synthesis of 1T-MoS_2 with perfect structural geometry is still challenging. In addition to that, the density functional theory (DFT) analysis on the structural stability showed that 2H-MoS_2 possesses more stability compared to 1T-MoS_2. The third type of MoS_2 which is known as 3R-MoS_2 is also based on the stacking order as well as the coordination of S atoms with respect to the central Mo atom. But the similarities in the crystallographic parameters, the distance between the atoms, thickness, and lattice constant of 2H and 3R also bring almost identical chemical, physical, and electronic properties. Besides experimental analysis, computational approaches are well utilized for the accurate electronic property predictions of different MoS_2 phases. It is well known that the closed-shell electronic configuration of the 4*d* orbitals of Mo atoms is mainly responsible for the semiconducting electronic properties of 2H-MoS_2. The 2H-MoS_2 is a semiconductor with a direct bandgap of ~1.9 eV, whereas the 1T-MoS_2 is metallic in nature. In consistent with the electronic properties, the ultraviolet–visible (UV) visible spectroscopy studies confirm that no visible absorption peaks are observed in metallic 1T-MoS_2.

Furthermore, the structural similarities including the similar prismatic units of 2H and 3R phases, almost comparable elastic constants, can also be observed for them. Both 2H and 3R phases exhibit a similar bulk modulus of 55 GPa, whereas the bulk modulus of 1T is 43 GPa. Besides, the mechanical property analysis showed that 2H-MoS_2 possesses superior mechanical stability compared to 1T-MoS_2 and exhibit anisotropic mechanical properties consistent with other 2D materials including graphene and h-BN. Considering the recent studies and understandings, it will be challenging for MoS_2-based applications to completely replace the current materials used in optoelectronic and photonics applications. Hence, a more detailed analysis has to be carried out to accelerate the realization of MoS_2-based applications in different fields.

References

1. Novoselov KS, Jiang D, Schedin F et al (2005) Two-dimensional atomic crystals. PNAS 102:10451–10453. https://doi.org/10.1073/pnas.0502848102
2. Zhang K, Feng Y, Wang F et al (2017) Two-dimensional hexagonal boron nitride (2D-hBN): synthesis, properties and applications. J Mater Chem C 5:11992–12022. https://doi.org/10.1039/C7TC04300G
3. Wang Z, Zhao K, Li H et al (2010) Ultra-narrow WS_2 nanoribbons encapsulated in carbon nanotubes. J Mater Chem 21:171–180. https://doi.org/10.1039/C0JM02821E
4. Wang Z, Li H, Liu Z et al (2010) Mixed low-dimensional nanomaterial: 2D ultranarrow MoS_2 inorganic nanoribbons encapsulated in quasi-1D carbon nanotubes. J Am Chem Soc 132:13840–13847. https://doi.org/10.1021/ja1058026
5. Ueno A, Fujita T, Matsue M et al (2006) Scanning tunneling microscopy study on a BC_3 covered NbB_2(0001) surface. Surf Sci 600:3518–3521. https://doi.org/10.1016/j.susc.2006.07.007
6. Kara A, Enriquez H, Seitsonen AP et al (2012) A review on silicene—new candidate for electronics. Surf Sci Rep 67:1–18. https://doi.org/10.1016/j.surfrep.2011.10.001
7. Naguib M, Mashtalir O, Carle J et al (2012) Two-dimensional transition metal carbides. ACS Nano 6:1322–1331. https://doi.org/10.1021/nn204153h
8. Naguib M, Kurtoglu M, Presser V et al (2011) Two-dimensional nanocrystals produced by exfoliation of Ti_3AlC_2. Adv Mater 23:4248–4253. https://doi.org/10.1002/adma.201102306
9. Lehnert T, Ghorbani-Asl M, Köster J et al (2019) Electron-beam-driven structure evolution of single-layer $MoTe_2$ for quantum devices. ACS Appl Nano Mater 2:3262–3270. https://doi.org/10.1021/acsanm.9b00616
10. Littlejohn AJ, Li Z, Lu Z et al (2019) Large metallic vanadium disulfide ultrathin flakes for spintronic circuits and quantum computing devices. ACS Appl Nano Mater 2:3684–3694. https://doi.org/10.1021/acsanm.9b00608
11. Lu Y, Chen T, Ryu GH et al (2019) Self-limiting growth of high-quality 2D monolayer MoS2 by direct sulfurization using precursor-soluble substrates for advanced field-effect transistors and photodetectors. ACS Appl Nano Mater 2:369–378. https://doi.org/10.1021/acsanm.8b01955
12. Belete M, Kataria S, Koch U et al (2018) Dielectric properties and ion transport in layered MoS2 grown by vapor-phase sulfurization for potential applications in nanoelectronics. ACS Appl Nano Mater 1:6197–6204. https://doi.org/10.1021/acsanm.8b01412
13. Feng T, Zhang D, Li X et al (2020) SnS2 nanosheets for Er-doped fiber lasers. ACS Appl Nano Mater 3:674–681. https://doi.org/10.1021/acsanm.9b02194
14. Jaiswal K, Girish YR, De M (2020) Influence of a tunable band gap on photoredox catalysis by various two-dimensional transition-metal dichalcogenides. ACS Appl Nano Mater 3:84–93. https://doi.org/10.1021/acsanm.9b01711
15. Hejazi D, Liu S, Ostadabbas S, Kar S (2019) Transition metal dichalcogenide thin films for precise optical wavelength estimation using bayesian inference. ACS Appl Nano Mater 2:4075–4084. https://doi.org/10.1021/acsanm.9b00489
16. Coleman JN, Lotya M, O'Neill A et al (2011) Two-dimensional nanosheets produced by liquid exfoliation of layered materials. Science 331:568–571. https://doi.org/10.1126/science.1194975
17. Wang K, Wang J, Fan J et al (2013) Ultrafast saturable absorption of two-dimensional MoS_2 nanosheets. ACS Nano 7:9260–9267. https://doi.org/10.1021/nn403886t
18. Lopez-Sanchez O, Lembke D, Kayci M et al (2013) Ultrasensitive photodetectors based on monolayer MoS_2. Nat Nanotechnol 8:497–501. https://doi.org/10.1038/nnano.2013.100
19. Magda GZ, Pető J, Dobrik G et al (2015) Exfoliation of large-area transition metal chalcogenide single layers. Sci Rep 5:14714. https://doi.org/10.1038/srep14714
20. Sun J, Li X, Guo W et al (2017) Synthesis methods of two-dimensional MoS_2: a brief review. Crystals 7:198. https://doi.org/10.3390/cryst7070198
21. Zhao X, Ning S, Fu W et al (2018) Differentiating polymorphs in molybdenum disulfide via electron microscopy. Adv Mater 30:1802397. https://doi.org/10.1002/adma.201802397

22. Zhao W, Pan J, Fang Y et al (2018) Metastable MoS_2: crystal structure, electronic band structure, synthetic approach and intriguing physical properties. Chem–Eur J 24:15942–15954. https://doi.org/10.1002/chem.201801018
23. Wang QH, Kalantar-Zadeh K, Kis A et al (2012) Electronics and optoelectronics of two-dimensional transition metal dichalcogenides. Nat Nanotechnol 7:699–712. https://doi.org/10.1038/nnano.2012.193
24. Fang Y, Pan J, He J et al (2018) Structure re-determination and superconductivity observation of bulk 1T MoS_2. AngewandteChemie 130:1246–1249. https://doi.org/10.1002/ange.201710512
25. Voiry D, Salehi M, Silva R et al (2013) Conducting MoS_2 nanosheets as catalysts for hydrogen evolution reaction. Nano Lett 13:6222–6227. https://doi.org/10.1021/nl403661s
26. Kim JS, Kim J, Zhao J et al (2016) Electrical transport properties of polymorphic MoS_2. ACS Nano 10:7500–7506. https://doi.org/10.1021/acsnano.6b02267
27. Gao G, Jiao Y, Ma F et al (2015) Charge mediated semiconducting-to-metallic phase transition in molybdenum disulfide monolayer and hydrogen evolution reaction in new 1T′ phase. J Phys Chem C 119:13124–13128. https://doi.org/10.1021/acs.jpcc.5b04658
28. Song I, Park C, Choi HC (2014) Synthesis and properties of molybdenum disulphide: from bulk to atomic layers. RSC Adv 5:7495–7514. https://doi.org/10.1039/C4RA11852A
29. Eda G, Fujita T, Yamaguchi H et al (2012) Coherent atomic and electronic heterostructures of single-layer MoS_2. ACS Nano 6:7311–7317. https://doi.org/10.1021/nn302422x
30. Guo X, Yang G, Zhang J, Xu X (2015) Structural, mechanical and electronic properties of in-plane 1T/2H phase interface of MoS_2 heterostructures. AIP Adv 5:097174. https://doi.org/10.1063/1.4932040
31. Bertolazzi S, Brivio J, Kis A (2011) Stretching and breaking of ultrathin MoS_2. ACS Nano 5:9703–9709. https://doi.org/10.1021/nn203879f
32. Jiao Y, Hafez AM, Cao D et al (2018) Metallic MoS_2 for high performance energy storage and energy conversion. Small 14:1800640. https://doi.org/10.1002/smll.201800640
33. Imani Yengejeh S, Liu J, Kazemi SA et al (2020) Effect of structural phases on mechanical properties of molybdenum disulfide. ACS Omega 5:5994–6002. https://doi.org/10.1021/acsomega.9b04360
34. Arif Khalil RM, Hussain F, Manzoor Rana A et al (2019) Comparative study of polytype 2H-MoS_2 and 3R-MoS_2 systems by employing DFT. Phys E 106:338–345. https://doi.org/10.1016/j.physe.2018.07.003
35. Kan M, Wang JY, Li XW et al (2014) Structures and phase transition of a MoS_2 monolayer. J Phys Chem C 118:1515–1522. https://doi.org/10.1021/jp4076355
36. Geng X, Sun W, Wu W et al (2016) Pure and stable metallic phase molybdenum disulfide nanosheets for hydrogen evolution reaction. Nat Commun 7:10672. https://doi.org/10.1038/ncomms10672
37. Mouhat F, Coudert F-X (2014) Necessary and sufficient elastic stability conditions in various crystal systems. Phys Rev B 90:224104. https://doi.org/10.1103/PhysRevB.90.224104
38. Thomas S, Ajith KM, Valsakumar MC (2016) Directional anisotropy, finite size effect and elastic properties of hexagonal boron nitride. J Phys: Condens Matter 28:295302. https://doi.org/10.1088/0953-8984/28/29/295302
39. Thomas S, Ajith KM, Lee SU, Valsakumar MC (2018) Assessment of the mechanical properties of monolayer graphene using the energy and strain-fluctuation methods. RSC Adv 8:27283–27292. https://doi.org/10.1039/C8RA02967A

Chapter 6
Phase Transition in MoS_2: Structure, Strategies and Applications

Payal Chauhan, C. K. Sumesh, and Dattatray J. Late

1 Introduction

In recent years, transition metal dichalcogenides (TMDs) have emerged as appealing possibilities for electrical devices, transistors, biosensors, as well as catalysis and energy storage devices [1–3].TMDs have piqued the interest of researchers due to their layered structure like graphite [4, 5]. TMDs have a 3-atom-thick unit layer in which the transition metal (M) atom is sandwiched between two chalcogens (X), yielding the MX_2 stoichiometry [6]. Interlayer bonds between two MX_2 slabs are normally van der Waals, whereas interlayer bonds are covalent [7]. In semiconducting, mostly the transition metal Mo or W (VI group) has gained great attention for applications including flexible electronics [8]. Among these TMDs, molybdenum diselenide ($MoSe_2$), molybdenum disulfide (MoS_2), tungsten diselenide (WSe_2), and others are potential members.

Molybdenum disulfide (MoS_2) has become a preferred material among the aforementioned TMDs due to its outstanding in-plane carrier mobility, strong catalytic performance, good stability, low cost, increased abundance, and advantageous mechanical features. The MoS_2 is most studied due to its tunable band gap, and unique chemical and physical properties. Each layer of the MoS_2 crystal is made of S-Mo-S′ triple atomic planes with substantial in-plane bonding, and the stacks of atomic layers are held together by van der Waals forces. Numerous theoretical and experimental researches have established that MoS_2 has a layer-dependent band gap

P. Chauhan (✉) · C. K. Sumesh
Department of Physical Sciences, P. D. Patel Institute of Applied Science, CHARUSAT Campus, Highway 139, Off. Nadiad-Petlad Road, Changa, Gujarat 388421, India
e-mail: payalnchauhan64@gmail.com

D. J. Late (✉)
Department of Physics, Federal University of Lavras, Campus Universitário, PO Box 3037, Lavras, Minas Gerais, Brazil
e-mail: datta099@gmail.com

D. J. Late and C. S. Rout (eds.), *Science and Technology of 2D MoS_2*,
Materials Horizons: From Nature to Nanomaterials,
https://doi.org/10.1007/978-981-97-7367-1_6

and it possesses layer-dependent bandgap characteristics. The bulk form of the material has an indirect band gap of 1.2 eV and a direct band gap of 1.8 eV [9]. Several studies on the physical and chemical treatment of MoS_2 to achieve a reversible transition between these two phases have recently been published.

In 2D materials, a wide variety of phase transitions have been investigated, including superconductivity, metal–insulator transitions, the quantum anomalous hall (QAH) effect, and magnetic phase transitions. In this chapter, firstly, we highlight recent significant research on MoS_2 material phase transitions and secondly discuss the key parameters which cause/affect the phase shifts in MoS_2. We also explain special characteristics of phase transitions that develop in MoS_2 materials and also discussed the MoS_2 for different applications.

2 Structure of Different Phases of MoS_2

2.1 1T, 2H, and 3R Phase

As seen in Fig. 1, two-dimensional (2D) MoS_2's single layer has the 0.7 nm-thick lamellar S-Mo-S structure [10]. Each MoS_2 2D crystal layer is made up of two S atoms in a trigonal prismatic shape and two S atoms in a hexagonal plane that are both modified by covalent interactions.

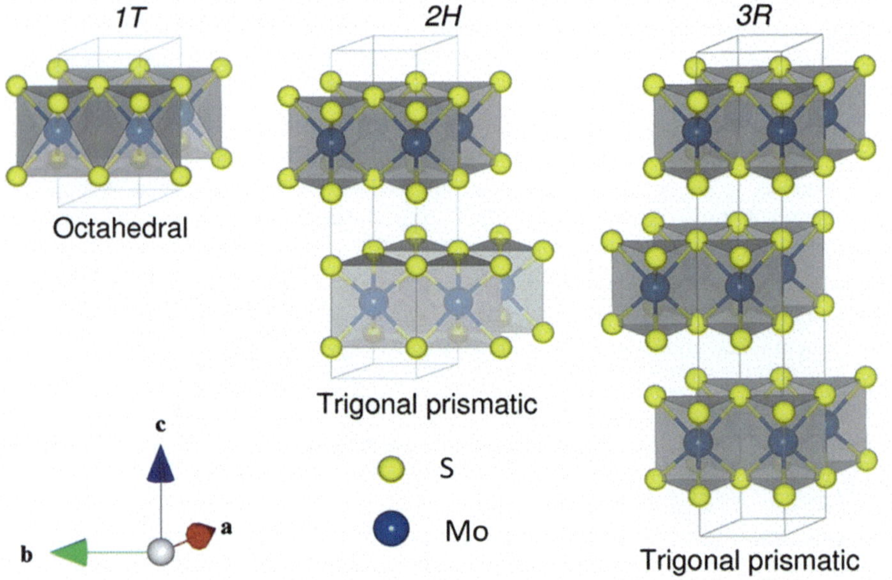

Fig. 1 Crystal structure of MoS_2: Octahedral (1T), trigonal prismatic (2H), and trigonal prismatic (3R) unit cell structures. Produced with permission [10]

Weak van der Waals forces connect adjacent atomic sandwich units. MoS_2 has different crystalline structures such as hexagonal (H), tetrahedral (T), and rhombohedra (r). It has 2H and 3R semiconducting phases and 1T metallic phase. Among these, the 1T phase is
more conductive and less stable compared to the 2H phase [10]. The 2H-MoS_2 (2H phase) typically has a band gap in the 1.3–2 eV, but the octahedral 1T phase has metallic characteristics and significantly lower charge transfer resistance [11, 12]. The unstable 1T-MoS_2 with the S-Mo-S framework tends to change into stable 2H-MoS_2 as a result of S-S van der Waals interaction. The d orbital breaks into three degenerate states during the 1H phase, with an energy gap of B1 eV between the dz2 and dx2y2xy orbitals. Whereas, for tetragonal symmetry of the 1T phase, the d-orbitals of the metal degenerate into dxy, yz, zx (t2g) and dx2y2, z2 (eg) orbitals. The phase transition from 2H-MoS_2 to 1T-MoS_2 and 1T-MoS_2 to 2H-MoS_2 occurs in certain conditions [13, 14]. Phase engineering can be utilized to modify both the materials' polytype and electrical characteristics. Both phases are prominent for different applications. The 2H-MoS_2 was wildly used for the field-effect transistor (FET), solar cells, and memory devices [15–18]. The d-orbitals in 2H-MoS_2 are divided by an energy gap into three bands: z2 d, x2-y2, xy d, and x z, yz d. As a result, 2H-MoS_2 displays semiconductor properties and has weak charge transfer abilities. Whereas the 1T-MoS_2 was most active for electrocatalyst and energy storage devices [19, 20] because the metallic phase removes the Schottky barrier and increases the charge transfer rate, which has clear benefits in electrochemical applications.

Figure 2a–c [21] show the monolayer-MoS_2 estimated band structures in the 1H, 1T, and 1T′ phases. As per the theoretical investigation, the *K*-*K* transition is observed in the d orbit of the transition metal. The transition metal d-orbitals contain two electrons, which are often coupled and fully occupied in their lowest dz2 state as a result of the hexagonal symmetry of its 1H phase. The STM analysis demonstrates the different layers ((single (SL), Bi (BL) and trilayer (TL)) with the honeycomb-like feature as shown in Fig. 2d. The optical spectra of MoS_2 are presented in Fig. 2e [21].

3 Phase Transition in MoS_2

3.1 Importance of Phase Transition in MoS_2

The properties of MoS_2 can be modified by the phase transition, which may be changed by varying the electron density within MoS_2. It is still challenging to control the phase transition in most two-dimensional materials. The utilization of MoS_2 may be improved with a greater understanding of the process and the circumstances that cause the phase change. The phase transition can improve the current application [22, 23]. Recently, K. Leng et al reported chemical lithiation of bulk MoS_2 distorts its

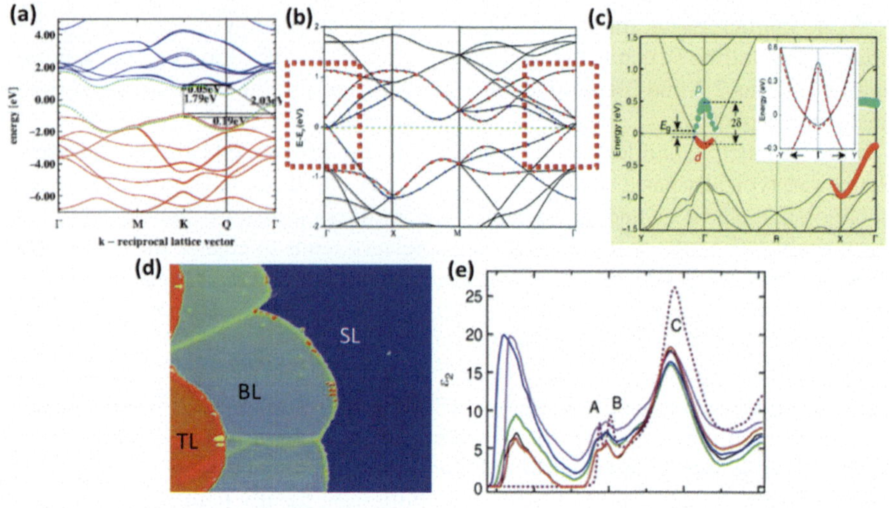

Fig. 2 The band structures (left side), total densities of states (DOSs), and estimated DOSs (right side) of the d orbital of Mo atom (Mo-d) and p orbital of S atom (S-p) for **a** 1H-MoS_2, **b** 1T-MoS_2, and **c** 1T′-MoS_2. The lines indicated by red and blue dot triangles represent the bands that contributed by the 4d-orbitals of Mo atoms and 3p-orbitals of S atoms, respectively. **d** STM image of single-layer, bilayer, and trilayer. (e) Optical spectra of MoS_2. Panel **a**–**e** reproduced from Ref. [14] with permission from Elsevier, copyright 2012

crystalline domains in three primary directions for energy storage which enhanced capacitance and stability, shown in Fig. 3a [24].

Metallic 1T-MoS_2 is highly stable compared to the semiconducting 2H phase. The 2H/1T-phase transition improved conductivity and HER activity with the lowest Tafel slope, as shown in Fig. 3b and c [25]. The transition from 2H to metallic 1T/1T′ has been extensively observed through metal ion intercalation, which is the ideal candidate for energy storage as it shows excellent specific capacitance, as shown in Fig. 3d [26]. Mostly, to break 2H-MoS_2 into three orientation variations, the phase transition (2H-1T′) is observed which results in uniform charge distribution and more homogenous phase conversion. Moreover, the phase transition can be utilized to promote the applications of surface Raman enhancement as shown in Fig. 3e. The 2H-to-1T phase transition enhanced the speed of charge transfer between material and environment. The phase transition gives chance to TMDs to fully utilize the advantages of all phases. As well as it provides TMDs with a large platform for several unique applications involving the alteration of electron structures.

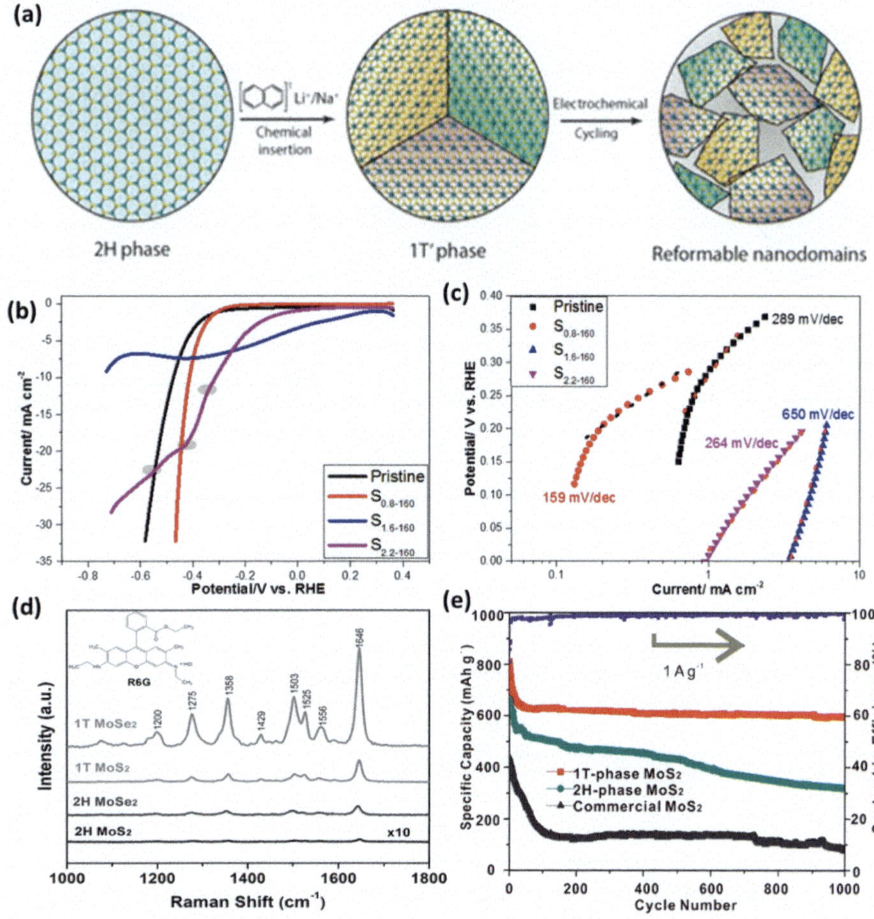

Fig. 3 **a** Schematic diagram in which domain size reduction after chemical Li intercalation obstructs the material to form nanocrystals during electrochemical Li intercalation, **b** Polarization curve, and **c** Tafel plot of MoS_2 nanosheet. **d** Raman spectra: The intensity of Raman spectra is enhanced after the charge transfer from 1T′ phase substrate to target molecules. **d** Cycling behaviors of these three electrodes at the current density of 1 A g^{-1} and the upper line corresponding to the Coulombic efficiency of 1T-phase MoS2 electrode during the cycles. Panel **a** reproduced with permission [24]. Panel **b** and **c** reproduced Copyright 2017, WILEY–VCH Verlag GmbH & Co. Panel **d** reproduced with permission [26]

3.2 Different Approaches to Prompt the Phase Transition

3.2.1 Doping

As per the previous study, the transition barrier between 2H and 1T can reduce by prominent doping [27]. Due to the higher energy difference between 2H and 1T in

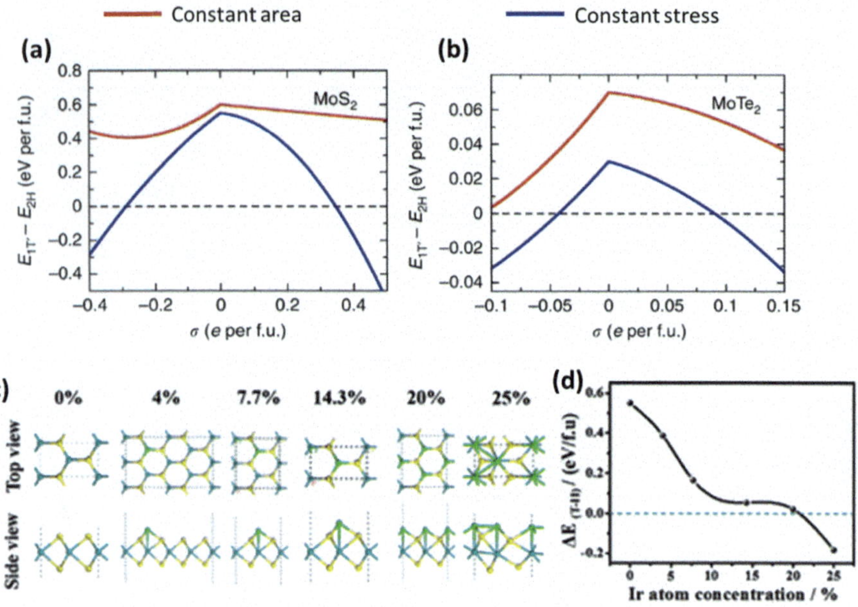

Fig. 4 **a–b** The energy difference between 2H and 1T′ phases in MoTe2 and MoS2 under constant stress and constant area conditions, respectively. Theoretical model **c** and energy differences **d** between Ir/1T-MoS2 and Ir/2H-MoS2 as a function of the atomic concentration of the adsorbed Ir. Panel **a** and **b** are reproduced under the terms of the Creative Commons Attribution 4.0 International License. [28] Copyright 2016, the Authors, published by Springer Nature. Panels **c** and **d** were produced with permission [29]

MoS_2, it is difficult to prompt phase transitions to compare to the other material [28].The MoS_2 exhibits an energy difference of 0.6 eV under area and 0.5 eV under constant stress without any doping. Whereas the energy difference in $MoTe_2$ is 0.03 and 0.07 eV, respectively, shown in Fig. 4a and b [28]. S. Wei et al. demonstrated that the energy difference between Ir/2H-MoS_2 and Ir/1T-MoS_2 slowly decreases with the concentration of Ir. After some concentration (20% Ir), the energy difference becomes negative which confirms that the 1T-MoS_2 is more stable than the 2H phase. Figure 4c and d [29] further confirms that the high-density Ir atoms may be adsorbed on the surface of MoS_2 and cause a phase transition of the material.

3.2.2 Heterostructures

The phase individual layer mainly depends on the d orbital electronic density of transition metal. Tuning the filling of the d orbital permits phase engineering. The phase modification in transition metal (MoS_2, WS_2, $MoSe_2$, etc.) has been already reported. As shown in Fig. 5a–d [30], Yazdani et al. examined the heterointerface on intercalation-induced phase transition demonstrating the more stable 1T phase

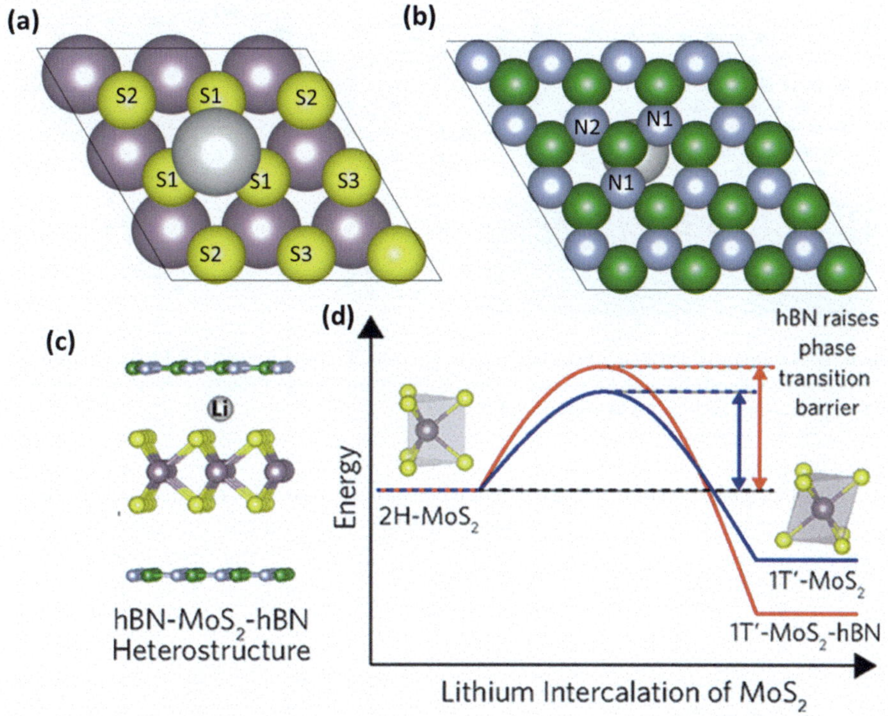

Fig. 5 Distribution of charge density and atoms near Li in an hBN-MoS_2-hBN heterostructure. Top view of **a** heterostructure presenting the Li (white) and its Nth-nearest-neighbor S atoms represented as first-S1, second-S2; **b** heterostructure with hBN layer above a Li atom presenting the first and second nearest-neighbor N atoms represented as N1 and N2, respectively; **c** schematic presentation of the formation of heterostructures; **d** band structure after lithium intercalation of MoS_2. Panel (a-d) reproduced from [30]

compared to the 2H phase when Li inserted on the surface. The interface between two materials lowers the energy barrier of phase transition. Sometimes, the exfoliated single-layer TMDs exhibit incomplete phase change, resulting in distortion in one of the phases. Z. Hu et al. observed the positive and negative energy difference in the case of phase transition of 1T-1T$''$ which happens in the zigzag shapes, and 2H-2H$'$, 1T-1T$'$ in armchair shape, respectively [31].

3.2.3 Plasma Induction

Phase transition using plasma treatment is one of the clean, most efficient, and facile methods. Recently, Qiu et al. reported the phase transition by inducing Ar plasma in MoS_2 which is limited up to 40% and N2 bombarded plasma which elevates the 2H-to-1T phase transition to a conversion of ca. 62% [32].

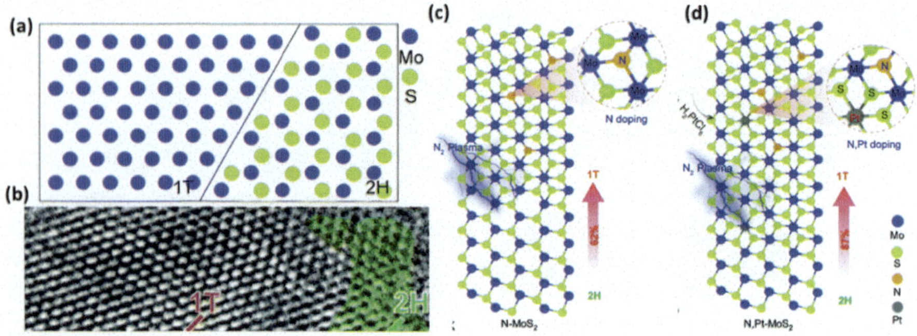

Fig. 6 **a** Schematic illustration of the lattice structures of 2H and 1T phases of MoS_2 nanosheets, **b** HRTEM images of 2H and 1T phases of MoS_2, Structural illustration of **c** N-MoS_2 and **d** N, Pt-MoS_2 nanosheets. Panel **a–d** reproduced from [32]

After the plasma induction, the surface becomes rough which is depicted in scanning electron microscopy (SEM) images of MoS_2 (Fig. 6a and b). Also, the Mo-S bond twisted due to N2 plasma treatment which results in the partial replacement of the S atom by the N atom, causing the phase transition from 2H to 1T, shown in Fig. 6c and d.

3.2.4 Synthesis Control

There are various methods used for the synthesis of TMDs. The phase formation can be triggered by the synthesis method by tuning parameters such as temperature, time, concentration, and procedure.

The temperature-dependent phase transition observed in MoS_2 by the microwave-assisted route is shown in Fig. 7a. The MoS_2 nanosheet exhibits 1T phase with metallic properties. The 1T phase will return to the stable 2H phase after surmounting the relevant energy barrier because of the 1T phase's metastability. Also, the reduction of Pt salt while microwave irradiation will foam the 2H-MoS_2/Pt NP composite. Further, Fig. 7b of absorption from the UV to NIR region of 1T-MoS_2 demonstrates the A-, B- and C-excitons of the 2H-MoS_2 after the few minutes of irradiation, suggesting the 1T-to-2H phase transition. Mostly, the MoS_2 mainly focused on the 2H and 1T phases. A 3R phase of MoS_2 was first synthesized for the first time [33]. Recently, the 3R-MoS_2 was fabricated in a one-step sulfuration process using an annealing treatment. The scanning electron microscopy (SEM) image indicates a 3R phase with flake-like structure and high-resolution transmission microscopy (HRTEM) displays that the lattice fringes are distributed uniformly on the surface of MoS_2, and the interplanar distance of 0.27 nm corresponds to (101) lattice plane of 3R-MoS_2, displayed in Fig. 8a–c [34].

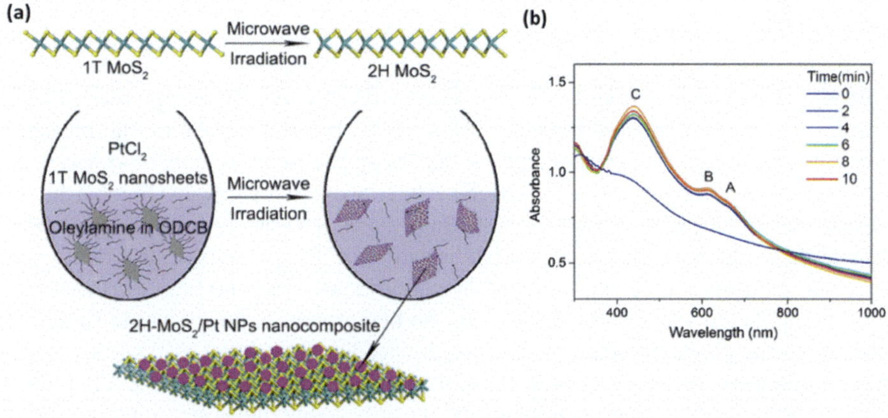

Fig. 7 **a** Phase conversion and the preparation of the 2H-MoS_2/Pt NP nanocomposite by microwave method, **b** Absorption spectra of MoS_2 nanosheets in a solution after microwave irradiation for different times. Panel **a–b** produced with permission [33]

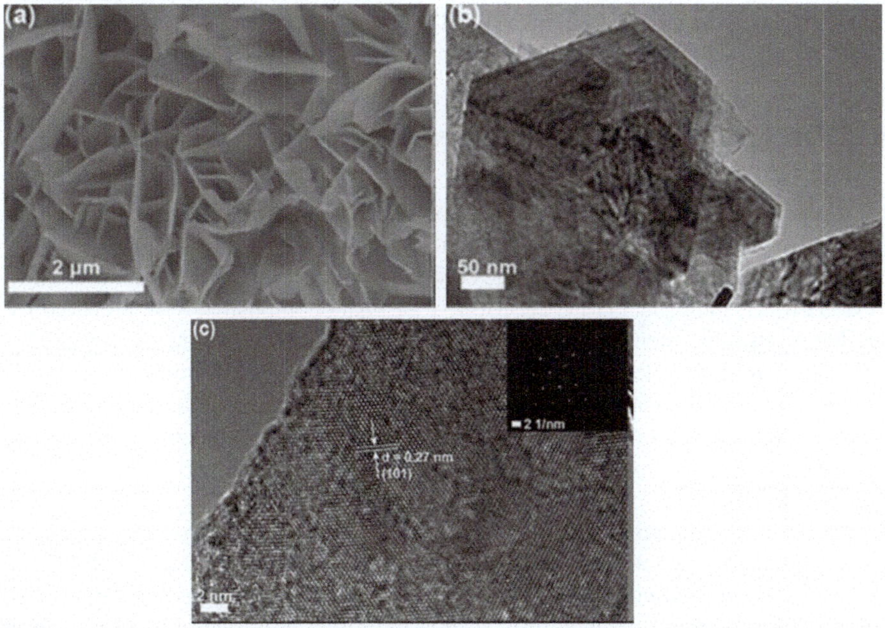

Fig. 8 **a** SEM image and **b** a TEM image of MoS_2-700. **c** An HRTEM image of 3R phase of MoS_2. Panel **a–c** produced with permission [34]

3.3 Strain

In MoS_2, both uniaxial and isotropic pressures can be triggered by compression which initiates the phase transition between 2H and 1T′. Besides the strain, the physical collision also causes the phase transition. Recently, the Ar plasma method was used to trigger phase transition 2H-1T′ in MoS_2. The low-energy Ar plasma bombardment influences the 2H phase in two ways: (i) introduces S vacancies and (ii) facilitates the sliding of the S layer. Large-scale fabrication of 2H-1T′ hybrid structures with particular patterns is made possible by the pattern mask (Fig. 9a) [35]. Furthermore, by utilizing the Ar plasma employed here, no heteroatom impurity would be added throughout the phase transition procedure. It was discovered that a 40 s low-energy plasma treatment can only convert up to 40% of the 2H phase into the 1T′ phase. Even after the weak contact, generated 1T′ phase is stable under ambient conditions due to the stabilization effect of S vacancies. The Raman spectra after seem to induce phase transitions, in good agreement with the mark of the 1T-ML-MoS_2, as shown in Fig. 9b. Also, it suggests the gradual increase of the 1T-phase percentage of MoS_2 with a growing loss of MoS_2 crystal symmetry. The PL spectra as displayed in Fig. 9c [35] confirm an increase in the metallic 1T-phase component.

4 Device Applications with Phase Transition

The change in the physical and structural properties of MoS_2 at the phase transition point has potential applications in many fields. For example, the floating gate transistor structure of MoS_2 has advantages toward the memory device application [36]. 1T@2H-MoS_2 nanosheet (1T@2H-MoS_2-PVP) nanocomposites were prepared for flexible nonvolatile memory devices which have low set/reset voltages, larger ON/OFF ratio, and long retention time [37]. Additionally, the phase transition in MoS_2 has played a key role in memory devices and photodetection [38]. The MoS_2 has potential applications in functional devices due to its different magnetic properties. Also, the MoS_2 transistor was fabricated by channel length scaling to the sub-10-nm regime (Fig. 10).

The fabricated metallic 1T-MoS_2 exhibits excellent electrocatalyst properties due to its conductivity and active sites. The facile and large-scale production of 1T-2H-MoS_2 nanosheet focused toward water splitting and energy conversion system [39]. The room temperature magnetoresistance (MR) effect is observed in MoS_2-based heterostructures [40]. Based on the spin direction, spin Hall magnetoresistance can be calculated. Thus, the phase transition MoS_2 enables new structure magnetoelectronic devices [41].

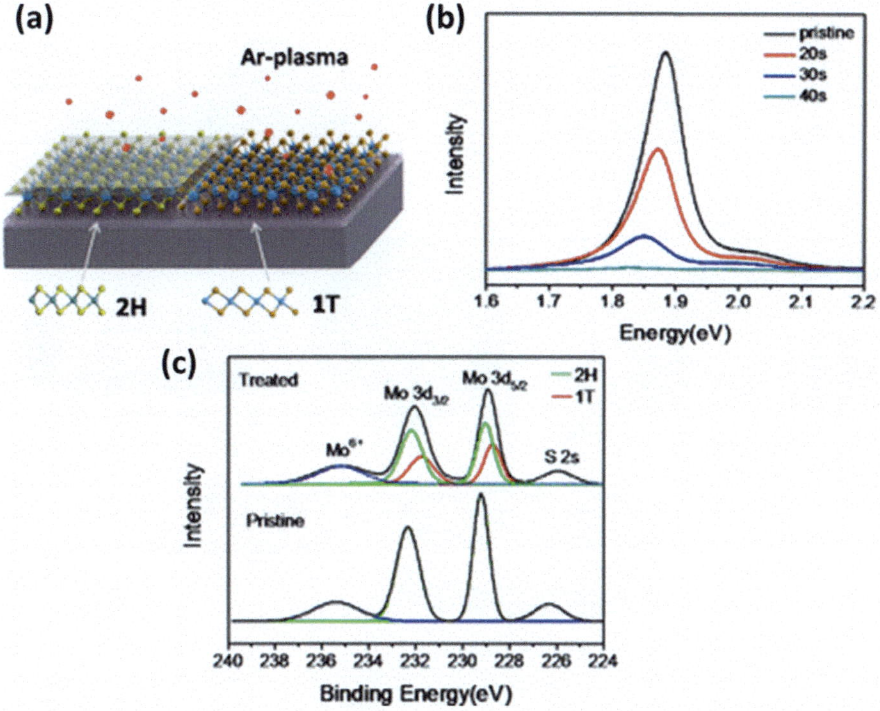

Fig. 9 1T-phase MoS_2 formation by plasma therapy. **a** Schematic diagram showing the plasma-treated process. **b** PL spectra of plasma-treated MoS_2 monolayer. **c** XPS spectra of pristine and plasma-treated MoS_2 showing Mo 3d and S 2s core-level peak regions. The green and red curves show the contributions of 2H and 1T phases to the Mo 3d peaks

5 Conclusion and Future Prospects

The properties of MoS_2 are determined based on the phase transition. So, modifying and controlling the phase of MoS_2 is important for future applications. In this section, we are focusing on the discussion on phase transitions in MoS_2. Phase engineering in MoS_2 provides chances for doing both basic and technologically significant research. The MoS_2 has a variety of phases in which 1T is not much stable. Therefore, the phase conversion will allow examinations of novel properties of the metallic phase. For practical applications, the MoS_2 by hydrothermal/solvothermal synthesis provides more active sites which boost the charge transfer mechanism. Therefore, controlling the phase transition by appropriate synthesis could be the key aspect in future. Though it is still challenging to synthesize large-scale high-quality film. Strain is another good factor to achieve a phase transition. The Ar bombardment causes phase transition. The doping and formation of heterostructures are also good choices to modulate phase transition.

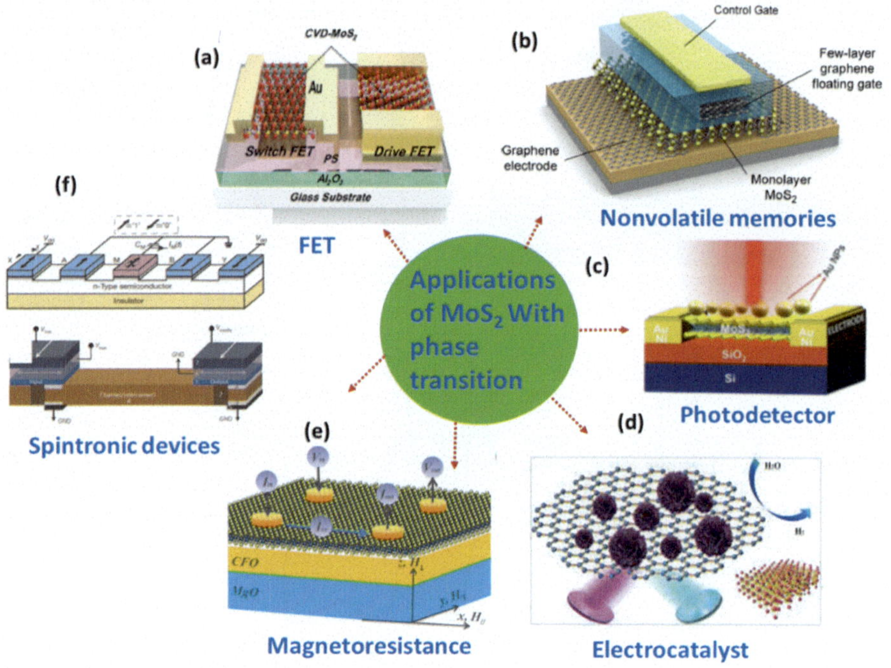

Fig. 10 Various applications of MoS_2-based devices. Panel **a** reproduced from Ref. [36], Panel **b** from Ref. [37]. Panel **c** reproduced from Ref. [38]. Panel **d** from Ref. [39]. Panels **e** and **f** were produced with permission [40, 41], respectively

Still, there are many issues related to phase transition. Firstly, for the communal finite group of the initial and final phases, a moral reversibility of phase transition is necessary [42]. The H and T phases of MoS_2 don't fulfill this condition, resulting in crystal distortion. The kinetics of phase transition which is less focused is the second point in a phase transition.

Here, we have discussed the best idea to improve phase transition. Alkali-ion intercalation offers clear benefits over alternative techniques that could permit phase change on a wide scale for energy storage applications. For other applications, doping and heterostructure formation are key strategies to prompt phase transition.

References

1. Fan X, Zhang G, Zhang F (2015) Chem Soc Rev 44:3023–3035
2. Zhang H (2015) ACS Nano 9:9451–9469
3. Chauhan P, Patel AB, Solanki GK, Machhi HK, Soni SS, Pathak VM, Patel V, Narayan S, Jha PK (2021) Adv Opt Mater 9:1–10
4. Choi W, Choudhary N, Han GH, Park J, Akinwande D, Lee YH (2017) Mater Today 20:116–130
5. Pandey M, Vojvodic A, Thygesen KS, Jacobsen KW (2015) J Phys Chem Lett 6:1527–1585

6. Patel AB, Machhi HK, Chauhan P, Narayan S, Dixit V, Soni SS, Jha PK, Solanki GK, Patel KD, Pathak VM, Appl ACS (2019) Mater Interfaces 11:4093–4102
7. Li HM, Lee D, Qu D, Liu X, Ryu J, Seabaugh A, Yoo WJ (2015) Nat Commun 6:1–9
8. Jiang D, Liu Z, Xiao Z, Qian Z, Sun Y, Zeng Z, Wang R (2022) J Mater Chem A 10:89–121
9. Li X, Zhu H (2015) J Mater 1:33–44
10. Han B, Hu YH (2016) Energy Sci Eng 4:285–304
11. Ambrosi A, Sofer Z, Pumera M (2015) Chem Commun 51:8450–8453
12. Tang Q, Jiang DE (2015) Chem Mater 27:3743–3748
13. Hu JQ, Shi XH, Wu SQ, Ho KM, Zhu ZZ (2019) Nanoscale Res Lett 14:1–13
14. Lin S, Li X, Wang P, Xu Z, Zhang S, Zhong H, Wu Z, Xu W, Chen H (2015) Sci Rep 5:1–9
15. Tong X, Ashalley E, Lin F, Li H, Wang ZM (2015) Nano-Micro Lett 7:203–218
16. Liao F, Guo Z, Wang Y, Xie Y, Zhang S, Sheng Y, Tang H, Xu Z, Riaud A, Zhou P, Wan J, Fuhrer MS, Jiang X, Zhang DW, Chai Y, Bao W, Appl ACS (2020) Electron Mater 2:111–119
17. Ma J, Bai H, Zhao W, Yuan Y, Zhang K (2018) Sol Energy 160:76–84
18. Sebastian A, Pendurthi R, Choudhury TH, Redwing JM, Das S (2021) Nat Commun 12:1–12
19. Sumesh CK (2020) Int J Hydrog Energy 45:619–628
20. Jayabal S, Wu J, Chen J, Geng D, Meng X (2018) Mater Today Energy 10:264–279
21. Yin X, Tang CS, Zheng Y, Gao J, Wu J, Zhang H, Chhowalla M, Chen W, Wee ATS (2021) Chem Soc Rev 50:10087–10115
22. Chhowalla M, Voiry D, Yang J, Shin HS, Loh KP (2015) MRS Bull 40:585–591
23. Chen W, Gu J, Liu Q, Luo R, Yao L, Sun B, Zhang W, Su H, Chen B, Liu P, Zhang D (2018) ACS Nano 12:308–316
24. Leng K, Chen Z, Zhao X, Tang W, Tian B, Nai CT, Zhou W, Loh KP (2016) ACS Nano 10:9208–9215
25. Gan X, Lee LYS, Wong KY, Lo TW, Ho KH, Lei DY, Zhao H, Appl ACS (2018) Energy Mater 1:4754–4765
26. A. Manuscript. https://doi.org/10.1039/xxxxxx
27. Li L, Li Z (2019) Mater Res Express 6:0–8
28. Li Y, Duerloo KAN, Wauson K, Reed EJ (2016) Nat Commun 7:1–8
29. Wei S, Cui X, Xu Y, Shang B, Zhang Q, Gu L, Fan X, Zheng L, Hou C, Huang H, Wen S, Zheng W (2019) ACS Energy Lett 4:368–374
30. Yazdani S, Pondick JV, Kumar A, Yarali M, Woods JM, Hynek DJ, Qiu DY, Cha JJ, Appl ACS (2021) Mater Interfaces 13:10603–10611
31. Voiry D, Yamaguchi H, Li J, Silva R, Alves DCB, Fujita T, Chen M, Asefa T, Shenoy VB, Eda G, Chhowalla M (2013) Nat Mater 12:850–855
32. Sun Y, Zang Y, Tian W, Yu X, Qi J, Chen L, Liu X, Qiu H (2022) Energy Environ Sci 15:1201–1210
33. Xu D, Zhu Y, Liu J, Li Y, Peng W, Zhang G, Zhang F, Fan X (2016) Nanotechnology 27:1–7
34. Fang B, Yao J, Zhang X, Ma L, Ye Y, Tang J, Zou G, Zhang J, Jiang L, Sun Y (2021) New J Chem 45:2488–2495
35. Zhu J, Wang Z, Yu H, Li N, Zhang J, Meng J, Liao M, Zhao J, Lu X, Du L, Yang R, Shi D, Jiang Y, Zhang G (2017) J Am Chem Soc 139:10216–10219
36. Nourbakhsh A, Zubair A, Sajjad RN, Amir TKG, Chen W, Fang S, Ling X, Kong J, Dresselhaus MS, Kaxiras E, Berggren KK, Antoniadis D, Palacios T (2016) Nano Lett 16:7798–7806
37. Bertolazzi S, Krasnozhon D, Kis A (2013) ACS Nano 7:3246–3252
38. Hu T, Zhang R, Li J-P, Cao J-Y, Qiu F (2022) Chip 1:100017
39. Gao B, Du X, Li Y, Ding S, Xiao C, Song Z, Appl ACS (2020) Mater Interfaces 12:877–885
40. Jie W, Yang Z, Zhang F, Bai G, Leung CW, Hao J (2017) ACS Nano 11:6950–6958
41. Ahn EC, NPJ 2D Mater Appl. https://doi.org/10.1038/s41699-020-0152-0
42. Bhattacharya K, Conti S, Zanzotto G, Zimmer J (2004) Nature 428:55–59

Chapter 7
Molybdenum Disulfide (MoS_2) as an Efficient Solid Lubricant

Shweta Sharma and Dattatray J. Late

1 Introduction

In the 1960s, increasing machine failures in heavy industry due to equipment wear led to high financial losses. In the subsequent years, with the rapid rise in technological advancement, the issue of wear, friction, and corrosion increased. However, in March 1966, Dr. H. Peter Jost, known as the "Father of Tribology," presented a report called "The Jost Report" that revolutionized the work in this field [1]. The report explained "the science and technology of interacting surfaces in relative motion". Dr. Jost also introduced the concept of "Green Tribology," which explained the tribological characteristics in the context of the ecological balance of the environmental and biological impacts [1, 2].

With the expansion in the economy, the use of two terminologies, friction and wear, is entirely on the rise. These terms are widely used in transportation, aerospace, machine manufacturing, and daily life [3, 4]. Statistically, the energy released by friction is around one-third of the total annual consumption of fossil fuels. Further, the associated wear considerably reduces the life cycle of the moving parts [5, 6]. So, researchers are conducting in-depth research in tribology to eliminate the adverse effects of friction by achieving high lubricity or low friction and wear in mechanical systems [7].

Lubricants are used for dealing with the issue of friction and wear, which forms a protective layer on the surface of the substance to prevent direct contact of the friction pairs. Liquid and solid lubricants are widely used to optimize tribological

S. Sharma
Schulich Faculty of Chemistry, Technion-Israel Institute of Technology (IIT), 32000 Haifa, Israel

D. J. Late (✉)
Department of Physics, Federal University of Lavras, Campus Universitário, PO Box 3037, Lavras, Minas Gerais, Brazil
e-mail: datta099@gmail.com

D. J. Late and C. S. Rout (eds.), *Science and Technology of 2D MoS_2*,
Materials Horizons: From Nature to Nanomaterials,
https://doi.org/10.1007/978-981-97-7367-1_7

properties [8]. Solid lubricant materials are employed as a powder or thin film to safeguard against damage when in relative motion and to reduce friction and wear. Some other commonly used terms include dry lubrication, solid-film lubrication, and dry-film lubrication. Liquid lubricants are added to optimize the lubricating proficiency of the base oils. Solid lubricants are preferred over liquid lubricants due to some advantages, such as permanent lubrication without leakages, reduced heat generated, and low startup torque. Further, solid lubricants provide extensive load-bearing tendency, stability in cryogenic temperatures, freedom for system designing, and exclusion of oil pumps compared to liquid lubricants [9, 10]. Archaeologists have found that Egyptians used gypsum (hydrated calcium sulfate) during the 3rd millennium BC to ease the movement of big construction stone blocks [11]. Some of the desirable characteristic properties of solid lubricants are

(i) to deliver low, persistent, and controlled friction among the two bearing surfaces;
(ii) to be chemically stable over the required temperature range;
(iii) to not deteriorate the bearing material;
(iv) adhere intensely to one or both bearing surfaces;
(v) have adequate resistance to wear to enrich it with a useful life; and
(vi) to be non-toxic and economical.

A half-century ago, solid lubrication was deliberated as an art only, but it has now emerged as an integral part of engineering and materials science. It can be implemented for space applications, textile and food industries, transportation industries, mining, construction, agriculture and aviation applications, etc. Considering the classical theory formulated by Bowden and Tabor, friction force (F) results from the shear strength of lubricant, $A_r t$, and actual contact area. The friction coefficient, μ, can be expressed by the following equation [12]:

$$\mu = \frac{F}{L} = \frac{A_r \cdot \tau}{L} = \frac{\tau}{P_H} = \frac{\tau_0}{P_H} + \alpha,$$

where τ denotes the interfacial shear strength and α denotes the pressure dependence of the shear strength. The constant α is the lowest acquired friction coefficient for a known friction couple. Thus, it can be concluded from the above equation that the friction coefficient is the ratio of shear strength to pressure if both forces are divided by the contact area [13]. Solid lubricants are used in the automobile and shipping industries, machining fluids, mining, and power generation. Friction reduction in the mentioned industries helps in energy savings [14]. Different applications in medical devices, where friction plays a significant role, have also been reported.

Significant categories and examples of most commonly employed lubricant materials presented in the literature are given below and are also compiled in Fig. 1:

(i) Compounds such as graphite, titanium disulfide, tungsten diselenide, molybdenum disulfide, tungsten disulfide, molybdenum diselenide, titanium diselenide, and calcium fluoride.

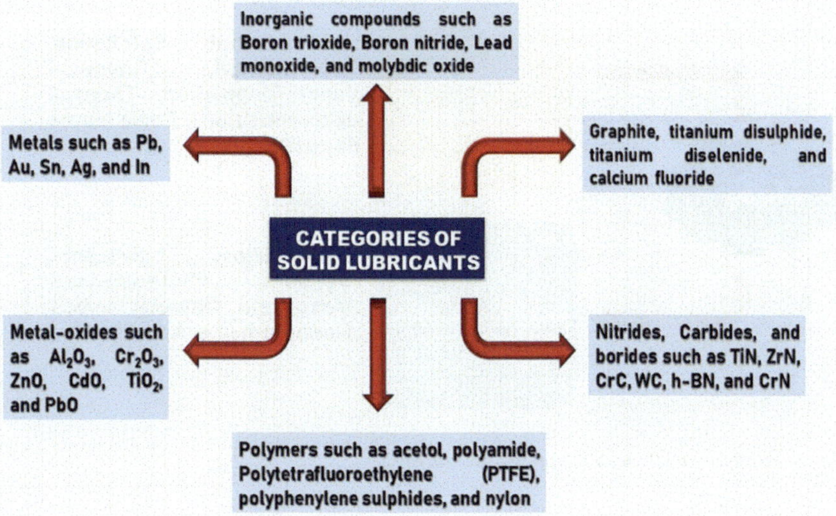

Fig. 1 Different categories and examples of solid lubricants

(ii) Polymers are also employed for this application, such as acetol, polytetrafluoroethylene (PTFE), polyphenylene sulfides, polyamide, and nylon.
(iii) Metals including Sn, In, Pb, Au, and Ag.
(iv) Inorganic compounds such as boron trioxide, boron nitride, lead monoxide, and molybdic oxide.
(v) Metal oxides such as Al_2O_3, Cr_2O_3, ZnO, CdO, TiO_2, and PbO, etc.
(vi) Nitrides, carbides, and borides such as TiN, ZrN, CrC, WC, hBN, and CrN.

In the recent few years, extensive research has been conducted on the lamellar two-dimensional (2D) materials such as layered double hydroxides (LDH), graphene, transition metal dichalcogenides (TMDCs), metal–organic frameworks, and transition metal carbides or nitrides (MXenes), etc. [15–17]. These materials exhibit superior structural and novel characteristics, broad utilization in biomedical technology, catalysis, energy generation/conversion, electronics, lubrication, optoelectronics, and catalysis [18–20]. Due to the variation in crystal configurations, various structural phases can also be observed at the molecular level. Lamellar compounds exhibit mainly three essential characteristics, which are (i) they deform at deficient shear stress levels, (ii) they adhere intensely to the worn surface, and (iii) there occurs no chemical degradation of the lamellar under the operating temperature [21]. Among the lamellar materials, TMDCs and graphene-like compounds have fascinated much research interest [22]. TMDCs are structural compounds with the chemical formula MX_2, where M represents the transition metal of groups IV-VII B (V, Mo, W, and Nb, etc.) and X denotes the chalcogen element (S, Se, and Te) [23]. The most common and extensively studied compound among TMDCs is molybdenum disulfide (MoS_2) which is a 2D material with extraordinary properties such

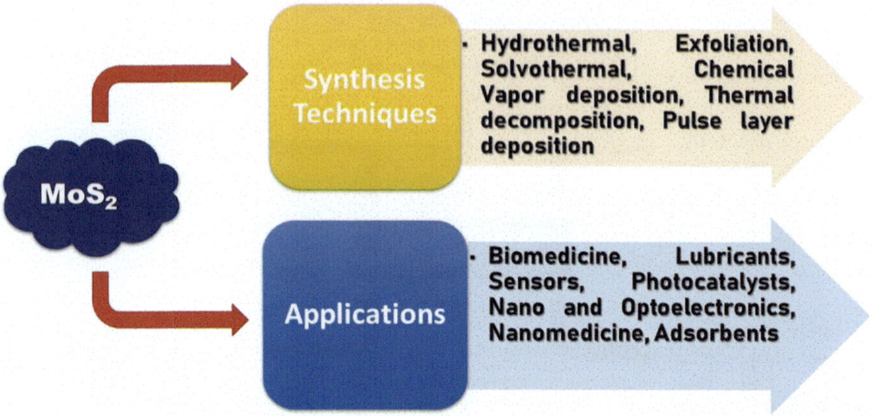

Fig. 2 Synthesis methods and applications of MoS_2

as low tendency to contaminate products, unique bandgap structure, exceptional mechanical strength, remarkable magnetic features, high mobility, low volatility, stability to radioactivity, high absorbance in NIR region, practically no tendency to flow, chemical inertness, and high surface area-to-volume ratio [23, 24]. It occurs naturally as the mineral molybdenite in the earth's crust with dark-blue coloration. It is chemically stable, insensible to radiation, and resistant to most acids. It exhibits semiconductor and diamagnetic properties in its purest form [11]. Possible synthesis techniques and numerous applications of MoS_2 are presented in Fig. 2.

Fascinated by these enormous promises, researchers are exploring the in-depth information about MoS_2. So, this chapter is prepared to deliver a brief introduction about fascinating material to readers of various backgrounds to aid them in understanding its characteristics and exploiting this material further for solid lubrication-based applications. In this chapter, we have comprehensively studied various synthesis techniques, structures, properties, and methodologies employed for MoS_2 coatings. In one of the sections, we briefly discuss the possible methods used for MoS_2 coating. We have briefly introduced various possible mechanisms responsible for heat production. Finally, the ability of MoS_2 to act as a solid lubricant is also supported by a proper literature survey.

2 Structure of MoS_2

The naturally occurring mineral of crystalline MoS_2 is molybdenite. MoS_2 consists of a single plane of atoms per sheet, which comprises a layer of Mo atoms sandwiched between two layers of S atoms. Strong covalent bonds hold Mo and S atoms together, and weak van der Waal's forces stack the layers. It exists in several polymorphic forms due to the disparity in the atomic stacking and symmetry, such as 1T, 2H, and

3R where T, H, and R denote tetragonal, hexagonal, and rhombohedral symmetries, respectively, and 1, 2, and 3 represent the number of layers present in an individual crystallographic unit cell [25, 26]. The single-sheet structure having trigonal prismatic coordination around Mo and having semiconducting properties is represented as 1H ("hexagonal") structure, and for octahedral bonding, it is designated as metallic 1T ("trigonal") phase (Fig. 3a). As per the density-functional theory (DFT) and X-ray diffraction (XRD) studies, 1T is quite an unstable form in the context of the distortion and is metallic (Fig. 3b) [27]. 1T' phase is the distorted 1T structure that triples the unit cell [28–30]. The thermodynamically stable and extensively explored polytype of MoS_2 is 2H, which is semiconducting but has negligible electrical conductivity [31]. The 3R polytype consists of three sheets per cell with ABC stacking in rhombohedral symmetry and is also semiconducting [32, 33]. It is commonly employed for industrial applications such as hydrodesulphurization catalysts, dry lubricants, and lithium-ion batteries [34].

The Mo-S bond length in all polytypes is approximately 2.4 Å [26], and the S–S bond length is 3.08 Å [35]. As per the DFT, 2H and 3R have comparatively similar energies [26]. The reduced surface energy and insensitivity to stacking of 2H polytype reduces its sliding barriers and, thus, exhibits low friction [36]. Although 2H and 3R polytypes are abundant in nature, each polytype can easily be synthesized by chemical vapor transport (CVT) in the laboratory. The UV–Vis absorption spectrum

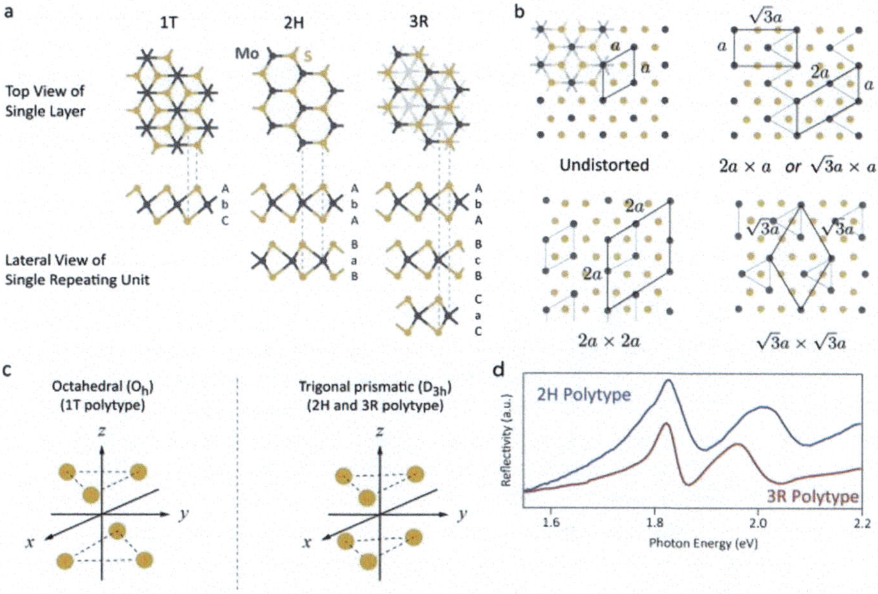

Fig. 3 **a** Schematic structures of the polytypes of MoS_2 with a top and lateral views, **b** Schematic presentation of 4 possible distortions of 1T polytype, **c** Splitting Mo 4d orbitals under octahedral (1T polytype) and trigonal prismatic ligand field (2H and 3R polytypes), and **d** Optical reflectivity of 2H- and 3R-MoS_2 polytype. Reprinted from ref. [34] Copyright (2015) RSC

of 2H and 3R exhibits major peaks at 1.9 and 2.1 eV, usually designated as the A1 and B1 exciton absorptions, respectively (Fig. 3d). Table 1 includes the lattice parameters and other essential characteristics of major polytypes of MoS_2.

The tribological properties of graphite and MoS_2 are assumed to be the same, considering the similarity in their crystal structure, as shown in Fig. 4. Graphite consists of a single plane of carbon atoms per sheet; however, MoS_2 consists of several sheets, and every sheet comprises a Mo atom plane sandwiched between two S atom planes. Both consist of strong bonds between the atoms, but weak interlayer strength is due to the high interlayer distance. However, this bond strength anisotropy is quite prominent in MoS_2 compared to graphite, as layers of the former slide easily over each other without causing structural damage [37]. Interestingly, it exhibits a constant friction coefficient value of 0.2 up to 800 °C, a temperature after which the decomposition of the MoS_2 occurs [38]. In addition, it offers high lubricity in vacuum environments, whereas graphite or graphene's lubrication ability fails in such conditions.

Table 1 Lattice parameters and characteristics of polytypes of MoS_2 [34]

Polytypes	Atoms per Conv. Cell	Stacking order	Lattice parameters	Properties	Common application
1T	9	AbC	a = 5.60 Å, c = 5.99 Å	Metallic, metastable	Intercalation
2H	6	AbABaB	a = 3.16 Å, c = 12.29 Å	Semiconducting, naturally occurring	Dry lubricants
3R	9	AbABcBCaC	a = 3.17 Å, c = 18.38 Å	Semiconducting, naturally occurring	Dry lubricants and non-linear optical devices

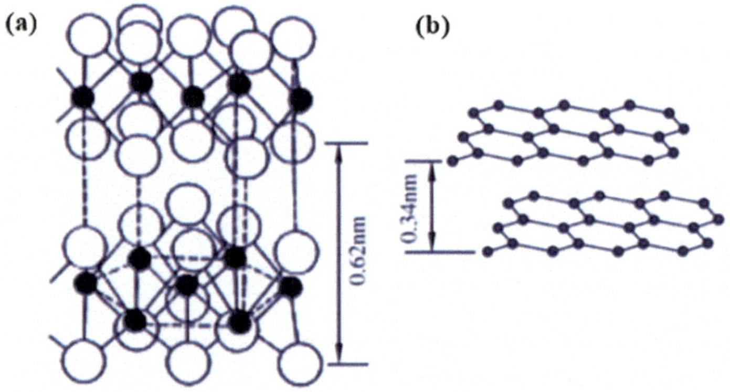

Fig. 4 Crystal structure of **a** MoS_2 and **b** graphite. Reprinted from ref. [39] Copyright (2015) Springer

3 Synthesis Techniques for MoS_2

The synthesis technique allows the manipulation of characteristic properties of the molecules, such as cross section, shape, size, porosity, stoichiometry, and surface arrangement. For the designing of MoS_2-based lubricants, prominently employed synthesis techniques are mechanical exfoliation, microwave-assisted, hydrothermal, and chemical vapor deposition (CVD), etc. [40].

3.1 *Mechanical Exfoliation*

Mechanical exfoliation, or simply "micromechanical cleavage," was first explained by *Novolselov* et al. in their experiment, where bulk crystals of 2D graphite were converted into a few layers and flakes by mechanical fragmentation [41]. It is also known as the "Scotch tape method," where bulk crystals are detached by rubbing against a solid surface. A gold film is commonly used as an intermediary substrate that improves mechanical exfoliation as it exhibits appreciable tolerance for chalcogens and efficiently overpowers the van der Waal's forces on the top layer and residue of the crystal. It is an effective way to create atomically thin nanosheets, pure and crystalline materials. Despite such effectiveness, this technique cannot be used for large-scale processing due to limited yield, which limits its practical utility [42].

3.2 *Electrochemical Exfoliation*

It is a simple technique carried out in mild conditions and suitable for large-scale production. This method is used for the exfoliation of bulk-layered MoS_2. In this method, metal foil of lithium and MoS_2 is used as cathodic and anodic poles, respectively. When a constant current is discharged, lithium ions get intercalated in the bulk MoS_2 structures that weaken the van der-Waal's forces between the layers. After washing, the intercalated compounds are ultrasonicated in water or acetone to exfoliate the MoS_2 sheets [42].

3.3 *Hydrothermal or Solvothermal*

Hydro/solvothermally synthesized MoS_2 has a homogeneous thickness of layers, controllable size, and high yield. It can be carried out in the presence of ionic liquids. In this method, Mo-salt and S-source reactions at high temperatures produce MoS_2. Pirarath and his coworkers in 2021 produced few-layered MoS_2 nanosheets using a surfactant-assisted hydrothermal system in which petal-like morphology was obtained [43]. Du et al. synthesized MoS_2 nanospheres using a solvothermal method using an ionic liquid. In dimethyl formamide/water mixed solvents, they employed the template of 1-ethyl-3-methylimidazolium bromide. The ratio of dimethyl formamide: H_2O exerts a crucial impact on the morphology and size of MoS_2 produced [44]. Several papers on MoS_2 synthesis by hydro/solvothermal method are reported in the literature [45–49].

3.4 Microwave-Assisted Synthesis

It is a simple, scalable, ultrafast, secure, time, and energy-efficient way of synthesizing MoS_2 [50]. It provides a uniform and instant heating of the reaction vessel and reduces the reaction time. In one study, the microwave method was used, and a reaction time of just 200 s yielded fullerene-like MoS_2 nanoparticles with double layers of MoS_2 [51]. In another study, this technique assisted the growth of MoS_2 directly on a graphene substrate under an inert environment [52]. This method produces smoother edges of the layers [53].

3.5 Chemical Vapor Deposition (CVD)

It is a bottom-up technique that allows the synthesis of layered MoS_2 with controlled morphology, reduced structural defects, and high crystallinity [54]. In this method, the substrate, preheated at 700 °C, is placed in a furnace containing precursor and background gas. Reacted species diffuse on the surface of the substrate and result in successful deposition of material on the surface of the substrate. This technique generally requires extreme temperature conditions (around 700 °C) for the material growth and substrate such as SiO_2 for the growth [55]. Chen and his coworkers synthesized MoS_2 flakes using the CVD method, and the diagrammatic presentation of MoS_2 synthesis is presented in Fig. 5a, b. Here, two different zones of varying temperatures and different precursors were used. The S was placed in zone 1, and the Mo foil was placed on the SiO2/Si substrate in zone 2 in a downward direction. The furnace was flushed with Ar flow for 30 min at room temperature before the experiment. The optical microscope of the resultant product showed MoS_2 domains having an equilateral triangular configuration (Fig. 5 (c)), which was further consistent with the corresponding AFM images, and the height profile data is presented in Fig. 5e [56].

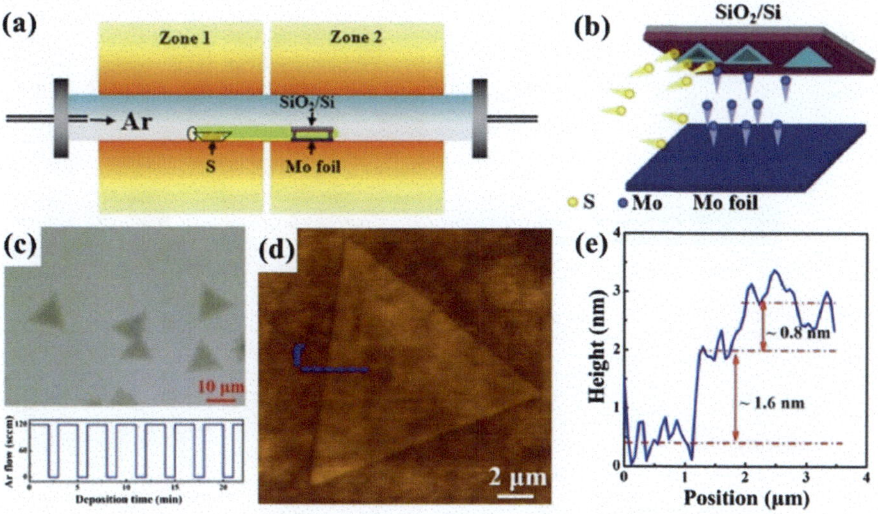

Fig. 5 **a** Diagrammatic presentation of equipment for the fabrication of MoS_2, **b** space-confined CVD growth of the MoS_2 samples, **c** optical microscope images of MoS_2 flakes and pulsed flow of Ar gas during the growth time, **d, e** AFM images and height profiles of MoS_2 flakes. Reprinted from ref. [56] Copyright (2020) Elsevier

3.6 Physical Vapor Deposition

In this method, high energy is used to convert pristine solid material to vapor form, solidifying on the substrate's surface under either a high vacuum or inert gases. This method also includes sputtering and pulsed laser deposition, in which sputtering is a commonly employed synthesis technique for MoS_2 films [57, 58]. A typical sputtering system consists of a power supply, a planar magnetic system, a pumping system, the source of substrate heating, and a gas inlet source [58].

Shahzad and his coworkers employed a two-step MoS_2 synthesis scheme employing sputtering and sulfurization (Fig. 6a, b). In the first step, a thin film of Mo was deposited on the SiO_2 substrate, and in the second step, sulfurization of Mo films was performed in a quartz tube furnace that resulted in MoS_2 films. Factors such as deposition time and sputtering parameters influenced the thickness of the formed MoS_2 sheets. AFM studies (Fig. 6c–h) suggested that Mo and MoS_2 synthesized by this method exhibited smooth and flat surfaces. Mono-, bi-, and multilayer MoS_2 samples were synthesized, and their corresponding thickness was found to be 0.76, 1.58, and 8.34 nm, respectively, and the corresponding Mo film thicknesses were 0.43, 0.95, and 4.05 nm, respectively [59].

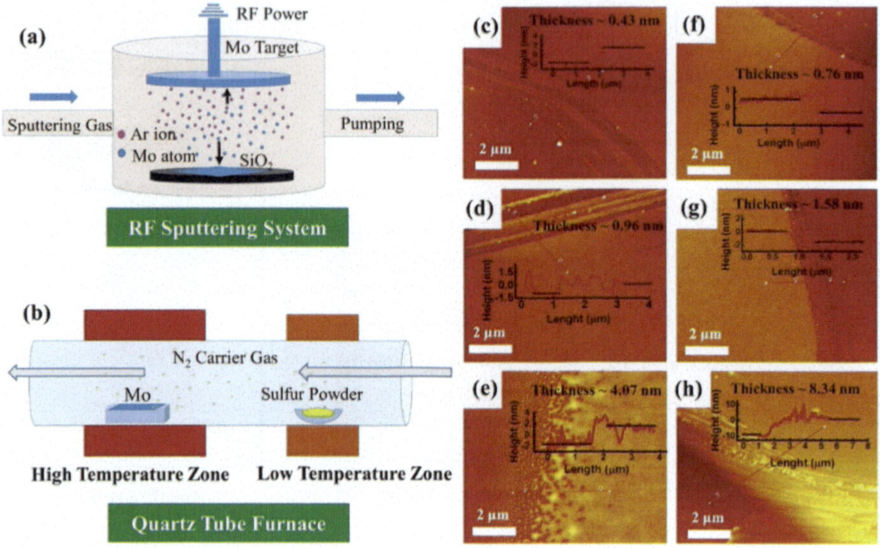

Fig. 6 Diagrammatic presentation of **a** sputtering system for deposition of atomic-scale Mo films; **b** sulfurization process of the pre-deposited Mo metal film in quartz tube furnace; AFM images with cross-sectional height profile insets: **c–e** sputter growth of Mo films deposited at different growth times and **f–h** corresponding mono-, bi-, and multilayer MoS_2 after sulfurization. Reprinted from ref. [59] Copyright (2017) Elsevier

4 Properties of MoS_2

The presence of oxygen and a humid environment can drastically reduce the friction ability of MoS_2. It can be employed under high vacuum and temperature as it doesn't require the absorbed vapors to act as a lubricant. The oxidation of MoS_2 starts at 350 °C and produces molybdic oxide (MoO_3) that also acts as a lubricant under extreme temperature conditions. Its characteristic properties change in the presence of strong alkali and oxidizing acids but are inert to most other chemicals; thus, it can be used where grease cannot be applied [60, 61]. Oxidation of MoS_2 significantly influences its properties, such as an upsurge in corrosion, increased friction, and reduced shelf life. Additionally, MoS_2 displays substantial instability and a high friction coefficient at 400 °C [62].

It has appreciable mechanical strength that is dependent on the thickness. For bulk MoS_2, Young modulus is found to be 0.24 TPa, which increases to 0.33 TPa, which increases with the reduction in the thickness of the flake. High elasticity and low pre-tension make it a compelling candidate for semiconductor composite films. Atomically thin sheets of MoS_2 can easily bear elastic deformations up to 25% before rupturing. Further, the bending modulus of MoS_2 is higher than graphene, with a single layer of S atom having 1.75 eV of bending modulus [62]. The average breaking strength of MoS_2 was $15 \pm 3\ Nm^{-1}$ [63].

For several centuries now, MoS_2 has been used as a solid lubricant, and its different shapes, such as nanotubes, nanoparticles, and platelets, have been studied [64]. Its low friction, high crystalline arrangement, and easy separation during synthesis make it a distinctive option. However, bulk MoS_2 should have three essential conditions to have low friction such as (i) the ability to slide against itself for which the transfer film needs to be made on its counterpart, (ii) its grains at the joint surface must align with the base plates parallel to the slip direction, and (iii) minimum contamination [65].

5 Deposition or Coating Methods of MoS_2

The durability of MoS_2 depends strongly on the method used for its deposition on the substrate. Strong adhesion between substrate and film is a fundamental criterion for prolonged survival of the lubricant film. Coated rolling bearings have been found to show a tenfold increase in fatigue lifetime and a 7-fold decrease in wear [66]. Techniques used for MoS_2 deposition include spray bonding, burnishing, vacuum coating, and particle incorporation in liquid or grease. The typical deposition method, properties, and applications of MoS_2 as a solid lubricant are presented in Fig. 7.

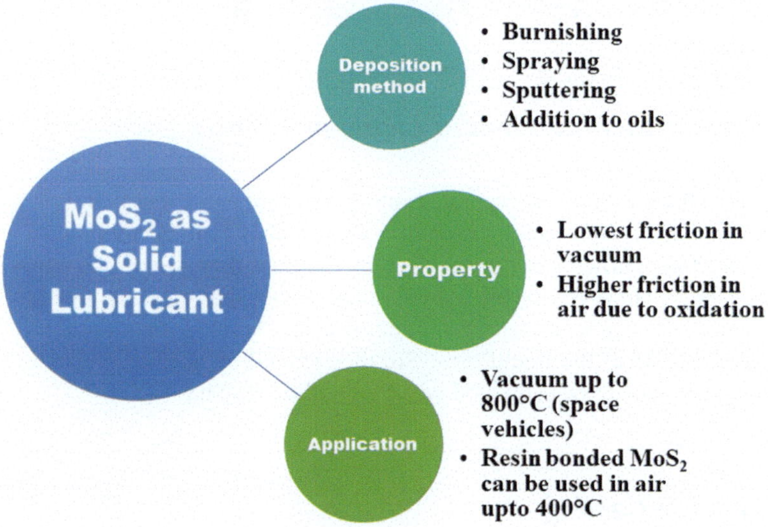

Fig. 7 Deposition method, properties, and applications of MoS_2 as a solid lubricant

5.1 Burnishing

It is one of the earliest methods to produce films of 0.1–10 μm thickness with cloth or brush. The films' thickness depends on the surface roughness, time, and relative humidity. The films produced by this method are significantly porous, contaminated, poorly adherent, and differ in thickness. However, this methodology is simple and cheap and can be easily employed in many non-critical applications [67, 68]. In this method, the metal surface is rubbed against MoS_2 powder, followed by its successive deposition, forming a thin lubricant film [60]. However, since this method does not use any binder, there is the possibility of low bondage of MoS_2 powder on the substrate.

5.2 Spray Bonding

It covers a wide range of methodologies in which the dispersion of a volatile fluid, a mixture of solid lubricant and binder that can be either an inorganic resin, such as borate, silicate, or organic resin, such as epoxy or acrylic, is used. The fluid is then sprayed on the surface to be lubricated, followed by drying. The binder forms a thick, composite-like film that advances the adhesion to the substrate and is usually mixed with the lubricant blend at 33% by volume [57, 69].

5.3 Particle Incorporation in Liquid or Greases

Solid lubricants are often incorporated into liquid lubricants that help in boundary lubrication. Usually, the contact geometry should be able to accommodate a large number of MoS_2 particles in the interface by the fluid flow, where the temperature should be high. However, recent studies have shown that graphite and MoS_2 are ineffective. For instance, under some circumstances, solid lubricant or a film made from them can affect the liquid flow. So, they should be used in oils and greases cautiously and need to be estimated on an individual case basis [57].

5.4 Vacuum Coating

This methodology is mainly used in every industry. Vacuum conditions during coating help exclude contaminants such as oxygen and moisture that significantly deteriorate the activity of the lubricants. Solid lubricants can be applied as plasma to the substrate, ensuring the deposition of solid lubricants on the surface as individual atoms and ions of highly high energy [69, 70]. Using this method, the crystal structure, life, and adhesion of the film can be improved significantly such as the life tenure of sputtering-coated MoS_2 film of 200 nm thickness was more than five times longer than that of resin-bonded MoS_2 film of thickness 13 μm. Further, this method uses a unique Ar ion bombardment cleaning process to eliminate the residual contaminants from the substrate surface [71, 72].

6 Possible Mechanisms of Heat Dissipation

The coefficient of friction and wear rate quantifies the extent of friction at a macro-cyclic scale [73]. Friction originates from the heat produced due to physical, chemical, and mechanical interactions at the interface of the sliding surfaces [74]. However, quantifying interactions is impossible due to the unavailability of a single theoretical model. Figure 8 shows various possible mechanisms responsible for heat dissipation, including wear, molecular deformation, thermal effect, bonding, and environmental factors.

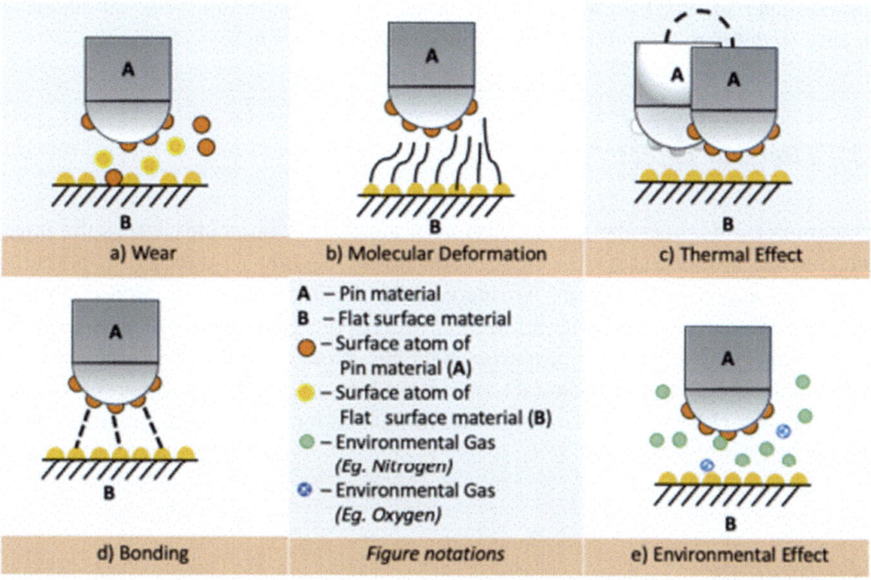

Fig. 8 Various possible mechanisms for frictional energy dissipation. Reprinted from ref. [73] Copyright (2021) MDPI

6.1 Wear

Wear is the physical destruction caused by the fracture in the solid-film materials that increases friction and makes the surface rough. Ductile deformation is more favorable for solid-film lubrication than brittle. Ductile materials can randomly undergo plastic deformation that can be easily scrutinized by modifying grain size and orientation. However, brittle fracture is promoted by crack initiation and propagation, which increase the wear rate [75]. During tribological studies, several wears can be produced, and a general wear sequence that helps identify the type of wear mechanism is presented in Fig. 9.

6.2 Molecular Deformation

Atomic corrugation is the deformation of 2D layers due to the adhesion of contaminant atoms, resulting in molecular deformation and heat dissipation. During the molecular contact of the sliding surface at the interface, the adhesion of atoms of the topmost layer of thin film to the tip of the counter body occurs [76]. 2D materials are easy to deform, considering the tip-layer adhesion strength.

6.3 Thermal Effect

Temperature influences the rate of motion of atoms and molecules across the interface, further influenced by the heat produced by friction or interaction potential between molecules and atoms. An increase in the interaction potential decreases the overall extent of heat generation [77]. Maximum temperature at which MoS_2 exhibits favorable lubricative properties depends strongly on the operating environment and the microstructural characteristics of MoS_2 coating [36]. Researchers focus on synthesizing MoS_2-based coatings with broad temperature range utilizations these days.

6.4 Bonding

Chemical interaction at the contact area strongly influences the friction rate. The formation of bonds between the rubbing surfaces leads to a decrease or increase in friction, which alters the bond properties. This strongly depends upon the bond energy [78].

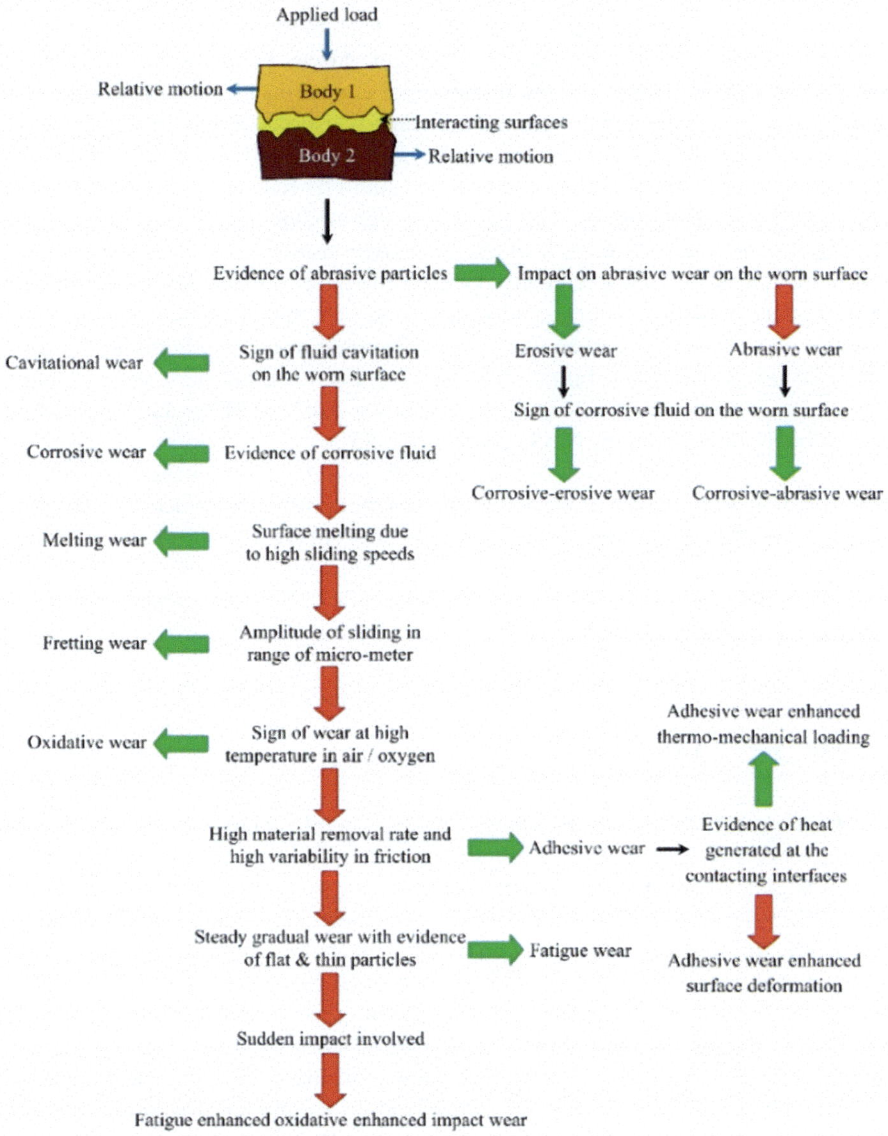

Fig. 9 General sequence of wear classification. Reprinted from ref. [1] Copyright (2015) Elsevier

6.5 *Environmental Effect*

Environmental factors such as humidity, temperature, vacuum, and gaseous environment influence the friction rate to a great extent. The physical or chemical interactions between film and the environment alter the performance of MoS_2 coatings

[79]. For example, with the rise in humidity, the friction of MoS_2 increases. In 1955, Ross studied the oxidation products of MoS_2 at various temperatures in the presence of water and air [80]. The accumulation of contaminants on the MoS_2 surface causes oxidation at the defect sites, which can be adsorbed atoms, sulfur dangling bonds, substitutional defects, and vacancies [31]. It is mainly initiated by the moisture and oxygen in the environment where O and OH radical groups replace S in Mo-S and form Mo-O bonds. This process is thermodynamically favorable as the Mo-O bonds are much stronger than Mo-S. Oxidation-induced etching starts preferably at the edges and then eventually breaches the flake's center. The oxidation rate on the surface of MoS_2 is accelerated with the rise in humidity and temperature [80]. Peterson and Johnson observed that with an increase in relative humidity, friction increased up to 65% [81]. Further, in space applications, the presence of atomic oxygen significantly alters the tribological characteristics of MoS_2 due to oxidation and loss of S atoms from the MoS_2 coating [82, 83].

It is crucial to consider the influence of MoS_2's oxidation or corrosion properties on the coating performance or the degradation of the substrate's surface. In 1968, Kay reported one of the initial studies indicating the effect of environmental factors on the corrosion of MoS_2-based coatings. They found that steel corrosion was augmented under high humidity when in contact with MoS_2 [84]. However, graphite showed a comparatively lower corrosion extent.

7 MoS_2 as a Lubricant

The unique structure of MoS_2 enriches it with the property of lubrication that provides enough power to be used even in extreme conditions of high temperature and vacuum. Its tribological characteristics can be easily altered by regulating synthesis procedures. MoS_2 coatings prepared by different synthesis techniques exhibit different properties; for example, MoS_2 prepared using magnetron sputtering was used for detector applications [85, 86], chemically deposited MoS_2 films for light-emitting devices [87], and chemical stripping for fireproofing materials [88]. Proper analysis of Moiré TEM patterns showed that the ultra-low friction of MoS_2 coatings is due to significant inconsistency in the rotational registry of single MoS_2 crystallites around the surface-normal direction (Fig. 10) which leads to a rotational structural mismatch in basal planes that cancels the atomic-scale forces [89].

As established, MoS_2 can exhibit varied properties by altering the structure and thickness of layers. The role of solid lubricants is to minimize the friction between the surfaces in critical environments, such as in extremely high or low temperatures, low/high pressure, and high shaft speed. They are a smart substitute for fluid lubricants to reduce friction and avoid seizing and galling, particularly in high/low-temperature conditions where fluids may freeze or vaporize. One of the practical utilizations of MoS_2 for space applications is the James Webb Space Telescope (JWST), in which it is employed as a solid lubricant in sensitive components such as Near-Infrared Spectrograph (NIR Spec) and Mid-Infrared Instrument (MIRI) [90].

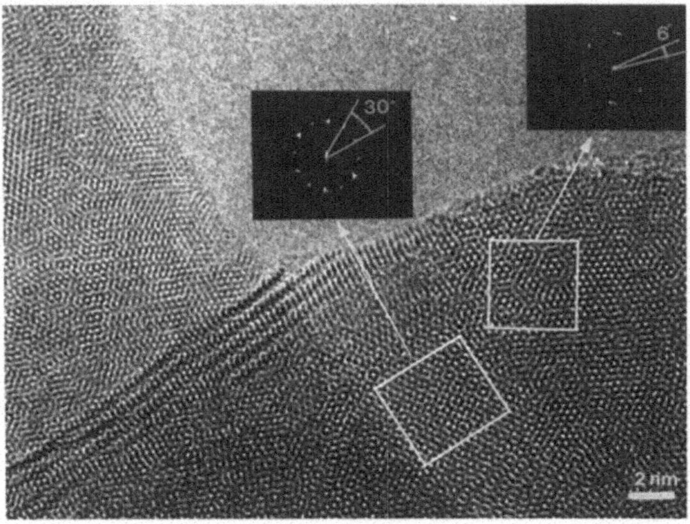

Fig. 10 HR-TEM image of MoS_2 signifying the presence of two different planes with different rotational registry. Reprinted from ref. [89] Copyright (1994) Elsevier

Bulk MoS_2 with a hexagonal lattice structure is a superb lubricant, considering its low friction and high adherence to materials. Further, high lattice spacing structures with lamellar structure, weak van der Waals attraction, and good thermal stability have raised its value as a lubricant. As explained in the above sections, the low friction of MoS_2 can be ascribed to the easy sliding of the layers over the other and the positively charged surface of the MoS_2 layer. During sliding, the crystalline layers of MoS_2 slide easily over each other and orientate parallel to the direction of the relative movement, enriching it with a lubricating effect (Fig. 11). The thickness of the lubricating film is usually 2–3 monolayers of MoS_2, with each monolayer having a thickness of 6.5 Å [91].

MoS_2 has been known for displaying a low coefficient of friction for a long time. There are three primary mechanisms used for explaining the frictional behavior of MoS_2, which are (a) the intrinsic crystal structure of MoS_2 where in its lamellar structure, the distance between S atoms of consecutive lamina layers is exceptionally high which leads to the cleavage of molybdenite; (b) the adsorbed layers of some foreign materials on its surface that weakens the structure; and (c) forces acting between individual crystallites and breakage along grain boundaries as opposed to the cleavage between the layers in an individual crystal.

It has been found that its chemical properties and topography significantly influence the tribological properties of MoS_2. However, the influence of crystallographic orientation on the tribological properties of MoS_2 was first studied by Bodik and his team in 2020 [92]. They synthesized vertically and horizontally aligned MoS_2 sheets using CVD, which showed varied tribological behavior on the macro- and nanoscales. They also found that the horizontally aligned MoS_2 sheets where the

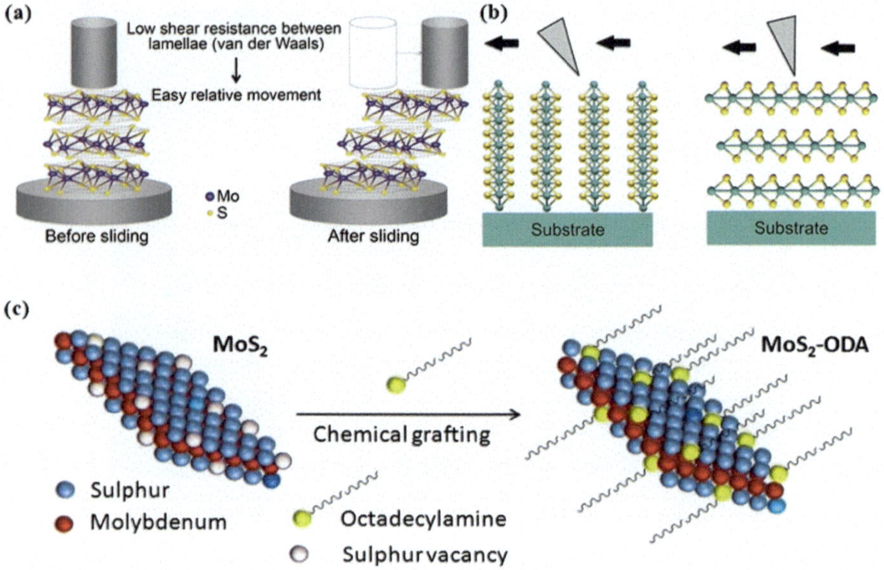

Fig. 11 **a** Sliding effect in MoS_2 [93]), **b** vertical and horizontal alignment of MoS_2 sheets [92], **c** grafting of octadecylamine directing to the structural defects and "S" vacancies of MoS_2 [94]. (a) (Reprinted from ref. [96] Copyright (2018) Elsevier (b) Reprinted from ref. [95] Copyright (2021) Elsevier (c) Reprinted from ref. [97] Copyright (2020) ACS

c-axis oriented along the surface typically favored the weak van der Waals forces significantly reduced the friction compared to the vertically aligned MoS_2 sheets (Fig. 11b).

Another interesting study by Wu and his coworkers showed the effect of synthesis on the tribological characteristics of MoS_2 [95]. They employed high-temperature expansion exfoliation followed by low-temperature freeze exfoliation to get ultrathin and small-sized 2D MoS_2 nanosheets. They tested the lubrication performance on the four-ball wear machine in base oil, where a significantly improved rate of the average friction coefficient (COF) and average wear scar diameter (AWSD) for 0.08 wt% MoS_2-based oil was reported to be 26.46% and 44.55% lower than that of the base oil.

To gain complete insight into the tribological properties of MoS_2 coatings in air, Dreva and team studied the effect of optimal conditions that contribute to low friction and wear [96]. They investigated the use of MoS_2 in aerospace applications in the air. They found that a slight temperature rise can be valuable for the tribological characteristics of MoS_2 in air. Around 75 °C, excess surface oxides are formed in small quantities on MoS_2, which do not alter its tribological characteristics. Water molecules hinder the proper alignment of MoS_2 layers at room temperature. Thus, desorption of H_2O molecules results in the re-orientation of MoS_2 layers and helps reinstate low shear strength, wear, and friction.

Thachnatharen and his team investigated the effect of incorporating microwave-synthesized MoS_2 nanoparticles into engine oil on the tribological properties. The performance was tested using the four-ball tribometer. In contrast to the engine oil, the MoS_2-based nanolubricant significantly improved the coefficient of friction and wear scar diameter by 10.25 and 10.6%, respectively. The formation of a protective layer called the tribofilm followed the tribo-mechanism related to reducing the friction coefficient [97].

Chen and his coworkers reported novel oil-soluble ultrathin MoS2 nanosheets prepared via a facile one-pot process where they modified its surface with oleylamine to improve its dispersion in oil and thus predicted a way of reducing the dependency on surfactant or dispersant [98]. They used a few atomic layer-thick and tens of nanometer-wide oil-soluble nanosheets of MoS_2 as lubricant additives. They reported that by adding just one wt% MoS_2 nanosheets at 120 °C, the load-bearing capacity of liquid paraffin increased tremendously from less than 50N to more than 2000N. Improved dispersion and ultrathin MoS_2 nanosheets aided in entering the contact areas of the opposite sliding surfaces and behaved as a protective film to avoid direct contact between them.

Kumari and her team presented an exciting study outlining the MoS2 nanosheets modified with octadecylamine [94]. Few layer MoS_2 nanosheets were obtained via the hydrothermal method, and nanosheets were then chemically functionalized with octadecylamine that targeted the "S" vacancies in the distorted MoS_2 surface resulting in grafting ((Fig. 11c). This modification showed improved dispersibility in lube base oil and was employed as an additive to mineral oil under an optimized dose of 0.1 mg/mL. Even with such low concentration, modified MoS_2 exhibited appreciably reduced value of wear (around 44%) and friction (around 48%).

Before concluding this chapter, it is worth mentioning that the composite formation can further improve the lubricative properties of MoS_2. A few of its drawbacks, such as the effect of atomic oxygen and humidity, can be easily tackled by MoS_2-composite-based lubricants. Various MoS_2-based composites have been synthesized to improve the tribological properties, such as Nb-MoS_2 [99], Pb-Ti/MoS_2 [100], 2D MoS_2-hBN hybrid coatings [60], Al/Al_2O_3/MoS_2 hybrid composite [101], WC–Co–Cu–MoS_2 [102], Sb_2O_3/Au-doped MoS_2 [103], Cu-Al/MoS_2 coatings [104], GNS/MoS_2 composite [105], Mo/MoS_2-Pb-PbS composite films [106], and MoS_2/Sb_2O_3/C composite films [107].

The practical utility of MoS_2 at the industrial scale is also rising. With the advancement of mechanical industries and the emergence of a distinctive industrial environment such as aerospace, research on lubricating materials has drastically increased. The efficiency and constancy of lubrication are two vital parameters for working any equipment. MoS_2 is rapidly used in such industries and is employed in various forms and concentrations, such as between 1 and 20% in grease formulations and 50–79% in pastes [108]. Steel and iron industries such as Freeport McMoRan (FCX) (USA), Jinduicheng Molybdenum Group Co. Ltd. (JDC) (China), and Thomson Creek Metals Company Inc. (USA) are its primary consumers. In India, Sunrise Ventures in New Delhi utilizes transition metal dichalcogenides as lubricant material. In particular, they are working on inorganic fullerene-tungsten disulfide (IF-WS_2)

that has exhibited various attractive properties in the context of excellent lubrication, improved cooling properties, reduced treatment rate, and reduction in disposal cost. Solid MoS_2, in combination with graphite, is often used as a raw material for grease in slow-moving and heavy-duty applications such as constant velocity joints, plain and roller bearings, and open gears [109].

8 Conclusions

Tribology can be considered a young scientific field where required analytical concepts have not yet been adequately developed to provide a clear guide to the tedious concepts of friction and wear. However, this chapter offers insights into the lubrication ability of MoS_2, which exhibits superior tribological, mechanical, and thermal properties. Significant observations of this chapter can be summarized as follows:

- Solid lubricants are highly desirable as compared to liquid lubricants.
- The tribological properties of MoS_2 depend strongly on the synthesis method and crystallographic arrangement.
- MoS_2's lubrication ability is susceptible to operating conditions such as humidity, gases, and atomic oxygen. Moisture and oxygen significantly deteriorate the tribological properties of pristine MoS_2 and reduce its utilization in terrestrial applications.
- Under vacuum conditions, the wear rate increases with the temperature rise, and the friction rate decreases with the decrease in temperature.
- MoS_2 is a superb solid lubricant, and efforts should be made to extend its working environment. In addition to tribological applications, its thermal and environmental stability should be improved for other applications such as photodetectors, photocatalysis, sensing, and semiconductor device fabrication.

References

1. Nirmal U, Hashim J, Ahmad MM (2015) A review on tribological performance of natural fibre polymeric composites. Tribol Int 83:77–104
2. Tzanakis I, Hadfield M, Thomas B, Noya S, Henshaw I, Austen S (2012) Future perspectives on sustainable tribology. Renew Sustain Energy Rev 16(6):4126–4140
3. Hod O, Meyer E, Zheng Q, Urbakh M (2018) Structural superlubricity and ultralow friction across the length scales. Nature 563(7732):485–492
4. Holmberg K, Erdemir A (2015) Global impact of friction on energy consumption, economy and environment. Fme Trans 43(3):181–185
5. Meng Y, Xu J, Jin Z, Prakash B, Hu Y (2020) A review of recent advances in tribology. Friction 8(2):221–300
6. Holmberg K, Siilasto R, Laitinen T, Andersson P, Jäsberg A (2013) Global energy consumption due to friction in paper machines. Tribol Int 62:58–77

7. Luo J, Zhou X (2020) Superlubricitive engineering—Future industry nearly getting rid of wear and frictional energy consumption. Friction 8(4):643–665
8. Gulzar M, Masjuki H, Kalam M, Varman M, Zulkifli N, Mufti R, Zahid R (2016) Tribological performance of nanoparticles as lubricating oil additives. J Nanopart Res 18(8):1–25
9. Martin P Jr (1972) A survey of solid lubricant technology Army Weapons Command Rock Island IL General Thomas J Rodman Lab
10. Kumar R, Hussainova I, Rahmani R, Antonov M (2022) Solid lubrication at high-temperatures—a review. Materials 15(5):1695
11. Miyoshi K (2019) Solid lubrication fundamentals and applications. CRC Press
12. Bowden F, Tabor D (1986) The friction and lubrication of solids. Clarendon Press, UK
13. Scharf T, Prasad S (2013) Solid lubricants: a review. J Mater Sci 48(2):511–531
14. Kaplan-Ashiri I, Tenne R (2016) On the mechanical properties of WS_2 and MoS_2 nanotubes and fullerene-like nanoparticles: In situ electron microscopy measurements. JOM 68(1):151–167
15. Marian M, Tremmel S, Wartzack S, Song G, Wang B, Yu J, Rosenkranz A (2020) Mxene nanosheets as an emerging solid lubricant for machine elements–towards increased energy efficiency and service life. Appl Surf Sci 523:146503
16. Liu L, Zhang Y, Qiao Y, Tan S, Feng S, Ma J, Liu Y, Luo J (2021) 2D metal-organic frameworks with square grid structure: a promising new-generation superlubricating material. Nano Today 40:101262
17. Zhou C, Li Z, Liu S, Zhan T, Li W, Wang J (2022) Layered double hydroxides for tribological application: recent advances and future prospective. Appl Clay Sci 221:106466
18. Anasori B, Gogotsi ÛG (2019) 2D metal carbides and nitrides (MXenes). Springer, Berlin
19. Kumbhakar P, Gowda CC, Mahapatra PL, Mukherjee M, Malviya KD, Chaker M, Chandra A, Lahiri B, Ajayan P, Jariwala D (2021) Emerging 2D metal oxides and their applications. Mater Today 45:142–168
20. Panchal H, Sadasivuni KK, Ahmed AAA, Hishan SS, Doranehgard MH, Essa FA, Shanmugan S, Khalid M (2021) Graphite powder mixed with black paint on the absorber plate of the solar still to enhance yield: an experimental investigation. Desalination 520:115349
21. Bailey AI, Courtney-Pratt J (1955) The area of real contact and the shear strength of monomolecular layers of a boundary lubricant. Proc R Soc Lond Ser A. Math Phys Sci 227(1171):500–515
22. Levita G, Molinari E, Polcar T, Righi MC (2015) First-principles comparative study on the interlayer adhesion and shear strength of transition-metal dichalcogenides and graphene. Phys Rev B 92(8):085434
23. Zhang X, Teng SY, Loy ACM, How BS, Leong WD, Tao X (2020) Transition metal dichalcogenides for the application of pollution reduction: a review. Nanomaterials 10(6):1012
24. Bartz WJ (1972) Some investigations on the influence of particle size on the lubricating effectiveness of molybdenum disulfide. Asle Trans 15(3):207–215
25. Song I, Park C, Choi HC (2015) Synthesis and properties of molybdenum disulphide: from bulk to atomic layers. RSC Adv 5(10):7495–7514
26. Braithwaite E (1969) MoS_2 second thoughts. Ind Lubr Tribol
27. Benavente E, Santa Ana M, Mendizábal F, González G (2002) Intercalation chemistry of molybdenum disulfide,. Coord Chem Rev 224(1–2):87–109
28. Hu T, Li R, Dong J (2013) A new (2× 1) dimerized structure of monolayer 1T-molybdenum disulfide, studied from first principles calculations. J Chem Phys 139(17):174702
29. Yu Y, Nam G-H, He Q, Wu X-J, Zhang K, Yang Z, Chen J, Ma Q, Zhao M, Liu Z (2018) High phase-purity 1T′-MoS_2-and 1T′-$MoSe_2$-layered crystals. Nat Chem 10(6):638–643
30. Wypych F, Schöllhorn R (1992) 1T-MoS_2, a new metallic modification of molybdenum disulfide. J Chem Soc Chem Commun 19:1386–1388
31. Joseph A, Vijayan AS, Shebeeb CM, Akshay K, Mathew KPJ, Sajith V (2023) A review on tailoring the corrosion and oxidation properties of MoS_2-based coatings. J Mater Chem A 11(7):3172–3209

32. Vazirisereshk MR, Martini A, Strubbe DA, Baykara MZ (2019) Solid lubrication with MoS_2: a review. Lubricants 7(7):57
33. Ding Q, Song B, Xu P, Jin S (2016) Efficient electrocatalytic and photoelectrochemical hydrogen generation using MoS_2 and related compounds. Chem 1(5):699–726
34. Chen X, Chen Z, Li J (2013) Critical electronic structures controlling phase transitions induced by lithium ion intercalation in molybdenum disulphide. Chin Sci Bull 58(14):1632–1641
35. Late DJ, Liu B, Matte HR, Rao C, Dravid VP (2012) Rapid characterization of ultrathin layers of chalcogenides on SiO_2/Si substrates. Adv Func Mater 22(9):1894–1905
36. Vazirisereshk MR, Martini A, Strubbe DA, Baykara MZ (2019) Solid lubrication with MoS_2: a review. Lubricants 7(7):57
37. Gardos MN (1988) The synergistic effects of graphite on the friction and wear of MoS_2 films in air. Tribol Trans 31(2):214–227
38. Rowe G (1960) Some observations on the frictional behaviour of boron nitride and of graphite. Wear 3(4):274–285
39. Miyake S, Wang M (2015) Nanoprocessing of layered crystalline materials by atomic force microscopy. Nanoscale Res Lett 10(1):1–16
40. Huang Y, Yao Q, Lu Z, Jiao L, Zhang S, Li Q, Meng Y (2018) Antiwear performance of monolayer MoS_2 modulated by residual straining. ACS Appl Nano Mater 1(12):7092–7097
41. Novoselov KS, Jiang D, Schedin F, Booth T, Khotkevich V, Morozov S, Geim AK (2005) Two-dimensional atomic crystals. Proc Natl Acad Sci 102(30):10451–10453
42. Abdel Maksoud M, Bedir AG, Bekhit M, Abouelela MM, Fahim RA, Awed A, Attia SY, Kassem SM, Elkodous MA, El-Sayyad GS (2021) MoS_2-based nanocomposites: synthesis, structure, and applications in water remediation and energy storage: a review. Environ Chem Lett 19(5):3645–3681
43. Pirarath R, Shivashanmugam P, Syed A, Elgorban AM, Anandan S, Ashokkumar M (2021) Mercury removal from aqueous solution using petal-like MoS_2 nanosheets. Front Environ Sci Eng 15(1):1–10
44. Du H, Liu D, Li M, Al Otaibi RL, Lv R, Zhang Y (2015) Solvothermal synthesis of MoS_2 nanospheres in DMF–water mixed solvents and their catalytic activity in hydrocracking of diphenylmethane. RSC Adv 5(97): 79724–79728
45. Ma L, Chen W-X, Li H, Zheng Y-F, Xu Z-D (2008) Ionic liquid-assisted hydrothermal synthesis of MoS_2 microspheres. Mater Lett 62(6–7):797–799
46. Kim D, Leem J-Y (2022) Synthesis of interface modified MoS_2/ZnO heterostructure via simple hydrothermal method and their enhanced UV photodetection characteristics with ultrafast photoresponse speed. Mater Res Bull 150:111767
47. Kour P, Yadav K (2022) Electrochemical performance of mixed-phase 1T/2H MoS_2 synthesized by conventional hydrothermal v/s microwave-assisted hydrothermal method for supercapacitor applications. J Alloy Compd 922:166194
48. Dharmalingam K, Ramasundaram S, Parthiban N, Kumar BA, Sriram S, Ramalingam G, Vijaibharathy P, Oh TH, Thangavel E (2022) Synthesis of $MoS_2/Mg(OH)_2/BiVO_4$ hybrid photocatalyst by ultrasonic homogenization assisted hydrothermal methods and its application as sunlight active photocatalyst for water decontamination. Chemosphere 308:136406
49. Yang H, Zhou J, Bao J, Ma Y, Zhou J, Shen C, Luo H, Yang M, Hou C, Huo D (2021) A simple hydrothermal one-step synthesis of 3D-MoS_2/rGO for the construction of sensitive enzyme-free hydrogen peroxide sensor. Microchem J 162:105746
50. Lee YB, Kim SK, Ji S, Song W, Chung H-S, Choi MK, Lee M, Myung S, Lim J, An K-S (2019) Facile microwave assisted synthesis of vastly edge exposed 1T/2H-MoS_2 with enhanced activity for hydrogen evolution catalysis. J Mater Chem A 7(8):3563–3569
51. Panigrahi PK, Pathak A (2011) A novel route for the synthesis of nanotubes and fullerene-like nanostructures of molybdenum disulfide. Mater Res Bull 46(12):2240–2246
52. Sarwar S, Nautiyal A, Cook J, Yuan Y, Li J, Uprety S, Shahbazian-Yassar R, Wang R, Park M, Bozack MJ (2020) Facile microwave approach towards high performance MoS_2/graphene nanocomposite for hydrogen evolution reaction. Sci China Mater 63(1):62–74

53. Solomon G, Mazzaro R, Morandi V, Concina I, Vomiero A (2020) Microwave-assisted vs. conventional hydrothermal synthesis of MoS_2 nanosheets: application towards hydrogen evolution reaction. Crystals 10(11):1040
54. Mouloua D, Kotbi A, Deokar G, Kaja K, El Marssi M, El Khakani MA, Jouiad M (2021) Recent progress in the synthesis of MoS_2 thin films for sensing, photovoltaic and plasmonic applications: a review. Mater Lett 14(12):3283
55. Gupta D, Chauhan V, Kumar R (2022) Sputter deposition of 2D MoS_2 thin films-a critical review from a surface and structural perspective. Inorg Chem Commun 144:109848
56. Chen F, Yao Y, Su W, Zhao S, Ding S, Fu L (2020) The synthesis of 2D MoS_2 flakes with tunable layer numbers via pulsed-argon-flow assisted CVD approach. Ceram Int 46(10):14523–14528
57. Savan A, Pflüger E, Voumard P, Schröer A, Simmonds M (2000) Modern solid lubrication: recent developments and applications of MoS_2. Lubr Sci 12(2):185–203
58. Gupta D, Chauhan V, Kumar R (2022) Sputter deposition of 2D MoS_2 thin films-a critical review from a surface and structural perspective. Inorg Chem Commun 109848
59. Shahzad R, Kim T, Kang S-W (2017) Effects of temperature and pressure on sulfurization of molybdenum nano-sheets for MoS_2 synthesis. Thin Solid Films 641:79–86
60. Joseph A, Gautham V, Akshay K, Sajith V (2022) 2D MoS_2-hBN hybrid coatings for enhanced corrosion resistance of solid lubricant coatings. Surf Coat Technol 443:128612
61. Shankara A, Menezes PL, Simha K, Kailas SV (2008) Study of solid lubrication with MoS_2 coating in the presence of additives using reciprocating ball-on-flat scratch tester. Sadhana 33(3):207–220
62. Krishnan U, Kaur M, Singh K, Kumar M, Kumar A (2019) A synoptic review of MoS_2: synthesis to applications. Superlattices Microstruct 128:274–297
63. Santhosh S, Madhavan AA (2019) A review on the structure, properties and characterization of 2D molybdenum disulfide. In 2019 advances in science and engineering technology international conferences (ASET), pp 1–5. IEEE
64. Choi MG, Belianinov A, Pawlicki A, Park S, Lee H, Ovchinnikova OS, Kim S (2021) Nanoscale friction of CVD single-layer MoS_2 with controlled defect formation. Surf Interfaces 26:101437
65. Serpini E, Rota A, Valeri S, Ukraintsev E, Rezek B, Polcar T, Nicolini P (2019) Nanoscale frictional properties of ordered and disordered MoS_2. Tribol Int 136:67–74
66. Holmberg K, Matthews A (2009) Coatings tribology: properties, mechanisms, techniques and applications in surface engineering, Elsevier
67. Rapoport L, Moshkovich A, Perfilyev V, Lapsker I, Halperin G, Itovich Y, Etsion I (2008) Friction and wear of MoS_2 films on laser textured steel surfaces, Surface and Coatings Technology 14:3332–3340
68. Lince JR (2020) Effective application of solid lubricants in spacecraft mechanisms. Lubricants 8(7):74
69. Stachowiak GW, Batchelor AW (2013) Engineering tribology. Butterworth-Heinemann
70. Gresham RM (1988) Solid-film lubricants: unique products for unique lubrication. Lubr Eng 44(2):143–145
71. Spalvins T (1970) Friction characteristics of sputtered solid film lubricants
72. Ohmae N, Tsukizoe T, Nakai T (1978) Ion-plated thin films for anti-wear applications. J Lubr Technol 100:129–135
73. Manu BR, Gupta A, Jayatissa AH (2021) Tribological properties of 2D materials and composites—a review of recent advances. Materials 14(7):1630
74. Bhushan B, Israelachvili JN, Landman U (1995) Nanotribology: friction, wear and lubrication at the atomic scale. Nature 374(6523):607–616
75. Aquaro D, Fontani E (2001) Erosion of ductile and brittle materials. Meccanica 36(6):651–661
76. Smolyanitsky A, Killgore JP (2012) Anomalous friction in suspended graphene. Phys Rev B 86(12):125432
77. Riedo E, Gnecco E, Bennewitz R, Meyer E, Brune H (2003) Interaction potential and hopping dynamics governing sliding friction. Phys Rev Lett 91(8):084502

78. Deng Z, Smolyanitsky A, Li Q, Feng X-Q, Cannara RJ (2012) Adhesion-dependent negative friction coefficient on chemically modified graphite at the nanoscale. Nat Mater 11(12):1032–1037
79. Gallagher P, Lee M, Amet F, Maksymovych P, Wang J, Wang S, Lu X, Zhang G, Watanabe K, Taniguchi T (2016) Switchable friction enabled by nanoscale self-assembly on graphene. Nat Commun 7(1):1–7
80. Ross S, Sussman A (1955) Surface oxidation of molybdenum disulfide. J Phys Chem 59(9):889–892
81. Peterson MB (1953) Friction and wear investigation of molybdenum disulfide I: effect of moisture
82. Gao X, Hu M, Sun J, Fu Y, Yang J, Liu W, Weng L (2015) Changes in the composition, structure and friction property of sputtered MoS_2 films by LEO environment exposure. Appl Surf Sci 330:30–38
83. Gao X, Fu Y, Jiang D, Wang D, Weng L, Yang J, Sun J, Hu M (2018) Responses of TMDs-metals composite films to atomic oxygen exposure. J Alloy Compd 765:854–861
84. Kay E (1968) The corrosion of steel in contact with molybdenum bisulphide. Wear 12(3):165–171
85. Pak SW, Chu D, Lee SK, Kim EK (2017) Enhancement of near-infrared detectability from InGaZnO thin film transistor with MoS_2 light absorbing layer. Nanotechnology 28(47):475206
86. Gong C, Xiao J, Zhu L, Qi M, Ma S (2019) Crystal structure and tribological properties of molybdenum disulfide films prepared by magnetron sputtering technology. Curr Appl Phys 19(12):1318–1324
87. Xue Y, Zhang Y, Liu Y, Liu H, Song J, Sophia J, Liu J, Xu Z, Xu Q, Wang Z (2016) Scalable production of a few-layer MoS_2/WS_2 vertical heterojunction array and its application for photodetectors. ACS Nano 10(1):573–580
88. Wang D, Song L, Zhou K, Yu X, Hu Y, Wang J (2015) Anomalous nano-barrier effects of ultrathin molybdenum disulfide nanosheets for improving the flame retardance of polymer nanocomposites. J Mater Chem A 3(27):14307–14317
89. Martin J, Pascal H, Donnet C, Le Mogne T, Loubet J, Epicier T (1994) Superlubricity of MoS_2: crystal orientation mechanisms. Surf Coat Technol 68:427–432
90. Krause O, Müller F, Birkmann S, Böhm A, Ebert M, Grözinger U, Henning T, Hofferbert R, Huber A, Lemke D (2010) High-precision cryogenic wheel mechanisms of the JWST/MIRI instrument: performance of the flight models. In: modern technologies in space-and ground-based telescopes and instrumentation, SPIE, pp 421–432
91. He Z, Que W (2016) Molybdenum disulfide nanomaterials: structures, properties, synthesis and recent progress on hydrogen evolution reaction. Appl Mater Today 3:23–56
92. Bodik M, Sojkova M, Hulman M, Tapajna M, Truchly M, Vegso K, Jergel M, Majkova E, Spankova M, Siffalovic P (2021) Friction control by engineering the crystallographic orientation of the lubricating few-layer MoS_2 films. Appl Surf Sci 540:148328
93. Furlan KP, de Mello JDB, Klein AN (2018) Self-lubricating composites containing MoS_2: a review. Tribol Int 120:280–298
94. Kumari S, Chouhan A, Sharma OP, Kuriakose S, Tawfik SA, Spencer MJ, Walia S, Sugimura H, Khatri OP (2020) Structural-defect-mediated grafting of alkylamine on few-layer MoS_2 and its potential for enhancement of tribological properties. ACS Appl Mater Interfaces 12(27):30720–30730
95. Wu P-R, Zhang W, Liu Z, Cheng Z-L (2019) A novel preparation method for MoS_2 nanosheets with good tribology performance by the combination of expansion and freeze exfoliation. Ceram Int 45(2):1730–1736
96. Dreva K, Morina A, Yang L, Neville A (2022) The effect of temperature on water desorption and oxide formation in MoS_2 coatings and its impact on tribological properties. Surf Coat Technol 433:128077
97. Thachnatharen N, Khalid M, Shahabuddin S, Anwar A, Sridewi N (2022) Tribological analysis of advanced microwave synthesized Molybdenum disulfide (MoS_2) as anti-friction additives in diesel engine oil for military vehicles. Mater Today: Proc

98. Chen Z, Liu X, Liu Y, Gunsel S, Luo J (2015) Ultrathin MoS_2 nanosheets with superior extreme pressure property as boundary lubricants. Sci Rep 5(1):1–7
99. Efeoglu I, Baran Ö, Yetim F, Altıntaş S (2008) Tribological characteristics of MoS_2–Nb solid lubricant film in different tribo-test conditions. Surf Coat Technol 203(5–7):766–770
100. Ren S, Li H, Cui M, Wang L, Pu J (2017) Functional regulation of Pb-Ti/MoS_2 composite coatings for environmentally adaptive solid lubrication. Appl Surf Sci 401:362–372
101. Kanthavel K, Sumesh K, Saravanakumar P (2016) Study of tribological properties on Al/Al_2O_3/MoS_2 hybrid composite processed by powder metallurgy. Alex Eng J 55(1):13–17
102. Yuan J, Zhu Y, Zheng X, Ji H, Yang T (2011) Fabrication and evaluation of atmospheric plasma spraying WC–Co–Cu–MoS_2 composite coatings. J Alloy Compd 509(5):2576–2581
103. Singh H, Mutyala K, Evans R, Doll G (2015) An investigation of material and tribological properties of Sb_2O_3/Au-doped MoS_2 solid lubricant films under sliding and rolling contact in different environments. Surf Coat Technol 284:281–289
104. Cao M, Zhao L, Wu L, Wang W (2018) Tribological properties of new Cu-Al/MoS_2 solid lubricant coatings using magnetron sputter deposition. Coatings 8(4):134
105. Song W, Yan J, Ji H (2019) Fabrication of GNS/MoS_2 composite with different morphology and its tribological performance as a lubricant additive. Appl Surf Sci 469:226–235
106. Han C, Li G, Ma G, Shi J, Liu Y, Li Z, Wang H (2021) Preparation and tribological properties of Mo/MoS_2-Pb-PbS composite films. Surf Coat Technol 405:126625
107. Zabinski J, Bultman J, Sanders J, Hu J (2006) Multi-environmental lubrication performance and lubrication mechanism of MoS_2/Sb_2O_3/C composite films. Tribol Lett 23(2):155–163
108. Mang T (2014) Encyclopedia of lubricants and lubrication. Springer, Berlin
109. Ordonez M, Farias M, Descartes S, Machado I, Souza RJTI (2021) Tribofilm formation during dry sliding of graphite-and MoS_2-based composites obtained by spark plasma sintering, 160:107035

Chapter 8
MoS_2 Heterostructures: Synthesis, Characterizations, and Applications

Rohit Thakur and H. S. S. Ramakrishna Matte

1 Introduction

Semiconductor junction devices are at the forefront of modern electronics. These junctions can be classified as homojunctions, where *n*- and *p*-type counterparts of same material are involved in interface formation, or heterojunctions, which involve an interface among two different semiconductors. In both cases, there is an inherent quasi-electric and magnetic field at the junction. However, in case of heterojunctions, the possible combinations of semiconductors are relatively huge which provides further flexibility in bandgap engineering. The experimental realization of heterojunctions revealed excellent carrier injection efficiencies relative to homojunctions which eventually won the Nobel prize in Physics, 2000. Semiconductor heterostructures have significantly improved the performance and economical footprint of devices resulting in a strong impact over various sectors like photovoltaics for space and terrestrial explorations, photodetectors, laser-based telecommunication systems, heterostructure bipolar transistors, low-noise HEMT for high-frequency applications, and semiconductor heterostructures which have been at the pinnacle of major technological breakthroughs [1].

The availability of vast number of semiconductors provides the opportunity to fabricate large variety of heterostructures. However, the realization is limited by factors like lattice mismatch, compatibility of the subsequent semiconductor with the other, etc. Lattice parameter mismatch is a substantial hurdle in realizing true potential of the heterojunctions formed. Larger the mismatch, poorer will be the charge carrier injection, optical and electronic confinement at the interface [2, 3]. Thus, the available library of semiconductors is restricted to just a few compatible

R. Thakur · H. S. S. Ramakrishna Matte (✉)
Centre for Nano and Soft Matter Sciences, Shivanapura, Bangalore 562162, India
e-mail: matte@cens.res.in
URL: https://www.cens.res.in/en/faculty/ramakrishna

D. J. Late and C. S. Rout (eds.), *Science and Technology of 2D MoS_2*,
Materials Horizons: From Nature to Nanomaterials,
https://doi.org/10.1007/978-981-97-7367-1_8

systems demonstrating optimum device performance. Layered materials are a class of materials distinct from conventional 3D materials in the nature of bonding along the *z*-axis. They possess weak van der Waals bond along this direction and strong covalent bonds along the *x*–*y*-plane. Upon exfoliation, such materials provide structures with naturally passivated surface due to absence of any dangling bonds, and do not require strong lattice matching with the materials to form heterostructures [2, 4, 5].

Owing to the vertical quantum confinement of states, conduction band and valence band edges possess sharp density of states which enables monolayer MoS_2 to interact strongly with light leading to absorption of nearly 10% of light despite being atomically thick [6–8]. However, the direct bandgap behavior leads to excellent absorption characteristics which are typical to a monolayer and are found to have drastic effect upon layer thickening, which is one of the major causes of limiting the photoresponsivity of the devices based on a particular layered material like MoS_2 [9, 10].

However, these materials do not require a strict lattice parameter matching, which enables formation of variety of heterojunctions from the large number of layered materials known [11, 12]. Furthermore, these junctions can be constructed from very few layers with atomically sharp interfaces and can be stacked in numerous sequences to obtain novel optical and electronic properties. Multidimensional heterostructures have also been investigated where excellent charge carrier mobilities of MoS_2 layers are employed to transport charge carriers generated by low-dimensional sensitizers like 0D QDs, 1D CNTs, and 2D and 3D films in order to modulate the bandgap, enhance light absorption, and generate a built-in electric field to accelerate interlayer charge transfer [4, 13]. For instance, graphene is a known material for FET fabrication but suffers with low ON/OFF ratios which upon forming a heterojunction with MoS_2 led to significant improvement in electron mobilities and ON/OFF ratios [14]. Similarly, forming a van der Waals hybrid of graphene and MoS_2 leverages from interfacial transparency among the two materials and high carrier mobilities of graphene to fabricate devices with highly sensitive photodetection ability for photon counting [15–17].

2 Synthesis

The significant augmentation in device figures of merits upon heterostructure formation asserts the importance of their fabrication. A typical heterostructure formation usually involves two steps. The first step is the obtaining layered materials from top-down or bottom-up approaches. In the second step, these layers are stacked together for heterointerface formation. First step can be achieved by protocols like chemical vapor deposition, mechanical exfoliation, and various solution routes like liquid phase exfoliation, solvothermal, and colloidal routes. Subsequent alignment of the formed layers can be achieved by methods like CVD growth over formerly grown layer; mechanical transfer and stacking of two materials; and solution-based route methods like spin coating, spray coating, or printing over grown or exfoliated layers.

Further, methods involving in situ formation of heterointerfaces via solvothermal route, modification in surface charge-induced self-assembly of layered materials are also employed. In the following sections, we will sequentially discuss these protocols in detail.

2.1 Mechanical Exfoliation

Micromechanical cleavage is the protocol employed in the groundbreaking discovery of graphene by Novoselov et al. in 2004 and the subsequent research that followed it. Square HOPG meshes were pressed against a thick layer of photoresist on a glass slide. Once the meshes were attached to the photoresist after baking, adhesive tape was used for thinning it down to a single layer. By dissolving the photoresist in acetone, the flakes are suspended. By dipping and washing in water and propanol, these flakes were transferred onto a solid substrate of SiO_2 (300 nm)/Si ($n+$ doped), Fig. 1a [18, 19]. As a principal control parameter, substrate to layer adhesion must be stronger than interlayer interaction energy of the cleaved material in order to achieve higher yields and high-quality monolayers.

The methodology has been at the forefront of probing fundamental optical, electronic, and physical phenomenon in layered materials and their heterostructures. It has been demonstrated that separately cleaved mono- and few layers can be stacked together to fabricate a heterostructure. For instance, a $PdSe_2$ and MoS_2 heterostructure was fabricated by this methodology to fabricate phototransistors, Fig. 1b [20]. Similarly, a few-layer WS_2/MoS_2 heterostructure was demonstrated as an IR photodetector (1020 nm) [21]. With same methodology, heterostructure formation of MoS_2 with different-layered materials like $MoTe_2$ [22], WSe_2 [23], and SnSe [24] been demonstrated with this approach. Predominantly, SiO_2/Si is the substrate used for transfer of mono- and few layers of layered materials, but other substrates have also been investigated as potential substrates. Gold Au (111) when adopted as a substrate for mechanical cleavage of MoS_2 from bulk crystal induces formation of a coherent heterostructure among 2H and 1T phases of MoS_2, Fig. 1d. Raman and STM/STS (scanning tunneling microscopy and spectroscopy) confirm the presence of a heterointerface among 2H and 1T phases. The dI/dV plot, inset of Fig. 1d, confirms metallic 1T phase at the red spot. Authors propose that due to strong coupling-induced charge transfer among Au and MoS_2, the activation barrier for 1T-to-2H phase transition increases [25]. The process of peeling and transfer usually exposes the cleaved layers to chemical environments such as polymers and solvents which hinder physico-chemical characteristics of the layers. In order to circumvent the challenge, van der Waals' adhesion methodology was demonstrated. In brief, PDMS/PVA strip was first employed to lift off a WSe_2 flake, followed by picking up graphene from the same strip and finally lifting off MoS_2 flake, ending up with a SiO_2/WSe_2/graphene/MoS_2 structure which was released at 70 °C [26].

Apart from 2D–2D-type heterostructure, 2D–1D and 2D–0D-type heterostructures have also been demonstrated. This can be achieved by solution protocols of thin

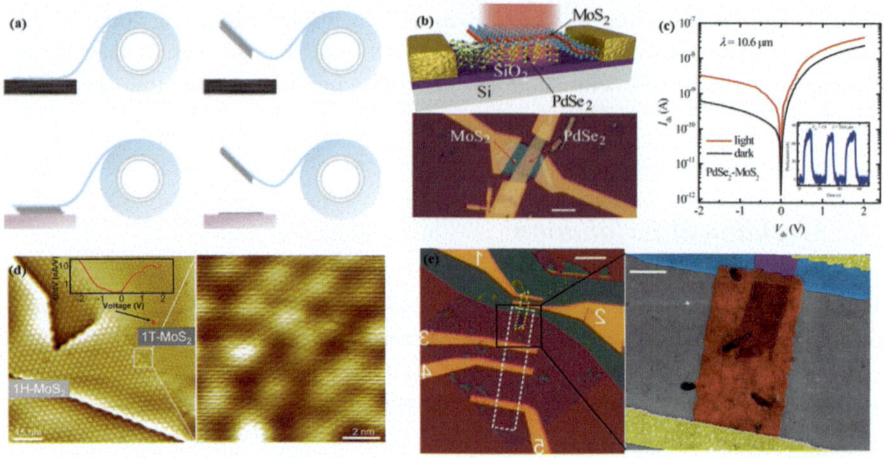

Fig. 1 **a** Schematic representation of a typical mechanical exfoliation process, where *Top row*: Topical few layers from graphite are cleaved with an adhesive tape. *Bottom left*: The adhesive with few layered flakes of graphite is pressed besides the substrate of choice. *Bottom right*: Upon stripping of the adhesive tape, some flakes stick onto the substrate [19]. **b** Schematic of the $PdSe_2$–MoS_2 device fabricated by micromechanical cleavage and transfer in Top panel, and Bottom panel shows an optical micrograph of the device, scale bar 5 μm. **c** Semilogarithmic output curve under dark and illuminated conditions. Inset: Temporal photoresponse of this device upon IR (10.6 μm) illumination at 1 V bias [20]. **d** STM image of a coherent 2H/1T lateral heterostructured region in a single layer of MoS_2 with zoomed-in view showing two different Moiré patterns. Inset: STS spectra at the red spot corroborating presence of metallic 1T phase [24]. **e** Microscopy and fabrication of the s-SWCNTs/ SL-MoS2 (*p-n*) heterojunction diode. Optical micrograph where white rectangle indicates presence of s-SWCNT film and yellow dashed boundary is the monolayer flake of MoS_2 (Scale bar, 10 μm). Zoomed-in region is a false-colored scanning electron micrograph of the heterostructure (Scalebar, 2.5 μm) where yellow regions are gold electrodes, violet region is MoS_2, pink represents the s-SWCNTs, and dark red region is the heterostructured region [29]

film formation, like spin coating, which has been demonstrated for ZnO nanowires, PbS QDs over mechanically exfoliated MoS_2 over flexible substrates [27, 28]. Alternatively, stamping method transfer of 1D nanostructures like SWCNTs has also been demonstrated to form a *p–n* junction with MoS_2, Fig. 1e [29].

Although micromechanical cleavage is ideally preferred to identify and investigate fundamental behavior of the layered materials and their heterostructures, characteristically, it has been restricted to lab-scale synthesis only.

2.2 *Chemical Vapor Deposition*

Chemical vapor deposition is a versatile technique imparting strong thickness and crystallinity control over 2D materials' growth. Plasma-enhanced, reactive, thermal [30], metallic seeding [31], and metal–organic chemical vapor deposition (MOCVD)

[32] are various methods of achieving high-quality thin films via chemical vapor deposition. These protocols enable to design and fabricate vertical as well as lateral heterostructures. Vertical heterostructures are the ones with layer-by-layer growth along the axis perpendicular to the substrate plane, whereas, in lateral heterostructures, the heterointerface is formed in the same plane.

2.2.1 Vertical Heterostructure

A vertical assembly can be realized by different approaches like CVD growth of material on already CVD-grown thin film; mechanical transfer onto a CVD-grown layer; and solution routes like spin coating, drop casting, and printing over CVD-grown layers. Furthermore, the dimensionality of materials being deposited can also be varied.

CVD over CVD

Vapor phase chalcogenization of MoO_2 was carried out to obtain MoS_2 monolayer on SiO_2 which was treated as substrate for vapor phase deposition of Bi_2Se_3 monolayers to fabricate the heterostructure. Interestingly, thermally induced reversible nature of interlayer coupling was exemplified by reemergence of the PL peaks [33]. Sequential CVD growth of wide library of layered materials onto CVD-grown MoS_2 has been demonstrated for single-layer graphene (SLGR) [34], β-In_2Se_3 [35], VS_2 [36], MoO_x [37], Sb_2Te_3 [38], Sb_2O_3 [39].

A two-step CVD process is usually employed to fabricate heterostructures of layered materials with graphene. Firstly, monolayer graphene is grown onto pretreated Cu foils. Pretreatment is a necessary step where the Cu substrate is exposed to high temperatures (~1000 ºC) under vacuum to get rid of copper oxides and to induce Cu grain formation. A similar approach was employed to demonstrate graphene heterostructures with hBN and WSe_2 as well [40]. The utility of this standard protocol in modern-day electronics is hindered by necessity of Cu pretreatment steps. In that context, it was shown that MoS_2 grown on epitaxially grown graphene possesses a significantly larger grain size when compared to the one grown on Cu substrates. The nature of the underlying substrate plays a crucial role in governing various physico-chemical characteristics of the sequential layer being grown. Thus, opting graphene as the substrate reduced the processing temperature to 400 °C as opposed to 580 °C in order to achieve equivalent quality of MoS_2 [41]. An atomically thin dual junction heterostructure was realized by sequential MOCVD of WSe_2 and MoS_2 onto epitaxial graphene (EG) monolayer. The methodology can be conveniently reversed in order to achieve an EG/MoS_2/WSe_2 dual junction heterostructure [42]. In order to assert the role of strain, thickness, and the buffer layer on the physico-chemical properties of heterostructure, normal EG and quasi-free standing epitaxial graphene (QFEG) were utilized as template for growth of MoS_2, WSe_2, and hBN. The EG template is achieved by treating (0001) SiC at 1700 °C for 15 min in Ar atmosphere, and further exposing EG to a 600 Torr pressure of H_2 at 1050 °C for 120 min led to preparation of QFEG. Growth temperatures for MoS_2 were brought down to 130 °C by incorporating QFEG as the substrate. Furthermore, when compared to EG as substrate, QFEG possessed prevalent and continuous coverage of MoS_2 monolayer, whereas, on EG, multilayer and discontinuous triangular stacks were observed. These observations were ascribed to various parameters, such as high density of nucleation site density in EG due to higher relative strain, attenuated reactivity of QFEG due to hydrogenation-induced deactivation of the buffer layer, and strain relaxation [43].

One can also physically or chemically modify the underlying substrate for enhanced quality of sequential layer. Selective region laser patterning of CVD-grown MoS_2 has been shown to cause formation of densely populated nucleation sites at the patterned region. HAADF-STEM and energy-dispersive X-ray (EDX) revealed that these sites suffer critical sulfur loss and end up with Mo-terminated surface, which induces selective growth of metallic layered material such as VSe_2 to fabricate vertical heterostructures [44]. Including a water vapor source in the growth process of MX_2 materials is another strategy to significantly change the growth process due to formation of metal oxyhydroxide. The excellent versatility of this approach is exemplified by demonstrating synthesizing wide variety of materials, ranging from monolayer, multilayer, rolled nanosheets, to wide area WS_2/MoS_2 heterostructures [45]. Capillary forces can be utilized to roll up CVD prepared heterostructures from the underlying substrate. This behavior has been demonstrated for a wide variety of hetero-, bi- and trilayers like MoS_2/WS2, SnS_2/MoS2, Ag nanowire/Al_2O_3/WSe2, Al_2O_3/SnS_2/WSe2, SnS2/WSe_2, Al_2O_3/WSe2, SnS_2/MoS_2/WS2, NbSe2/$MoSe_2$, $MoSe_2$/WSe_2, In_2Se_3/WSe_2, $SnSe_2$/WSe_2, and Cr_5Te_8/WSe_2 superlattices. In a typical procedure, ethanol:water (1:3) solution is dropped onto the planar heterostructure for intercalation which separated the structure from the substrate and the capillary forces cause the favorable roll up of the layers [46]. One can also incorporate MOCVD-grown MoS_2 monolayers as the seeding layer to provide nucleation centers for successive growth of TMDs like WSe_2 and WS_2 with respective gaseous metal precursors. The in-plane orientation of the basal planes of the first monolayer is an essential prerequisite for the proposed single layer-by-layer growth and alternate nine single layers of MoS_2 and WS_2 were demonstrated [47].

Sputtered Mo and W seeding layers were demonstrated to form atomically clean interface at the MoS_2/WS_2 heterojunction [48]. The same protocol was utilized to fabricate up to 8 such heterointerfaces across large areas (1 cm^2). Interestingly, formation of an alloy, superlattice, or a three-dimensional mixture of the components was found to be strongly correlated to the processing temperature, when various other parameters like. In the range of 400–800 °C, the grain size increased from 20 to 60 nm and transformed from atomically sharp superlattices at 400 °C to interfacially mixed layers at 600 °C to alloys at 800 °C, Fig. 2a–c [31]. This technique is not restricted to heterostructures with same chalcogen as it has been shown to form MoS_2/WSe_2 heterointerface, where Mo was sputtered over a WSe_2 thin film and then sulfurized to form the vertically aligned heterointerface [49]. Interestingly, metallic seed layer thickness can be used to direct a vertical crystal growth over horizontal growth. It was identified that a thickness of >3 nm preferentially promotes vertical growth, whereas thinner films compel horizontal crystal growth [50]. A large area

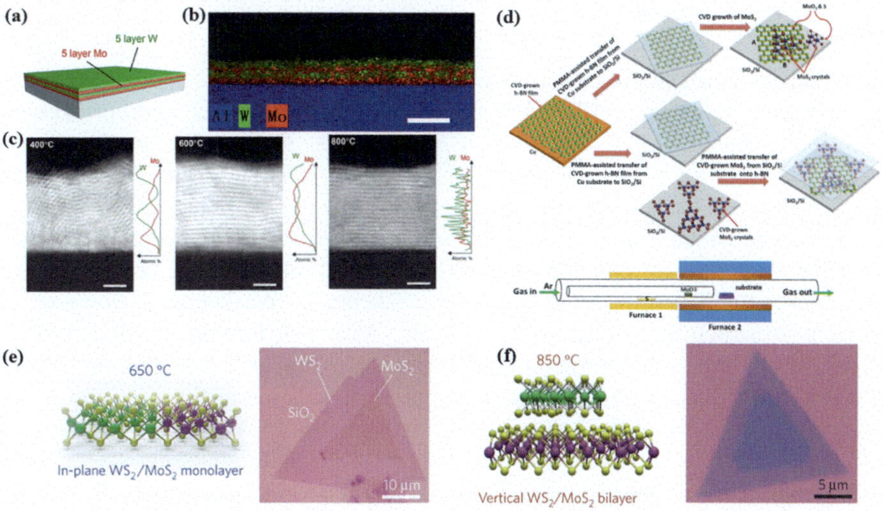

Fig. 2 Synthesizing $MoSe_2$/WSe_2 superlattices. **a** Alternate metal seeding layers as a schematic. **b** EDS map of the metal seed layer heterostructure with (Scalebar, 4 nm). **c** STEM and EDS spectra depicting temperature-dependent evolution of heterostructure, annealing at 400, 600, and 800 °C (Scalebar, 2 nm) [31]. Comparative study among mechanical transfer and CVD growth of hBN over MoS_2. **d** Top schematic illustrating the two synthesis procedures and bottom schematic portraying the CVD system utilized for MoS_2 growth [52]. Impact of annealing temperature over nature of the heterostructure forming. **e** At low temperature, kinetic product, lateral heterostructure is preferred, whereas **f** at elevated temperature, the thermodynamic product, vertical heterostructure is favored [58]

2D/1D heterostructure among MoS_2 nanosheets and tellurium (Te) nanowires was demonstrated via two-step CVD approach for telecommunication wavelength range photodetection [51].

Mechanical transfer over CVD-Grown layers

In this method, mono- and few layers are mechanically transferred to a CVD-grown layer to form heterointerface. A comparative study of the heterostructure properties formed by mechanical transfer of MoS_2 over hBN and CVD growth of MoS_2 over hBN was carried out. Raman and PL analysis revealed fundamental differences among the two like variation in lattice strain and doping levels of hBN Fig. 2d [52]. This protocol has been demonstrated for heterointerface formation of MoS_2 with $MoTe_2$ and SnS as well [53, 54].

Solution over CVD

Alternatively, utilization of solution-based protocols like spray coating, drop casting, etc. over a CVD-grown thin film provides a facile approach to heterostructure formation. This has been demonstrated for $CsPbBr_3$ which was drop casted onto CVD-grown MoS_2 to fabricate optoelectronic device [55].

2.2.2 Lateral Heterostructures

So far, we discussed synthesis of vertical heterostructures, which underlies layer-by-layer stacking by strategies like CVD growth, mechanical transfer, appropriate seed layer chalcogenization, and solution-based protocols. There exists another directional possibility for sequential material growth, lateral growth, wherein edge sites of one material act as the seed for epitaxial growth of another material. As we have discussed in previous sections, in case of vertical heterostructures borne out of layered materials, the lattice mismatch does not play a crucial role in tuning heterostructure formation as the governing interactions are weak van der Waals forces. However, in lateral heterostructures, strong covalent linkages have to be formed, which makes the process strongly dependent upon lattice mismatch. Another challenge with growth of lateral heterostructure is ambient air-induced passivation of the edge sites after first growth step. To address this, a home-built thermal CVD protocol was realized, which enables in situ replacement of reactants facilitating lateral growth of desired materials in a predestined sequence. Lateral interfaces among $MoS_2/MoSe_2$ and WS_2/WSe_2 were fabricated with this method. The in situ replacement of reactants facilitates retention of unsaturated dangling bonds which then act as the nucleation front for successive layers [56]. However, this protocol permits heterostructure formation among TMDs with same metals, leaving formation of intermetallic lateral heterostructure a challenge. In order to address that, a two-step chemical vapor deposition method, wherein WSe_2 grown onto a sapphire is utilized as the substrate for MoS_2 growth. Interestingly, the relative MoO_3 to S vapor ratio during growth provides the ability to orchestrate vertical or lateral, with excess MoO_3 led to vertical growth, whereas excessive S content consequently formed lateral heterointerface [57]. Temperature control is another parameter direct lateral growth. For WS_2/MoS_2, low temperature growth (650 °C) predominantly results in lateral heterostructure, higher temperature exposure (850 °C) led to vertical heterostructure formation, and at intermediate temperature (750 °C) a combination of both geometries was observed. The lateral heterostructure being the kinetic product forms at lower temperatures, whereas, upon providing enough thermal energy, the thermodynamically favored vertical heterostructures are formed, Fig. 2 e, f [58]. An ultra-low resistance (~0.5 kΩ·μm) 1D heterojunction was demonstrated among MoS_2 and VS_2, which was due to preferential nucleation of MoS_2 at the VS_2 vertices rather than edges [59]. Metallic seeding layer method has also been demonstrated to fabricate lateral heterostructures. Exposing lithography-assisted Mo and W seed layers to the chalcogen gas led to the formation of heterointerfaces along the substrate plane [60].

2.3 *Solution Routes*

The protocols discussed so far require expensive techniques such as lithography and clean rooms and are not suitable for mass production of thin films. Self-assembly protocol, wherein distinctly prepared materials are brought together and assembled intimately by tuning the chemical and physical interactions among the two, is a favorable substitute. Materials, obtained from top-down methods, such as liquid phase exfoliation, ball milling, and shear mixing, or bottom-up approaches, like hydrothermal, colloidal synthesis, and chemical vapor deposition, can be inculcated into solution routes for heterostructure formation. Stacking methodologies, like manual stacking, surface charged base solution phase assembly, sequential solvothermal-assisted synthesis, etc., can be employed for interface fabrication. Additionally, utilizing functional inks of exfoliated materials [61–63] can be utilized for heterostructure formation with the help of relatively inexpensive protocols such as vacuum filtration, inkjet printing, spray coating, roll-to-roll processing, and spin coating [64].

It has been demonstrated that the liquid phase exfoliated inks of 2D materials depict almost similar characteristics when pitched against micromechanically cleaved layered materials [62, 63]. A comparative study of three protocols, drop casting, inkjet printing, and vacuum filtration, was portrayed as possible means of thin film formation from 2D material inks to fabricate layer-by-layer assembly for graphene/X (X = hBN, WS_2, MoS_2) heterostructures up to three layers. In Fig. 3a [65], all-inkjet-printed Gr/MoS_2 heterostructures over a large area on numerous substrates such as paper, polymers, and programmable logic devices have also been realized, Fig. 3b [66].

Another approach for self-assembling independently obtained nanomaterials is tuning the zeta potential of involved species to assist spontaneous interface formation. For instance, poly (diallyl dimethylammonium chloride) (PDDA) was utilized to alter the naturally negative zeta potential of MoS_2 nanosheets (~−23.8 mV) to a positive value (+36.5 mV) which upon mixing with $ZnIn_2S_4$ nanosheets (zeta potential −

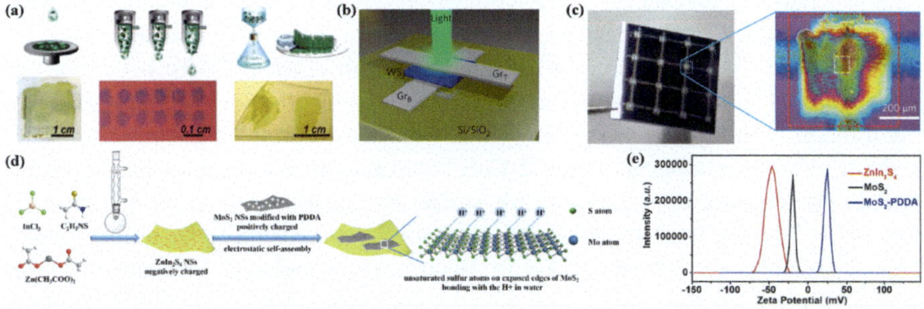

Fig. 3 **a** Schematic depiction (top) of deposition methods for thin-film formation and optical micrographs of fabricated films (bottom): from left to right: drop casting, inkjet printing, and vacuum filtration [65]. Entirely inkjet-printed heterostructures. **b** Schematic illustration of the printed GrB/WS_2/GrT heterostructure. **c** Optical micrograph of a 4 × 4 array of printed heterostructures, and Raman map of 2D band and A_{1g} mode in the marked region (500 × 500 μm) [66]. Electrostatic self-assembly by zeta-potential alteration. **d** Schematic illustration of the synthesis and self-assembly protocol. **e** Modification in zeta potential upon PDDA functionalization, facilitating self-assembly [68]

45.7 mV) self-assemble spontaneously in an aqueous solution to form heterostructure for efficient charge separation, Fig. 3d, e [67].

Mechanical mixing followed by sonication of independently prepared MoS_2 and CdS nanosheets has been demonstrated to form heterointerface [68]. Separately, ball milling was employed to self-assemble ultrasonically exfoliated nanosheets of black phosphorus and MoS_2. HAADF-STEM confirmed the formation of heterojunction and XPS analysis revealed that phosphorus–sulfur interaction governs the formation of heterostructure among the nanomaterials [69]. Hydrothermally synthesized Au/Bi_2WO_6 nanosheet composites dispersed in NMP when mixed with MoS_2 led to S–O bond formation, eventually forming heterostructures. The S–O bond formation was ascertained via XPS and HR-TEM analysis [70].g-C_3N_4 powder obtained from high-temperature polymerization of dicyanamide was mechanically mixed with commercial bulk MoS_2 and subjected to ball grinding, leading to formation of g-C_3N_4/MoS_2 heterostructure [71, 72].

Solvothermal process is another protocol for heterostructure formation among different nanomaterials. This method provides the ability to tune morphological composition, active surface area, and crystallinity. In this protocol, nanomaterials are crystallized at high pressures (1 MPa to 1 GPa) and temperature (100–250 °C) from a solution comprised of solvent and inorganic precursors. Another advantage of the protocol is that the obtained nanomaterials can be redispersed in appropriate solvent to develop inks and dispersions for large-scale thin-film fabrication.

One can employ solvothermal synthesized nanomaterials as substrate for MoS2 growth or vice versa. MoS_2 was deployed as the seed layer to grow ZnO nanorods [73]. Alternatively, Si-NWs have been utilized as substrate for subsequent growth of MoS_2 [74] Similarly, WS_2 nanowires and CeO_2 nanocubes were exploited as substrates for hydrothermal growth of MoS_2 [75, 76] Thermodynamically unstable 1T-MOS_2 was successfully synthesized forming a heterojunction with rGO. Incorporating rGO and successive heating with citric acid led to the selectivity for 1T phase as in their absence only 2H phase was observed [77]. To overcome the aggregation of nanoparticles upon synthesis, montmorillonite (Mt), an abundant clay mineral with layered morphology was chosen as a promising support for one-pot hydrothermal synthesis of MoS2-CdS heterostructure for photocatalytic applications. The equilibrium charge on Mt surface is negative, which electrostatically attracts Cd^{2+} cation from the precursor's solution, whereas molybdenum precursor (MoO_4^{2-}) cannot be adsorbed due to electrostatic repulsion. The sulfur source (thioacetamide), on the other hand, adsorbs on the surface of Mt via H-bonding, which upon hydrothermal synthesis decomposes and leads to nucleation and in situ growth across the exposed Mt surface to form the Mt@MoS_2/CdS [78].

One-step hydrothermal synthesis is also reported, wherein all precursors are mixed in the same chamber and treated to form heterostructures. Such method has been demonstrated to synthesize $CdIn_2S_4$/MoS_2 [79] and MoS_2/Al_2O_3/g-C_3N_4 [80] heterostructures. A facile and effective synthesis strategy for TMDs for HER was reported, wherein 1T-MoS_2/GO heterostructure was designed with intimate interface contact among MoS_2 basal plane and GO via one-pot hydrothermal method with content of 1T-MoS_2 reaching as much as 76.7%. Further, chemical interaction among MoS_2 and GO enables GO to act as a surface for 1T phase growth. The in situ prepared heterostructure augments the charge transfer kinetics, 1T phase assists in enhancing the number of active sites and electrical conductivity of the catalyst [81].

3 Characterization

Identifying and validating formation of heterostructures is an intricate process and cannot be ascertained by any solitary technique. In fact, complementing evidences are required to ascertain the presence of heterostructure. Technically, it is more feasible to identify presence of lateral heterostructures than vertical, owing to the encapsulation of the heterointerface by individual material layer. The characterization can be classified into **microscopic** and **spectroscopic** protocols. Microscopic techniques can be further sub-classified into **electron**, probe, and **optical** microscopy. Raman and photoluminescence are the spectroscopic techniques widely used for heterostructure characterization. We will discuss these approaches and employ them in order to strengthen the evidence for heterostructure formation and understand their optical, electronic, chemical, and physical properties.

3.1 Microscopy Protocols

Microscopy is a technique which is employed to assess samples invisible to the otherwise unaided eye. There are multiple modes of microscopy, ranging from optical to laser, infra-red, electron, probe, photoacoustic, X-ray, etc. However, in the context of heterostructures, optical, probe, and electron microscopy are the techniques widely employed. These techniques differ from each other in resolution, ease of operation, cost of equipment with optical microscopy being the most facile and easy to operate but suffers in resolution. Probe-and-electron microscopy, on the other hand, delivers resolutions at the atomic scale but requires specialized and expensive instrumentation and is intricate to operate.

Optical Microscopy

An optical microscope utilizes lenses to focus the visible light transmitting or reflecting from the sample to deliver a magnified view of the sample. It enables us to identify the location, shape, size, and layer thickness of associated materials constituting the heterostructure [82–84]. For instance, $WSe_2/MoSe_2$ heterostructure was characterized primarily by optical microcopy over a large area (>100 um). Furthermore, to assert the importance of cleaning during layer transfer, annealing was employed to depict emergence of various artifacts like formation of polymer residue pockets of mechanically transferred layers forms pockets of polymer residue, adversely affecting the interlayer interactions. The emergence of impurity pockets was recorded with help of optical microscope, Fig. 4a–d, where the white arrows represent the impurity packets in annealed samples [83]. A protocol for Artificial Neural Network (ANN) was established which portrayed the ability to distinguish various kinds of layered materials and provide intricate information about characteristics of the interface in vertical as well as lateral heterostructures. The protocol employed red, green, blue (RGB) and hue, saturation, view (HSV), six primary colors in pixels of the optical micrograph, Fig. 4e [85]. Upon utilizing this formulation for material and corresponding thickness detection, they report the ability to identify MoS_2/WS_2 heterostructures.

However, accuracy being a strong function of the constituents' color contrast, optical microscopy suffers from low accuracy. Furthermore, since, visible light is used as the reflecting and transmitting medium, lateral and axial resolutions are restricted to 200 and 500 nm, respectively. This limits its widespread utility in characterizing atomically pure interfaces.

Electron Microscopy

Lateral or axial resolution of a microscope is strongly impacted by wavelength of the source employed. In an electron microscope, instead of visible light, electron beam is utilized with significantly smaller which augments the resolution. Scanning Electron Microscopy (SEM) and Tunneling Electron Microscopy (TEM) are the

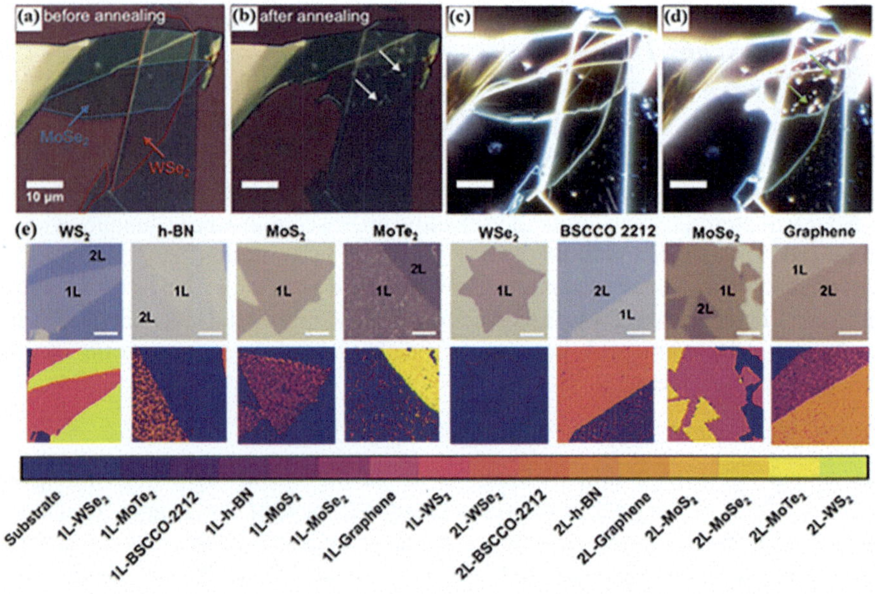

Fig. 4 Observing thermal annealing-induced changes in $MoSe_2/WSe_2$ heterostructure with help of optical microscope. **a, c** Bright and dark-field micrographs. **d, f** Bright and dark-field images of the same structure post-annealing under vacuum at 120 °C. White arrows display contamination pockets (Scale bars, 10 μm) [83]. ANN-assisted identification of 2D materials and their heterostructures. **e** Optical micrographs of various 2D materials (Scale bars, 10 μm). **f** Results of material types and number of layers [85]

primary forms of electron microscopy. SEM is limited to morphology determination or occasional reports of heterointerface identification in large area heterostructures, whereas TEM and its extended counterparts like High-Resolution Transmission Electron Microscopy (HR-TEM), High Angle Annular Dark Field Imaging (HAADF), Selected Area Electron Diffraction (SAED), Energy-Dispersive Spectroscopy (EDS), and Electron Energy Loss Spectrometry (EELS) have the ability to resolve the structure down to Angstrom (Å) scale and can reveal crucial material characteristics like chemical composition, grain boundaries, grain orientation, layer stack sequence, atomic, and structural defects [86–90].

HAADF-STEM selectively suppresses mass/thickness contrast and enhances the selectivity of atomic number contrast, enabling investigations of intricate systems. When employed collectively, HAADF-STEM (cross-sectional and plane view), EDX intensity profiling, and bright field STEM assert their viability in heterostructure characterization [31, 35, 52, 57, 58, 91–93]. This has been demonstrated in Fig. 5, wherein various modes of TEM like a focused ion beam (FIB) cross-sectional TEM, HAADF-STEM, EDS line profiling, and HR-TEM help corroborate formation of a clean $PdSe/Bi_2Se_3$ heterointerface [86]. Formation of a uniform seven-layered consecutive stack of WS_2 and MoS_2 monolayers was identified with the help of W-Lα and Mo-Kα edges from EDS in cross-sectional HAADF-TEM images. Furthermore, bright field STEM images exposed the presence of incoherent stacks ($<$10% of overall coverage) with dissimilar interlayer distance (0.647 nm) than the coherent stack (0.606 nm) [47].

Utilization of SAED to verify formation of a heterostructure over WSe_2 is shown in Fig. 6a and b, where upon heterostructure formation with VSe_2, two sets of identically orientated hexagonally arranged diffraction spots are observed which could be indexed individually to WSe_2 and VSe_2. STEM has also been demonstrated to reveal the apparent Moiré superlattice structure, Fig. 6c, and to extract lattice constants for VSe_2 (3.36 Å) and WSe_2 (3.29 Å) which sufficiently match with database values. Cross-sectional STEM and elemental map further corroborate the results from SAED and STEM images, Fig. 6d–h. Moreover, ideal vdW interface at the heterojunction was confirmed by determining metal-to-metal (V-to-W) and chalcogen-to-chalcogen (Se-to-Se) distance with the help of cross-sectional HAADF-STEM, Fig. 6i [44].

In this manner, electron microscopy assists extensive characterization of intricate structures at high resolutions. However, it is restricted to structural characterizations only and does not provide insights into electro-optic behavior of the system.

Probe Microscopy

In order to extract additional information, like spatio-temporal characterization of structural and electronic properties at sub-nm resolutions, probe microscopy is employed. In ***probe microscopy*** techniques like Scanning Tunneling Microscopy (STM) and Atomic Force Microscopy (AFM), a physical probe is scanned atop the surface of materials to investigate materials at resolutions on par with electron microscopy.

In STM, the local density of states at the surface is extracted from tunneling current recorded at a particular bias voltage between the probe and the surface,

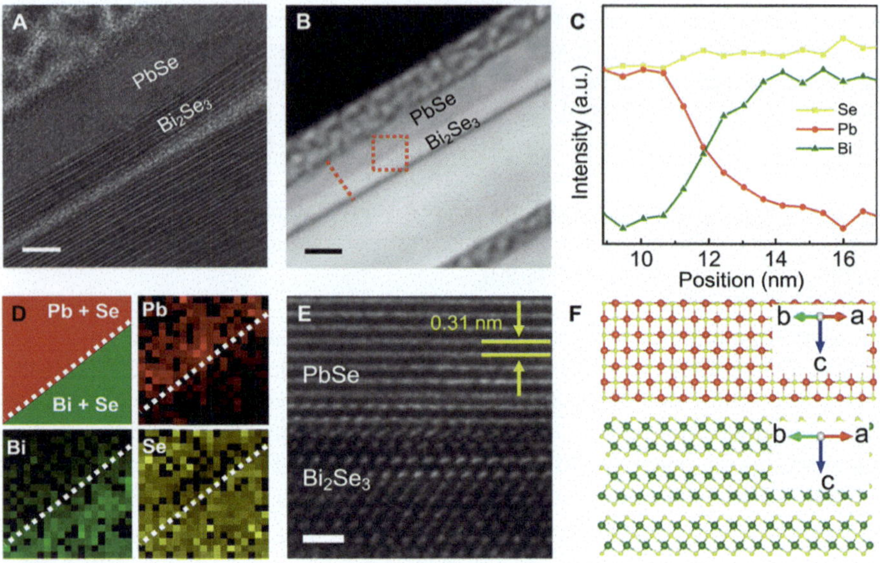

Fig. 5 Electron diffraction and Moiré pattern analysis of the $PbSe/Bi_2Se_3$ heterostructure. **a** A focused ion beam (FIB) cross-sectional TEM image of the heterostructure (Scale bar, 5 nm). **b** Cross-sectional HAADF-STEM image of the heterostructure (Scale bar, 10 nm), and the corresponding EDS line scan profile along the red line. **d** Elemental maps in the red square region of **b**. **e** Cross-sectional HR-TEM image of the interface (Scale bar, 1 nm). **f** Schematic diagram of the heterointerface based on atomic structure [86]

which is projected in the form of an image [94]. Furthermore, Scanning Tunneling Spectroscopy (STS), an extension of STM, can be employed to reveal information regarding localization of electron density among the available electronic states with ability to successfully study Moiré patterns as well [95, 96]. In order to utilize STM correctly, one should keep the following in mind. Firstly, in order to allow enough electron tunneling among the tip and the surface, near-surface region placement of the 2D electron gas (2DEG) is must. Secondly, the surface roughness needs to be smaller than the charge density-induced height deviations, thus a sub-angstrom scale roughness is recommended for high-quality plane-view images. These techniques were employed to study the surface of an InAs/GaAs heterostructure, Fig. 7a and b, to reveal in-plane height variation. Region A represents a triangular pit with step height representing a monolayer (~0.1 nm), B is an adsorbate species which was observed in multiple positions on the substrate, C to D represent a chain of empty states. Feature E has a depth of ~0.05 nm and is invariant with applied bias, thus confirming the artifact is not due to any electronic effects [94]. Carrying out an FFT analysis in selected region of STM topography image can be utilized to extract the lattice constants at distinct locations. This has been demonstrated for a graphene/hBN heterostructure, where Fourier analysis is carried out in different regions of STM images, Fig. 7c, d. The Fourier analysis, Fig. 7e, is then employed to extract relative lattice constants of the sample, Fig. 7f [95]. A similar approach was employed for characterizing MoS2/

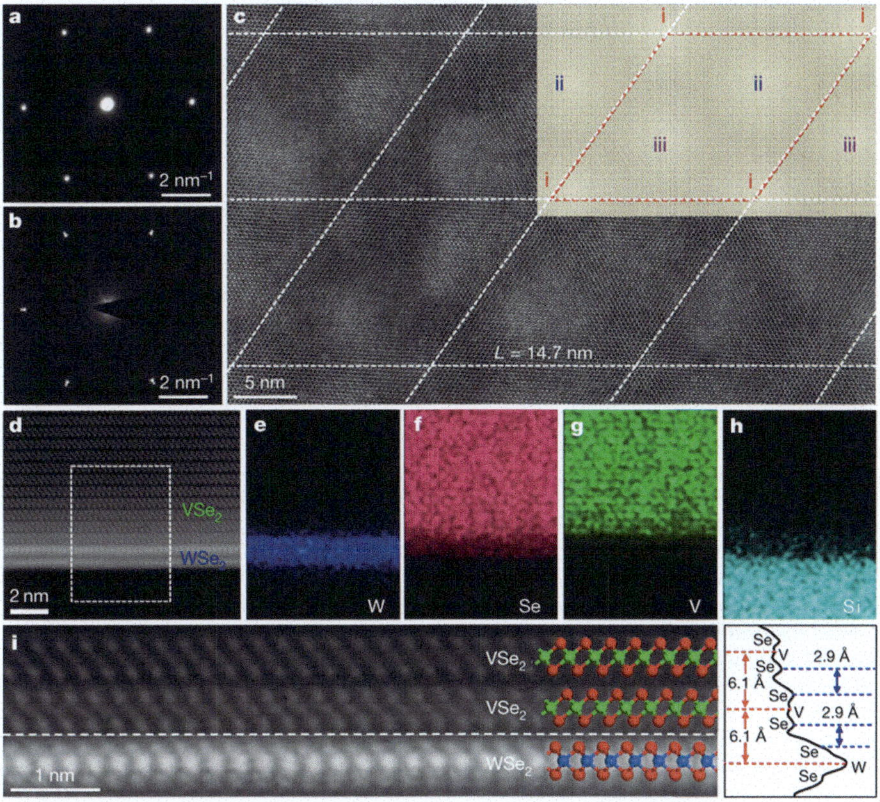

Fig. 6 Electron microscopy characterizations of a WSe_2/VSe_2 heterostructure. **a** and **b** are the SAED patterns from WSe_2 and the heterostructure, respectively, **c** is the STEM image of heterointerfaced region displaying the apparent Moiré superlattice structure, where white dashed lines portray the periodic ($L = 14.7$ nm) repetition of Moiré cells, and inset showing a simulation of atomic structure in a period. The simulation revealed three types of local stacking, (i) V atom over V, (ii) V over Se, and (iii) V over W. **d** Cross-sectional STEM image of the heterointerface and **e–h** corresponding elemental maps in the marked region of **d**. (i) is a cross-sectional STEM image and the associated atomic model with the intensity profile (right), where the heterointerface is marked by the white dashed line [44]

WS2 heterostructure as well, Fig. 7g. Furthermore, distinguishing materials based on their electronic properties can be carried out with the help of STS. The evolution of tunneling behavior from a monolayer of MoS_2 and WS_2 to their heterostructure was recorded, Fig. 7h. This technique not only facilitates validating formation of a heterostructure but also identifies band alignment at the interface [97].

AFM is another versatile probe-induced method which can reveal morphology, layer orientation, defects, adhesion energies, friction parameters, and elastic moduli and that too with atomic-scale resolution. A schematic of the nanoindentation protocol using an AFM tip is shown in Fig. 8a, using which the elastic moduli of suspended layers of different materials or their heterostructures can be calculated.

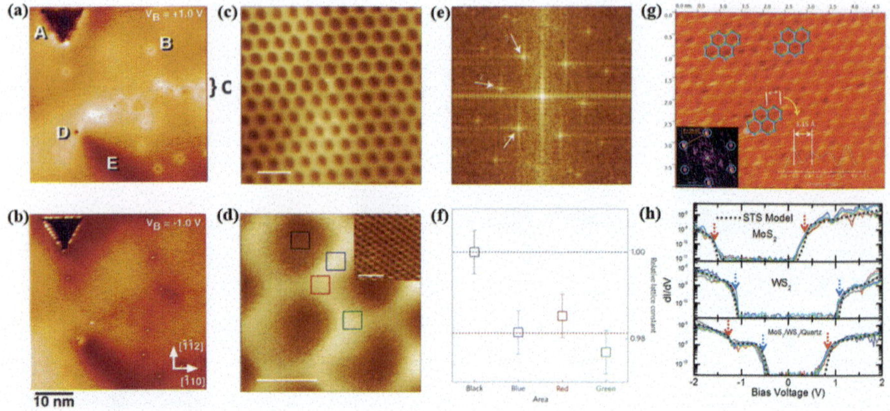

Fig. 7 A topography image of 8-nm-thick InAs thin film grown over (111) GaAs substrate, generated from STM operating under constant-current mode, with a bias voltage of **a** +1.0 V and **b** −1.0 V [94] STM measurements of aligned graphene-on-hBN heterostructures. **c** STM topography image under sample bias and tunneling current of −0.1 V and 300 pA, respectively, showing a clear Moiré pattern (Scale bar, 30 nm). **d** Zoomed-in view of **c** (Scale bar, 10 nm). Inset shows area marked by black square, clearly showing the atomic structure (Scale bar, 1 nm). Interatomic distance is computed by Fourier transformation at the regions marked by colored squares (3 nm). **e** Fourier transformation of atomically resolved structure in red square. **f** Relative lattice constants computed at different regions (colored squares) with respect to region under black square [95]. **g** STM image of WS_2 monolayer acquired at a bias of +1.5 V and constant current of 100 pA. As a guide for the eye, illustrative honeycomb lattices are overlaid and used to extract apparent lattice constant, which correspond to WS_2, inset on the right. Further, a Fourier transform of the real-space image, inset on the lower left, displays the expected periodicity, which could be indexed to WS_2. **h** Averages over multiple random measurements in STS data are shown for individual materials and their heterostructure as solid-colored lines, whereas fits generated from a model are represented by dotted lines. To designate band onsets, red and blue dotted arrows are employed for MoS_2 and WS_2, respectively [97]

For MoS_2/WS_2 and MoS_2/Gr heterostructures, this technique revealed that individual monolayers have higher elastic moduli relative to a heterostructure comprised of monolayers, Fig. 8b [98]. Furthermore, in lateral heterostructures, the interfacial buckling-induced relative strain has been probed by AFM [96]. Extensions of AFM like Kelvin probe force microscopy (KPFM) have been utilized to spatially identify heterointerfaces by analyzing the current variations. This was demonstrated for a $MoTe_2$/Gr heterostructure [99]. Similarly, it was employed to locate $MoSe_2$ in a $WSe_2/MoSe_2$ heterostructure due to considerable gradient in lateral dimension of the two layers (~300 nm for $MoSe_2$ and ~2 um for WSe_2) [42, 99, 100]. KPFM also enables probing work function of surface of interest to deduce Fermi level, eventually revealing band alignment at the heterointerface, as shown in Fig. 8c–h [101, 102].

Despite these abilities, probe microscopic protocols have challenges associated with their deployment as a general heterostructure characterization tool. For instance, scanning tunneling microscopes require ultra-high vacuum and near-absolute zero temperatures for optimum performance. Secondly, the limitation on thickness of

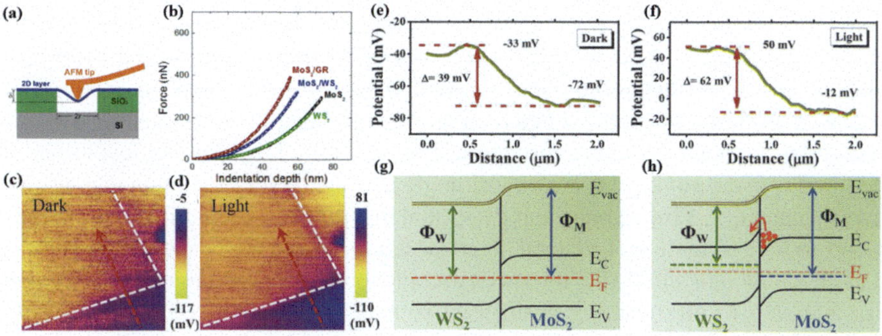

Fig. 8 Elastic properties of layered materials and their heterostructures. **a** Schematic illustration of the indentation experiment. **b** Force–displacement curves of individual monolayers and heterostructures. The white dashed lines are fitted for the force–displacement relationship [98]. Characterizing WS_2/MoS_2 heterostructure with KPFM. **c** and **d** are potential images obtained from KPFM of the heterostructure. **e** and **f** display the potential profiles across marked region in **c** and **d** under dark and illuminated conditions, respectively. Energy band alignment between WS_2 and MoS_2 under dark **g** illuminated conditions **h** [99]

heterostructure limits its use in characterization of solution-processed or printed heterostructures which tend to be composed of few layers of individual materials. Finally, sample preparation is a great hurdle onto which the sub-Angstrom surface roughness requirement restricts its deployment of wide variety of heterostructures.

3.2 *Spectroscopic Techniques*

Techniques such as Raman and PL demonstrate excellent ability to identify and differentiate heterostructures with important advantages over microscopy protocols like data collection takes just minutes as opposed to hours and no need for intricate sample preparation. Raman spectroscopy relies on variation in vibrational markers of individual materials upon formation of heterostructures [103]. Variation in vibrational markers can be a result of changes in crystal defects, lattice strain [104, 105], doping [106], or chemical composition [106, 107] which usually ensue heterostructure formation. Furthermore, Raman spectrum varies if the heterostructure formation entails interlayer coupling, which is primarily due to the increment in electronic density of states at the interface [108, 109]. This can be used to affirm heterostructure formation.

Low-frequency Raman spectroscopy has been used to illustrate the interlayer coupling in vertical heterostructure stacks of different TMDs, wherein the disappearance of shearing mode implies the heterojunction formation, Fig. 9a [110, 111]. This is ascribed to the significant reduction in shear force of a lattice mismatched heterostructure [112]. On the other hand, high-frequency Raman spectra, due to its ability to identify strain deviations, have been depicted as a tool to characterize lateral heterostructures by tracking strain variations at the heterostructure interface [57, 113]. When coupled with PL, evolution of interlayer optical and electronic mechanisms at the heterointerface is revealed. Tracing the extended lifetimes of interlayer excitons for vertical heterostructures has been demonstrated as a venue to corroborate heterostructure formation, Fig. 9b [114, 115]. Further, evolution of the interlayer excitonic peak with twist angle and interlayer separation can also be investigated [115, 116]. Raman and PL spatial mapping has been utilized in synergy to corroborate formation of a lateral heterostructure among MoS_2 and WS_2, Fig. 9c–f [117] Even though simplistic, the spectroscopic techniques are limited by laser spot size and step width, resulting in diffused signals at the interface [115, 117–119]. Utility of PL is further limited by the bandgap of materials involved in heterojunction formation. The PL signals for indirect bandgap semiconductors are weak and cannot be characterized within an acceptable margin of error [120]. For MoS_2, nature of the bandgap is strongly thickness dependent which restricts its employability for samples prepared with solution-processed routes.

In summary, one has to rely on various characterization strategies to identify, characterize, and investigate electrical as well as optical properties and their evolution upon heterostructure formation. Optical microscopy and spectroscopic methods are simplistic tools for optical as well as electrical characterizing heterostructures, with compromises to be made in resolution. The electron microscopy and probe microscopy, on the other hand, are relatively more time-consuming, require expensive tools, and are complex to record and analyze information but enables one to study the structures at atomic-scale resolution, with excellent accuracy and versatility in providing information about optical, electrical, and structural properties of the sample of interest. Thus, one has to recognize the information to be studied and compromises to be made in order to choose appropriate characterization tool.

4 Applications

Photodetectors

Photodetectors are electronic devices whose electronic response varies upon interacting with electromagnetic radiation. These devices are largely incorporated in modern-day electronics to sense the nature of incident electromagnetic radiations. The charge carriers generated upon absorption of incident light can follow two pathways to disburse absorbed energy. Either they can travel under the influence of applied bias and collected at the associated current collector or they can recombine and release absorbed energy via radiative or non-radiative routes. The recombination route is counterproductive and hampers the important figures of merit of a device. Primary consequence of a heterostructure formation is spontaneous injection of a charge carrier from photoactive layer to another, thereby reducing recombination losses.

Thus, optoelectronic response of a device can be greatly enhanced by incorporating a photosensitizer with selective absorption characteristics and utilizing MoS_2 as the transport layer for the optically generated charge carriers. Usually, 0D or 1D nanostructures are employed as photosensitizers due to their synergistic role in augmenting optoelectronic FOMs relative to MoS_2 alone [121, 122]. The tunability aspect emerges in the selection and engineering of the sensitizer employed for desired application. PbS quantum dots have been demonstrated as an excellent sensitizer source with large carrier density of electrons and holes generated upon illumination, wherein the electrons are injected into the underlying MoS_2 layer under the influence of in-built field. This scheme of charge carrier separation depicts elevated responsivity of 1400 AW^{-1} at 1.8 μm and D* of 2.8×10^{12} jones, Fig. 10a and b [27]. Furthermore, inclusion of a TiO_2 buffer layer enhances the efficiency of photogenerated electron injection into MoS_2 generating responsivities as high as ~10^5 AW^{-1}. The principle behind this approach is justified by the perceived band alignment at the two heterojunctions [123]. Similarly, HgTe QDs with a TiO_2 buffer layer are reported as well [124].

A 2D/1D-like heterostructure among Te nanowires and MoS_2, Fig. 10c, has been reported to enhance the optoelectronic performance in near-IR range (1,550 nm) with responsivities as high as 10^3 AW^{-1}, Fig. 10d [51]. On the other hand, to extend the response to near-UV range, heterostructure between MoS_2 and ZnO nanowires is reported with high EQE (8.28×10^3%) and responsivity (~20 A W^{-1}) at 365 nm [125] Apart from efficient charge carrier separation and absorption enhancement [126], heterostructures with MoS_2 involving other van der Waals materials bring auxiliary benefit of enabling bandgap manipulation, which assists in designing PD depicting excellent FOMs [127–131]. Materials such as $MoTe_2$ [22], WS_2 [101],

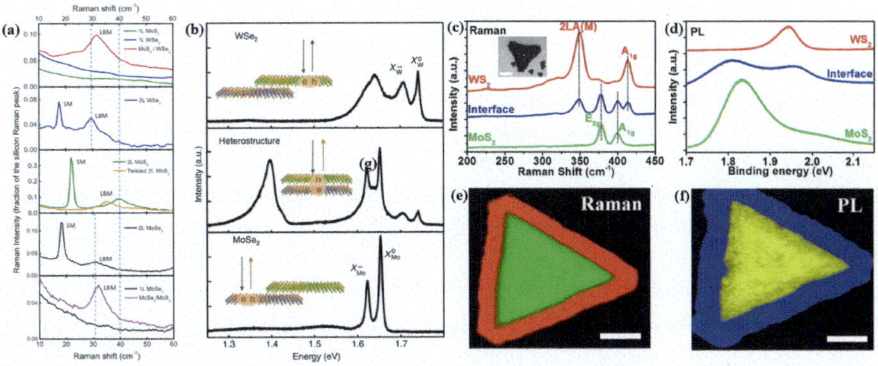

Fig. 9 Raman and PL spectra for vertical and lateral heterostructure characterizations **a** Low-frequency Raman spectra for individual materials and its evolution upon heterostructure formation [111]. **b** PL spectra of individual monolayers and resulting heterostructure upon excitation at 1.88 eV [115] **c** Raman spectra and **d** PL spectra of WS_2, MoS_2 and laterally heterointerfaced domains. **e** Raman intensity map for A_{1g} mode of WS_2 (red) and E_{2g} mode of MoS_2 (green), and **f** image of PL intensity map for characteristic PL peaks of MoS_2 and WS_2 at 677 (yellow) and 638 nm (blue), respectively (Scale bar, 10 μm) [117]

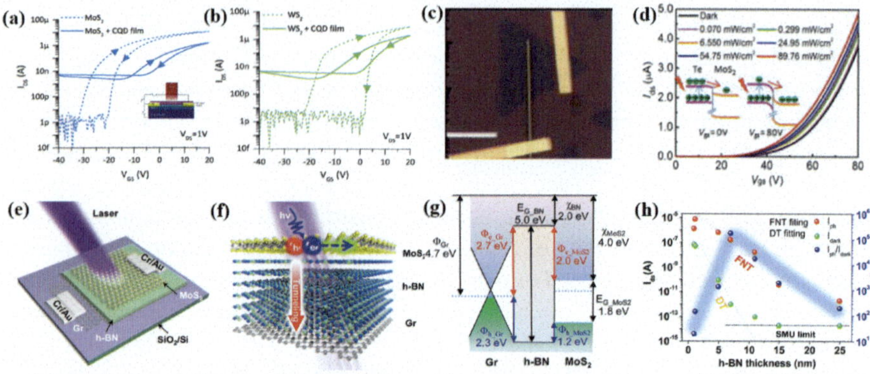

Fig. 10 Hybrid photodetector devices from TMDs and PbS QDs. Transfer curves under dark conditions for dark for MoS_2 device (**a**) and WS_2 device (**b**), with inset showing schematic of the device structure [27]. **c** Optical micrograph of the 2D–1D-type MoS_2/Te nanowire heterojunction, (Scale bar, 5 μm). **d** Transfer curves of the device upon illumination with 1550 nm laser at altered power densities ($V_{ds} = 2$ V), the inset displays the band alignment at the heterointerface at $V_{gs} = 0$ V and 80 V [51] Tuning photoresponse of MoS_2/Gr heterostructure with an intermediate hBN layer. **e** Schematic of the fabricated heterostructure device assembly. **f** Illustration of electron–hole pair generation under illumination and selective hole tunneling through hBN. **g** Band alignment for MoS_2/hBN/graphene heterostructure with flat band model. Here, χ is electron affinity, E_G is the bandgap; Φ_e is electron tunneling barrier, and Φ is the hole tunneling. **h** Evolution of dark current (I_{dark}), photocurrent (I_{ph}), and their I_{ph}/I_{dark} ratio with thickness of intermediate hBN layer [136]

CdTe [132], β-In_2Se_3 [35], WSe_2 [133], $BaTiO_3$ [134], MoO_x [37], $SnSe_2$ [135], Sb_2Te_3 [38], Sb_2O_3 [39], SnSe [24], and $PdSe_2$ [20] have been demonstrated as high-performance photodetectors across UV, visible, and near-IR spectrum.

The optoelectronic behavior of a heterostructure is strongly dependent upon the interlayer coupling which can be tuned by inculcating an additional layer of appropriate material within a heterostructure stack, Fig. 10e. Upon illumination, the dynamics of generated charge carrier is schematically illustrated in Fig. 10f, and the band alignment at flat band potential is depicted in (g). The thickness of hBN intermediate layer strongly governs photoresponse of the device by efficiently suppressing the dark-carrier tunneling and improving the photocarrier tunneling. These ultrathin devices demonstrate ultra-high FOMs like detectivity of ~10^{13} Jones, responsivity of 180 AW^{-1}, and high photo/dark current ratio of >10^5, Fig. 10h [136]. Multiple reports have been made for such hetero-stacks like Gr/MoS_2/WS_2 [137], WSe_2/Gr/MoS_2 [26], and WSe_2/hBN/MoS_2 [138]. These reports exemplify the tunability of interlayer electronic and optical coupling by varying layer thickness, stack angle orientation, etc.

Photovoltaic

Apart from photodetectors, charge carrier separation induced by heterojunction formation is of utility in photovoltaic devices as well. Ultrathin PV devices fabricated from monolayers of MoS_2 and WS_2 depict photovoltaic effect-driven electronics with incident light to power conversion efficiencies (PCE) of ~1% and an impressive power density of 2.5 MW/kg, Fig. 11a and b [139, 140] When pitched against already reported ultrathin film solar cells, these come out virtuous with about three orders of improvement in important FOMs for PV devices. Further improvement in PCE up to ~1.5% was achieved by modulating electrical doping state with help of back gate junction. However, the performance indices like carrier diffusion timescale and interlayer charge carrier lifetimes are still poor and limited by low in-plane charge carrier mobility of the TMDs. In order to overcome stunted mobility, inclusion of graphene electrodes in heterostructure has been demonstrated leading to about five orders of enhancement in photoresponse due to efficient charge extraction into the graphene layer [116].

A Gr/MoS_2/Gr double heterostructure was reported where substrate-induced doping of the bottom contact graphene led to different Schottky barrier heights, resulting in a device with EQE of ~27% [116, 141]. Similarly, individual heterostructures like Gr/MoS_2 and Gr/WS_2 have also been reported [142]. Interestingly, a Gr/MoS2/WSe2/Gr heterostructure is reported where the sandwiching of the *p–n* junction within graphene layers enhances the photogenerated charge carrier collection, Fig. 11c, d [116].

Integration with already upscaled Si manufacturing process is a great opportunity for rapid introduction of MoS_2 into the PV industry. In fact, it has been reported that a monolayer MoS_2 with *p*-type-doped silicon forms a Type-II heterojunction resulting

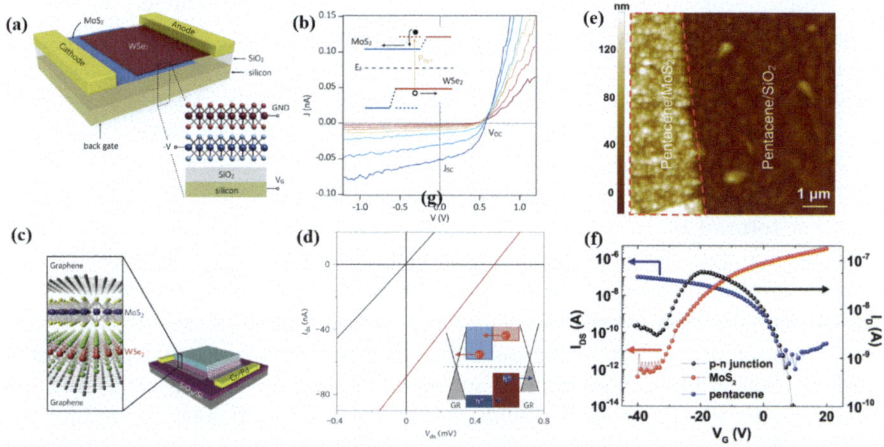

Fig. 11 **a** Schematic illustration of the device assembly. **b** *J-V* characteristics under illumination power (P_{opt}) of 180, 400, 670, 1100, 1800, 4000, and 6400 W/m^2. Inset illustrates the photovoltaic effect at the heterointerface [140]. Van der Waals *p-n* heterojunction sandwiched within graphene electrodes. **c** Schematic illustration of the device assembly and layer orientation. **d** *J-V* characteristics of the device under dark (black) and illuminated (532 nm) conditions (red). Inset shows the band alignment and charge carrier collection processes by arrows. The dashed line represents the Fermi level [116]. Organic–Inorganic *p-n* heterostructure for photovoltaics. **e** AFM image of pentacene/MoS2 heterostructure. The heterostructure region is marked with red dashes. **f** Semilogarithmic *J-V* characteristics of individual materials and their respective heterostructure [146]

in enhanced optically induced charge carrier separation over large areas with PCE of ~5.23% [143].

Plasma techniques have been reported to modulate the electrical density of states in MoS_2, which then can be employed to form a homojunction among *n*-type and *p*-type MoS_2. Such a device depicted decent performance parameters with EQE values of about 80% at 700 nm and PCE of ~3% [144]. The charge carrier transport in heterojunction of a photodiode constructed from GaTe and MoS_2 was identified to be limited by Shockley–Read–Hall (SRH) recombination mechanism. The photodiode managed to acquire a high EQE of ~62% [145]. Interfacing inorganic materials with organic semiconductors is another approach to form photovoltaic systems. Such a device has been demonstrated with *p*-type pentacene and *n*-type MoS_2, wherein transient absorption spectroscopy revealed domination of MoS_2 in exciton formation and one order of magnitude larger recombination lifetimes when compared to previously reported 2D material heterostructures, Fig. 11e, f [146].

Transistors

The advantage of layered materials is the wide class of materials available to select from with very different optical and electrical properties with same processability and device fabrication protocols. For instance, layered materials have semi-metals with 0 bandgap like graphene; large number of semiconductors, such as TMDs, group III-VI compound materials; and insulators such as hBN, which can be manufactured from same experimental setups by modifying simple processing parameters and precursors, which speaks to upscalability of these materials. Heterostructure formation further enhances the device parameters. For instance, graphene is a good candidate to demonstrate rapid response FETs but are limited by low ON/OFF ratios. This can be overcome by inclusion of TMDs as the channel materials. An all-2D-material FET, schematically shown in Fig. 12a, was reported, where the device portrayed excellent electron mobilities of about 30 cm^2/Vs and high ON/OFF ratio of $\sim 10^7$, Fig. 12b, c [14]. Devising a structure with vertical transport across the heterojunction is another form of FETs based on 2D materials. Here, basic structure is Gr/TMD/metal. In such a device, appropriate metal contact with the TMD facilitates Ohmic transport of charge carriers with the graphene interface enabling tunability on the Schottky barrier front. Another concession of opting a vertical device structure is miniaturization of device area and simultaneous increment of current density, which asserts to their utility in high-density integrated circuits. In that regard, a Gr/MoS_2 heterostructure was demonstrated with current densities up to 5000 A/cm^2 and 10^3 for ON/OFF ratios [147]. Further enhancement in the ON/OFF ratios (up to 10^5) and current densities (10^4 A/cm^2) was demonstrated by employing a graphene/MoS_2/metal-type geometry, Fig. 12d, e. Effect of number of MoS_2 layers (N) upon the ON and OFF currents was also recorded, Fig. 12f [148]. A heterojunction among p-type Ge and n-type MoS2 vertically stacked tunnel FET revealing the venue for employment in low-power and high-performance electronics, Fig. 12g. The device exhibits *p–n* junction characteristics in both forward and reverse bias conditions and the electrical characteristics trend toward negative differential resistance (NDR), asserting band-to-band tunneling, Fig. 12h [149].

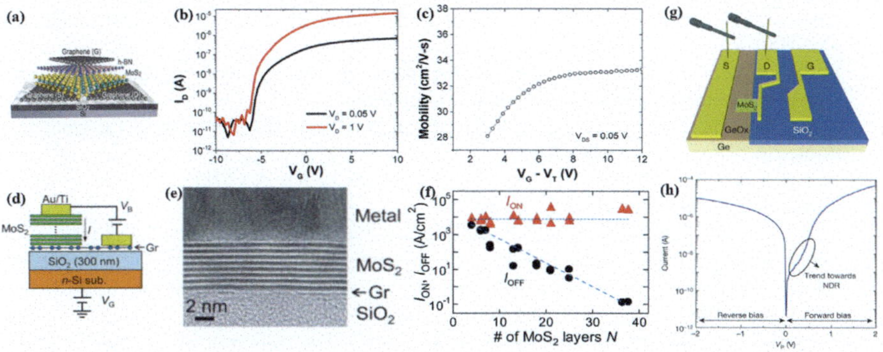

Fig. 12 All-2D FET with MoS_2 as the channel material, bilayer and multilayer graphene as the source–drain and gate electrodes, respectively, and few-layer hBN as the gate dielectric. **a** Schematic illustration of the device assembly. **b** Transfer characteristics and **c** calculated mobility values as a function of V_G-V_T. [14]. A graphene/MoS_2/metal vertical heterostructure. **d** Schematic illustration of device assembly and measurement circuit. **e** Cross-sectional TEM image of the device, and **f** the number of MoS_2 layers *(N)* dependence of ON and OFF currents [148]. **g** Schematic illustration of device assembly. **h** Current as a function of bias applied (V_P) to Ge with MoS_2 contact being grounded [149]

Apart from these, the employment of MoS2-based heterostructures is vast with utility in biosensors [150], catalysis [151–156], light-emitting diodes [157–160], and optical modulator [161–163].

5 Conclusions

Heterostructures of MoS_2 have been at the forefront of 2D materials research in case of synthesis, properties, and applications. Lab-scale investigations of various properties show great promise and assert the importance of upscaling the devices. In order to convince industry players to invest in the upscaling practice, synthesis protocols have to be prepared with higher yields and sustainable processes. CVD approaches have been demonstrated for large-scale prodution of layered materials. Liquid-phase exfoliation is a great tool for obtaining nanomaterial dispersion which can be further utilized for that large-scale protocols like printing, roll-to-roll processing.

References

1. Alferov ZI (2001) Nobel lecture: the double heterostructure concept and its applications in physics, electronics, and technology. Rev Mod Phys 73:767–782
2. Duong DL, Yun SJ, van der Lee YH (2017) Waals layered materials: opportunities and challenges. ACS Nano 11:11803–11830

3. Alferov ZhI (1998) The history and future of semiconductor heterostructures. Semiconductors 32:1–14
4. Li M-Y, Chen C-H, Shi Y, Li L-J (2016) Heterostructures based on two-dimensional layered materials and their potential applications. Mater Today 19:322–335
5. Zhou, X. *et al.* 2D Layered Material-Based van der Waals Heterostructures for Optoelectronics. *Adv Funct Mater* **28**, (2018).
6. Radisavljevic B, Radenovic A, Brivio J, Giacometti V, Kis A (2011) Single-layer MoS2 transistors. Nat Nanotechnol 6:147–150
7. Liu H et al (2014) Dense network of one-dimensional Midgap metallic modes in monolayer <math display="inline"> <mrow> <msub> <mrow> <mi>MoSe</mi> </mrow> <mrow> <mn>2</mn> </mrow> </msub> </mrow> </math> and their spatial undulations. Phys Rev Lett 113:066105
8. Zeng H, Dai J, Yao W, Xiao D, Cui X (2012) Valley polarization in MoS2 monolayers by optical pumping. Nat Nanotechnol 7:490–493
9. Lopez-Sanchez O, Lembke D, Kayci M, Radenovic A, Kis A (2013) Ultrasensitive photodetectors based on monolayer MoS2. Nat Nanotechnol 8:497–501
10. Long M, Wang P, Fang H, Hu W (2019) Progress, challenges, and opportunities for 2D material based photodetectors. Adv Funct Mater 29
11. Zhang W, Wang Q, Chen Y, Wang Z, Wee ATS (2016) Van der Waals stacked 2D layered materials for optoelectronics. 2d Mater 3:022001
12. Liu Y et al (2016) Van der Waals heterostructures and devices. Nat Rev Mater 1:16042
13. Hu T, Zhang R, Li J-P, Cao J-Y, Qiu F (2022) Photodetectors based on two-dimensional MoS2 and its assembled heterostructures. Chip 1:100017
14. Roy T et al (2014) Field-effect transistors built from all two-dimensional material components. ACS Nano 8:6259–6264
15. Konstantatos G et al (2012) Hybrid graphene–quantum dot phototransistors with ultrahigh gain. Nat Nanotechnol 7:363–368
16. Roy K et al (2013) Graphene–MoS2 hybrid structures for multifunctional photoresponsive memory devices. Nat Nanotechnol 8:826–830
17. Britnell L et al (2013) Strong light-matter interactions in heterostructures of atomically thin films. Science 1979(340):1311–1314
18. Novoselov KS et al (2004) Electric field effect in atomically thin carbon films. Science 1979(306):666–669
19. Konstantin Novoselov. Konstantin Novoselov – Nobel Lecture.
20. Long M et al (2019) Palladium diselenide long-wavelength infrared photodetector with high sensitivity and stability. ACS Nano acsnano.8b09476. https://doi.org/10.1021/acsnano.8b09476
21. Wang G et al (2019) Interlayer coupling induced infrared response in WS_2/MoS_2 heterostructures enhanced by surface plasmon resonance. Adv Funct Mater 28
22. Chen Y et al (2018) High-performance photovoltaic detector based on $MoTe_2$/MoS_2 Van der Waals heterostructure. Small 14
23. Wang F et al (2016) Configuration-dependent electrically tunable Van der Waals Heterostructures based on $MoTe_2$/MoS_2. Adv Funct Mater 26:5499–5506
24. Yang S et al (2017) Enhanced electrical and optoelectronic characteristics of few-layer type-II SnSe/MoS_2 van der Waals heterojunctions. ACS Appl Mater Interfaces 9:42149–42155
25. Wu F et al (2020) Formation of coherent 1H–1T heterostructures in single-layer MoS_2 on Au(111). ACS Nano 14:16939–16950
26. Long M et al (2016) Broadband photovoltaic detectors based on an atomically thin heterostructure. Nano Lett 16:2254–2259
27. Özdemir O, Ramiro I, Gupta S, Konstantatos G (2019) High sensitivity hybrid PbS CQD-TMDC Photodetectors up to 2 μm. ACS Photonics 6:2381–2386
28. Zhang J et al (2020) High-performance ultraviolet-visible light-sensitive 2D-MoS_2/1D-ZnO heterostructure photodetectors. Chem Sel 5:3438–3444

29. Jariwala D et al (2013) Gate-tunable carbon nanotube–MoS_2 heterojunction p-n diode. Proc Natl Acad Sci 110:18076–18080
30. Liu L, Zhai T (2021) <scp>Wafer-scale</scp> vertical van der <scp>Waals</scp> heterostructures. InfoMat 3:3–21
31. Motala MJ et al (2023) Synthesis of two-dimensional van der Waals superlattices, heterostructures, and alloys from conversion of sequentially layered sub-nanometer metal films. Mater Today Nano 22:100319
32. Bonaccorso F et al (2012) Production and processing of graphene and 2d crystals. Mater Today 15:564–589
33. Hennighausen Z et al (2019) Oxygen-induced in situ manipulation of the interlayer coupling and exciton recombination in Bi_2 Se_3/MoS_2 2D heterostructures. ACS Appl Mater Interfaces 11:15913–15921
34. Biroju RK, Pal S, Sharma R, Giri PK, Narayanan TN (2017) Stacking sequence dependent photo-electrocatalytic performance of CVD grown MoS_2/graphene van der Waals solids. Nanotechnology 28:085101
35. Zou Z et al (2020) Epitaxial synthesis of ultrathin β-In $_2$ Se_3/MoS_2 heterostructures with high visible/near-infrared photoresponse. Nanoscale 12:6480–6488
36. Yu SH et al (2019) In situ hybridizing MoS_2 microflowers on VS_2 microflakes in a One-Pot CVD process for electrolytic hydrogen evolution reaction. ACS Appl Energy Mater 2:5799–5808
37. Im H, Liu N, Bala A, Kim S, Choi W (2019) Large-area MoS2-MoOx heterojunction thin-film photodetectors with wide spectral range and enhanced photoresponse. APL Mater 7
38. Liu H et al (2019) Van der Waals epitaxial growth of vertically stacked Sb2Te3/MoS2 p–n heterojunctions for high performance optoelectronics. Nano Energy 59:66–74
39. Ye K et al (2020) High-performance broadband photodetectors of heterogeneous 2D inorganic molecular Sb_2 O_3 /monolayer MoS_2 crystals grown via chemical vapor deposition. Adv Opt Mater 8
40. Shi Y et al (2012) Van der Waals epitaxy of MoS $_2$ layers using graphene as growth templates. Nano Lett 12:2784–2791
41. Hoang AT et al (2020) Epitaxial growth of wafer-scale molybdenum disulfide/graphene heterostructures by metal-organic vapor-phase epitaxy and their application in photodetectors. ACS Appl Mater Interfaces 12:44335–44344
42. Lin Y-C et al (2015) Atomically thin resonant tunnel diodes built from synthetic van der Waals heterostructures. Nat Commun 6:7311
43. Lin Y-C et al (2014) Direct synthesis of van der Waals solids. ACS Nano 8:3715–3723
44. Li J et al (2020) General synthesis of two-dimensional van der Waals heterostructure arrays. Nature 579:368–374
45. Zhao Y, Jin S (2020) Controllable water vapor assisted chemical vapor transport synthesis of WS_2–MoS_2 heterostructure. ACS Mater Lett 2:42–48
46. Zhao B et al (2021) High-order superlattices by rolling up van der Waals heterostructures. Nature 591:385–390
47. Jin G et al (2021) Heteroepitaxial van der Waals semiconductor superlattices. Nat Nanotechnol 16:1092–1098
48. Woods JM et al (2016) One-step synthesis of MoS_2/WS_2 layered heterostructures and catalytic activity of defective transition metal dichalcogenide films. ACS Nano 10:2004–2009
49. Yu JH et al (2015) Vertical heterostructure of two-dimensional MoS_2 and WSe_2 with vertically aligned layers. Nano Lett 15:1031–1035
50. Jung Y et al (2014) Metal seed layer thickness-induced transition from vertical to horizontal growth of MoS_2 and WS_2. Nano Lett 14:6842–6849
51. Li N et al (2019) Strongly coupled van der Waals heterostructures for high-performance infrared phototransistor. Appl Phys Lett 114
52. Wang S, Wang X, Warner JH (2015) All chemical vapor deposition growth of MoS 2:h-BN vertical van der Waals heterostructures. ACS Nano 9:5246–5254

53. Zhang K et al (2016) Interlayer transition and infrared photodetection in atomically thin type-II $MoTe_2/MoS_2$ van der Waals heterostructures. ACS Nano 10:3852–3858
54. Olding JN et al (2019) Charge separation in epitaxial SnS/MoS_2 vertical heterojunctions grown by low-temperature pulsed MOCVD. ACS Appl Mater Interfaces 11:40543–40550
55. Song X et al (2018) Boosting two-dimensional MoS_2 /$CsPbBr_3$ photodetectors via enhanced light absorbance and interfacial carrier separation. ACS Appl Mater Interfaces 10:2801–2809
56. Duan X et al (2014) Lateral epitaxial growth of two-dimensional layered semiconductor heterojunctions. Nat Nanotechnol 9:1024–1030
57. Li M-Y et al (2015) Epitaxial growth of a monolayer WSe_2-MoS_2 lateral p-n junction with an atomically sharp interface. Science 1979(349):524–528
58. Gong Y et al (2014) Vertical and in-plane heterostructures from WS2/MoS2 monolayers. Nat Mater 13:1135–1142
59. Leong WS et al (2018) Synthetic lateral metal-semiconductor heterostructures of transition metal disulfides. J Am Chem Soc 140:12354–12358
60. Jung Y, Shen J, Sun Y, Cha JJ (2014) Chemically synthesized heterostructures of two-dimensional molybdenum/tungsten-based dichalcogenides with vertically aligned layers. ACS Nano 8:9550–9557
61. Blake P et al (2008) Graphene-based liquid crystal device. Nano Lett 8:1704–1708
62. Hernandez Y et al (2008) High-yield production of graphene by liquid-phase exfoliation of graphite. Nat Nanotechnol 3:563–568
63. Coleman JN et al (2011) Two-Dimensional Nanosheets Produced by Liquid Exfoliation of Layered Materials. Science 1979(331):568–571
64. Torrisi F et al (2012) Inkjet-printed graphene electronics. ACS Nano 6:2992–3006
65. Withers F et al (2014) Heterostructures produced from nanosheet-based inks. Nano Lett 14:3987–3992
66. McManus D et al (2017) Water-based and biocompatible 2D crystal inks for all-inkjet-printed heterostructures. Nat Nanotechnol 12:343–350
67. Huang L et al (2019) Ultrathin 2D/2D ZnIn2S4/MoS2 hybrids for boosted photocatalytic hydrogen evolution under visible light. J Alloys Compd 798:553–559
68. Xiong M, Yan J, Chai B, Fan G, Song G (2020) Liquid exfoliating CdS and MoS2 to construct 2D/2D MoS2/CdS heterojunctions with significantly boosted photocatalytic H2 evolution activity. J Mater Sci Technol 56:179–188
69. Huang Y et al (2021) Synthesis and photocatalytic performance of MoS2/Polycrystalline black phosphorus heterojunction composite. Int J Hydrog Energy 46:3530–3538
70. Wan J et al (2020) Effective charge kinetics steering in surface plasmons coupled two-dimensional chemical Au/Bi2WO6-MoS2 heterojunction for superior photocatalytic detoxification performance. J Hazard Mater 384:121484
71. Li W et al (2019) Fabrication of an ultrathin 2D/2D C3N4/MoS2 heterojunction photocatalyst with enhanced photocatalytic performance. J Alloys Compd 808:151681
72. Fu W et al (2016) Strong interfacial coupling of MoS2/g-C3N4 van de Waals solids for highly active water reduction. Nano Energy 27:44–50
73. Kim D, Leem J-Y (2022) Synthesis of interface modified MoS2/ZnO heterostructure via simple hydrothermal method and their enhanced UV photodetection characteristics with ultrafast photoresponse speed. Mater Res Bull 150:111767
74. Ben Naceur J, Benabderrahmane Zaghouani R, Amlouk M, Zaabi MA, Chtourou R (2023) Hydrothermal synthesis of MoS2-decorated silicon nanowires heterostructure with enhanced performance of photocatalytic activity under visible light. Inorg Chem Commun 147:110270
75. Chaudhary N, Khanuja M (2019) Architectural design of photodetector based on 2D (MoS_2 nanosheets)/1D (WS_2 nanorods) heterostructure synthesized by facile hydrothermal method. J Electrochem Soc 166:B1276–B1285
76. Chaudhary N, Kumar A, Imtiyaz S, Khanuja M (2021) Architectural design of electrode material for supercapacitor application based on a MoS_2/CeO_2 heterostructure synthesized by facile hydrothermal technique. ECS J Solid State Sci Technol 10:053005

77. He H-Y, He Z, Shen Q (2018) Efficient hydrogen evolution catalytic activity of graphene/metallic MoS2 nanosheet heterostructures synthesized by a one-step hydrothermal process. Int J Hydrog Energy 43:21835–21843
78. Peng K et al (2019) One-step hydrothermal growth of MoS2 nanosheets/CdS nanoparticles heterostructures on montmorillonite for enhanced visible light photocatalytic activity. Appl Clay Sci 175:86–93
79. Zhang B et al (2020) A novel S-scheme MoS $_2$ /CdIn $_2$ S $_4$ flower-like heterojunctions with enhanced photocatalytic degradation and H $_2$ evolution activity. J Phys D Appl Phys 53:205101
80. Vattikuti SVP, Byon C (2017) Hydrothermally synthesized ternary heterostructured MoS 2 / Al 2 O 3 /g-C 3 N 4 photocatalyst. Mater Res Bull 96:233–245
81. Lv Y et al (2022) One-pot hydrothermal approach towards 2D/2D heterostructure based on 1 T MoS2 chemically bonding with GO for extremely high electrocatalytic performance. Chem Eng J 428:132072
82. He H-K et al (2020) Multi-gate memristive synapses realized with the lateral heterostructure of 2D WSe_2 and WO_3. Nanoscale 12:380–387
83. Alexeev EM et al (2017) Imaging of interlayer coupling in van der Waals heterostructures using a bright-field optical microscope. Nano Lett 17:5342–5349
84. Gong Y et al (2015) Two-step growth of two-dimensional WSe_2/$MoSe_2$ heterostructures. Nano Lett 15:6135–6141
85. Zhu L et al (2022) Artificial neuron networks enabled identification and characterizations of 2D materials and van der Waals heterostructures. ACS Nano 16:2721–2729
86. Lin Z et al (2016) Scalable solution-phase epitaxial growth of symmetry-mismatched heterostructures on two-dimensional crystal soft template. Sci Adv 2
87. Yang M-Q et al (2017) Ultrathin nickel boron oxide nanosheets assembled vertically on graphene: a new hybrid 2D material for enhanced photo/electro-catalysis. Mater Horiz 4:885–894
88. Liu X et al (2017) Self-assembly of electronically abrupt borophene/organic lateral heterostructures. Sci Adv 3
89. Lin J et al (2019) Defect-rich heterogeneous MoS_2/NiS_2 nanosheets electrocatalysts for efficient overall water splitting. Adv Sci 6
90. Tao L et al (2017) Centimeter-scale CVD growth of highly crystalline single-layer MoS_2 film with spatial homogeneity and the visualization of grain boundaries. ACS Appl Mater Interfaces 9:12073–12081
91. Song X et al (2018) Boosting two-dimensional MoS_2/$CsPbBr_3$ photodetectors via enhanced light absorbance and interfacial carrier separation. ACS Appl Mater Interfaces 10:2801–2809
92. Shi Y et al (2012) Van der Waals epitaxy of MoS_2 layers using graphene as growth templates. Nano Lett 12:2784–2791
93. Liu Y et al (2021) Moiré superlattices and related moiré excitons in twisted van der Waals heterostructures. Chem Soc Rev 50:6401–6422
94. Hiroshi Yamaguchi HY, Yoshiro Hirayama YH (1998) Imaging of local charge density in an InAs/GaAs two-dimensional heterostructure by scanning tunneling microscopy. Jpn J Appl Phys 37:L899
95. Woods CR et al (2014) Commensurate–incommensurate transition in graphene on hexagonal boron nitride. Nat Phys 10:451–456
96. Naylor CH et al (2017) Synthesis and physical properties of phase-engineered transition metal dichalcogenide monolayer heterostructures. ACS Nano 11:8619–8627
97. Hill HM, Rigosi AF, Rim KT, Flynn GW, Heinz TF (2016) Band Alignment in MoS_2/WS_2 transition metal dichalcogenide heterostructures probed by scanning tunneling microscopy and spectroscopy. Nano Lett 16:4831–4837
98. Liu K et al (2014) Elastic properties of chemical-vapor-deposited monolayer MoS 2, WS 2, and their bilayer heterostructures. Nano Lett 14:5097–5103
99. Yu W et al (2017) Near-infrared photodetectors based on $MoTe_2$/graphene heterostructure with high responsivity and flexibility. Small 13:1700268

100. Zhou N et al (2018) P-GaSe/N-MoS_2 vertical heterostructures synthesized by van der waals epitaxy for photoresponse modulation. Small 14
101. Wang G et al (2018) Interlayer coupling induced infrared response in WS_2/MoS_2 heterostructures enhanced by surface plasmon resonance. Adv Funct Mater 28
102. Yuan Z et al (2019) Ultrathin black phosphorus-on-nitrogen doped graphene for efficient overall water splitting: dual modulation roles of directional interfacial charge transfer. J Am Chem Soc 141:4972–4979
103. Ding L et al (2018) Understanding Interlayer Coupling in TMD-hBN Heterostructure by Raman Spectroscopy. IEEE Trans Electron Devices 65:4059–4067
104. Neumann, C. *et al.* Line shape of the Raman 2D peak of graphene in van der Waals heterostructures. *physica status solidi (b)* **253**, 2326–2330 (2016).
105. Xu C et al (2017) Strongly coupled high-quality graphene/2D superconducting Mo_2 C vertical heterostructures with aligned orientation. ACS Nano 11:5906–5914
106. Zhang Y et al (2018) Edge-epitaxial growth of 2D NbS_2-WS_2 lateral metal-semiconductor heterostructures. Adv Mater 30
107. Zhang C et al (2017) Interlayer couplings, Moiré patterns, and 2D electronic superlattices in MoS_2/WSe_2 hetero-bilayers. Sci Adv 3
108. Tan Y, Ma L, Gao Z, Chen M, Chen F (2017) Two-dimensional heterostructure as a platform for surface-enhanced Raman scattering. Nano Lett 17:2621–2626
109. Seo J et al (2020) Ultrasensitive plasmon-free surface-enhanced Raman spectroscopy with femtomolar detection limit from 2D van der Waals heterostructure. Nano Lett 20:1620–1630
110. Esaki L, Chang LL (1974) New Transport Phenomenon in a Semiconductor 'Superlattice.' Phys Rev Lett 33:495–498
111. Lui CH et al (2015) Observation of interlayer phonon modes in van der Waals heterostructures. Phys Rev B 91:165403
112. Huang S et al (2016) Low-frequency interlayer Raman modes to probe interface of twisted bilayer MoS_2. Nano Lett 16:1435–1444
113. Ling X et al (2016) Parallel stitching of 2D Materials. Adv Mater 28:2322–2329
114. Cheng R et al (2014) Electroluminescence and photocurrent generation from atomically sharp WSe_2/MoS $_2$ heterojunction *p–n* DIODES. Nano Lett 14:5590–5597
115. Rivera P et al (2015) Observation of long-lived interlayer excitons in monolayer MoSe2–WSe2 heterostructures. Nat Commun 6:6242
116. Lee C-H et al (2014) Atomically thin p–n junctions with van der Waals heterointerfaces. Nat Nanotechnol 9:676–681
117. Chen K et al (2015) Electronic properties of MoS_2–WS_2 heterostructures synthesized with two-step lateral epitaxial strategy. ACS Nano 9:9868–9876
118. Ceballos F, Bellus MZ, Chiu H-Y, Zhao H (2014) Ultrafast charge separation and indirect exciton formation in a MoS_2–$MoSe_2$ van der Waals heterostructure. ACS Nano 8:12717–12724
119. Liu K et al (2014) Evolution of interlayer coupling in twisted molybdenum disulfide bilayers. Nat Commun 5:4966
120. Li B et al (2016) Direct vapor phase growth and optoelectronic application of large band offset SnS_2/MoS_2 vertical bilayer heterostructures with high lattice mismatch. Adv Electron Mater 2
121. Qiu F et al (2016) An investigation of exciton behavior in type-II self-assembled GaSb/GaAs quantum dots. Nanotechnology 27:065602
122. Schornbaum J et al (2014) Epitaxial growth of PbSe quantum dots on MoS_2 nanosheets and their near-infrared photoresponse. Adv Funct Mater 24:5798–5806
123. Kufer D, Lasanta T, Bernechea M, Koppens FHL, Konstantatos G (2016) Interface engineering in hybrid quantum dot–2D phototransistors. ACS Photonics 3:1324–1330
124. Huo N, Gupta S, Konstantatos G (2017) MoS $_2$ –HgTe quantum dot hybrid photodetectors beyond 2 μm. Adv Mater 29
125. Zhang J et al (2020) High-performance ultraviolet-visible light-sensitive 2D-MoS_2/1D-ZnO heterostructure photodetectors. ChemistrySelect 5:3438–3444

126. Liu Y, Huang Y, Duan XV (2019) Van der Waals integration before and beyond two-dimensional materials. Nature 567:323–333
127. Winther, K. T. & Thygesen, K. S. Band structure engineering in van der Waals heterostructures via dielectric screening: the GΔW method. *2d Mater* **4**, 025059 (2017).
128. Chiu M-H et al (2015) Determination of band alignment in the single-layer MoS2/WSe2 heterojunction. Nat Commun 6:7666
129. Qi T et al (2020) Interlayer transition in a vdW heterostructure toward ultrahigh detectivity shortwave infrared photodetectors. Adv Funct Mater 30
130. Li B et al (2019) Probing van der Waals interactions at two-dimensional heterointerfaces. Nat Nanotechnol 14:567–572
131. Boschker JE et al (2015) Coincident-site lattice matching during van der Waals epitaxy. Sci Rep 5:18079
132. Wang Y et al (2018) A room-temperature near-infrared photodetector based on a MoS $_2$ / CdTe p–n heterojunction with a broadband response up to 1700 nm. J Mater Chem C Mater 6:4861–4865
133. Lee HS, Ahn J, Shim W, Im S, Hwang DK (2018) 2D WSe2/MoS2 van der Waals heterojunction photodiode for visible-near infrared broadband detection. Appl Phys Lett 113
134. Ying H et al (2020) Band structure engineering in MoS_2 based heterostructures toward high-performance phototransistors. Adv Opt Mater 8
135. Zhou X et al (2018) Vertical heterostructures based on $SnSe_2/MoS_2$ for high performance photodetectors. 2d Mater 4:025048
136. Vu QA et al (2017) Tuning carrier tunneling in van der Waals heterostructures for ultrahigh detectivity. Nano Lett 17:453–459
137. Xiong Y, Chen J, Lu Y, Xu F (2019) Broadband optical-fiber-compatible photodetector based on a graphene-MoS_2-WS_2 heterostructure with a synergetic photogenerating mechanism. Adv Electron Mater 5
138. Fang H et al (2014) Strong interlayer coupling in van der Waals heterostructures built from single-layer chalcogenides. Proc Natl Acad Sci 111:6198–6202
139. Bernardi M, Palummo M, Grossman JC (2013) Extraordinary sunlight absorption and one nanometer thick photovoltaics using two-dimensional monolayer materials. Nano Lett 13:3664–3670
140. Furchi MM, Pospischil A, Libisch F, Burgdörfer J, Mueller T (2014) Photovoltaic effect in an electrically tunable van der Waals heterojunction. Nano Lett 14:4785–4791
141. Yu WJ et al (2013) Highly efficient gate-tunable photocurrent generation in vertical heterostructures of layered materials. Nat Nanotechnol 8:952–958
142. Shanmugam M, Jacobs-Gedrim R, Song ES, Yu B (2014) Two-dimensional layered semiconductor/graphene heterostructures for solar photovoltaic applications. Nanoscale 6:12682–12689
143. Tsai M-L et al (2014) Monolayer MoS $_2$ heterojunction solar cells. ACS Nano 8:8317–8322
144. Wi S et al (2014) Enhancement of photovoltaic response in multilayer MoS_2 induced by plasma doping. ACS Nano 8:5270–5281
145. Wang F et al (2015) Tunable GaTe-MoS $_2$ van der Waals p–n Junctions with novel optoelectronic performance. Nano Lett 15:7558–7566
146. Kim J-K et al (2016) Trap-mediated electronic transport properties of gate-tunable pentacene/MoS2 p-n heterojunction diodes. Sci Rep 6:36775
147. Yu WJ et al (2013) Vertically stacked multi-heterostructures of layered materials for logic transistors and complementary inverters. Nat Mater 12:246–252
148. Moriya R et al (2014) Large current modulation in exfoliated-graphene/MoS2/metal vertical heterostructures. Appl Phys Lett 105
149. Sarkar D et al (2015) A subthermionic tunnel field-effect transistor with an atomically thin channel. Nature 526:91–95
150. Loan PTK et al (2014) Graphene/MoS_2 heterostructures for ultrasensitive detection of DNA hybridisation. Adv Mater 26:4838–4844

151. Zhang J et al (2016) Crystal face regulating MoS2/TiO2(001) heterostructure for high photocatalytic activity. J Alloys Compd 688:840–848
152. Lu X et al (2016) Controllable synthesis of graphitic C_3N_4/ultrathin MoS_2 nanosheet hybrid nanostructures with enhanced photocatalytic performance. Dalton Trans 45:15406–15414
153. Meng X, Li Z, Zeng H, Chen J, Zhang Z (2017) MoS 2 quantum dots-interspersed Bi 2 WO 6 heterostructures for visible light-induced detoxification and disinfection. Appl Catal B 210:160–172
154. Ha E et al (2017) Cu2ZnSnS4/MoS2-reduced graphene oxide heterostructure: nanoscale interfacial contact and enhanced photocatalytic hydrogen generation. Sci Rep 7:39411
155. Tan Y-H, Yu K, Li J-Z, Fu H, Zhu Z-Q (2014) MoS2@ZnO nano-heterojunctions with enhanced photocatalysis and field emission properties. J Appl Phys 116
156. Biroju RK, Pal S, Sharma R, Giri PK, Narayanan TN (2017) Stacking sequence dependent photo-electrocatalytic performance of CVD grown MoS_2/graphene van der Waals solids. Nanotechnology 28:085101
157. Nikam RD, Sonawane PA, Sankar R, Chen Y-T (2017) Epitaxial growth of vertically stacked p-MoS2/n-MoS2 heterostructures by chemical vapor deposition for light emitting devices. Nano Energy 32:454–462
158. Lopez-Sanchez O et al (2014) Light Generation and Harvesting in a van der Waals heterostructure. ACS Nano 8:3042–3048
159. Withers F et al (2015) Light-emitting diodes by band-structure engineering in van der Waals heterostructures. Nat Mater 14:301–306
160. Li D et al (2015) Electric-field-induced strong enhancement of electroluminescence in multilayer molybdenum disulfide. Nat Commun 6:7509
161. Seyler KL et al (2015) Electrical control of second-harmonic generation in a WSe2 monolayer transistor. Nat Nanotechnol 10:407–411
162. Rigosi AF, Hill HM, Li Y, Chernikov A, Heinz TF (2015) Probing interlayer interactions in transition metal dichalcogenide heterostructures by optical spectroscopy: MoS_2/WS_2 and $MoSe_2$/WSe_2. Nano Lett 15:5033–5038
163. Hong X et al (2014) Ultrafast charge transfer in atomically thin MoS2/WS2 heterostructures. Nat Nanotechnol 9:682–686

Chapter 9
MoS_2 for Nanoelectronic Device Applications (Transistor, Sensor, Photodetector)

S. Reshmi and Dattatray J. Late

1 Introduction

The miniaturization of devices in accordance with the current world demands led to the search for materials that could substitute conventional semiconductors that were previously employed in electronic devices. The exciting properties displayed by MoS_2 made them a promising candidate for such device applications. The exotic properties of MoS_2 have grabbed eyes of scientists and are found to be promising materials to replace semiconductor industry. MoS_2 can be a promising material to substitute conventional materials as we move toward miniaturization of devices. These materials show favorable electronic and quantum properties while we go from bulk to 2D regime [1]. It can be used to solve the issues related to the existing devices [2–5]. MoS_2 has a large bandgap which modifies with layer numbers, thus is superior over zero-bandgap graphene [6]. The absence of dangling bonds and high mobility makes them ideal candidate for thin-film transistors [7] while the covalent bonds between Mo and S as well as the van der Waal's bond between layers make them ideal for sensor applications.

A major hurdle in Si-based devices is related to the contact metal–Si interface. The metal contacts are usually highly dense and narrow which causes an increase in contact resistance. Electron tunneling though the oxide layers, and thermal instability, interface incompatibility, reduced mobility, etc. associated with the high K dielectric materials are very common in conventional devices. MoS_2, however, as substitute

S. Reshmi
Institute of Physics, Bhubaneswar, Sachivalaya Marg, Gajapati Nagar, Bhubaneswar 751005, Odisha, India

D. J. Late (✉)
Department of Physics, Federal University of Lavras, Campus Universitário, PO Box 3037, Lavras, Minas Gerais, Brazil
e-mail: datta099@gmail.com

D. J. Late and C. S. Rout (eds.), *Science and Technology of 2D MoS_2*,
Materials Horizons: From Nature to Nanomaterials,
https://doi.org/10.1007/978-981-97-7367-1_9

for Si in devices provides easier and simpler synthesis and fabrication techniques. [8]

There are a wide variety of fields in which MoS_2 can be used like biosensing, virus detection, optical sensors, etc. [9, 10]. The superiority of MoS_2 in comparison to graphene and Si is its compatibility with human body [11]. The applications of MoS_2 materials are not limited to devices. It can serve as an excellent lubricant and is an essential part of hydrogen evolution reactions [12].

In this chapter, we will give a review on the applications of MoS_2 and MoS_2-based materials in the field of FET, sensors as well as photodetectors. The biomedical applications of MoS_2 are also discussed in length in this chapter.

2 FET Applications of MoS_2

The low atomic level thickness and absence of dangling bonds have naturally attracted scientists to incorporate 2D materials including graphene [13], black phosphorous, [14] and TMDs [15, 16] for device applications especially in field-effect transistors (FETs). Even though the carrier mobility is high, lower on–off ratio of graphene-based FETs makes them inapplicable in electronics. The high reactivity in normal atmosphere makes BP FET unstable and unsuitable for devices. The family of TMDs including MoS_2, WS_2, WSe_2, etc. has high on–off ratio, good mobility, etc. Thus, these materials can be used in low-power electronic applications. Particular interest is existing on MoS_2 FETs with high mobility, lower subthreshold swing, in comparison to other TMD materials. MoS_2 is a promising material for channel materials in FETs which can help in downscaling the size, thickness, high sensitivity, etc. of devices (Fig. 1).

The performance of MoS_2 FETs lies in the optimizing contact nature [17], regulating the channel [18] and selection of suitable dielectric layer [19]. Knowledge of the physics of metal contact with the 2D material is key to eliminate fermi level pinning effect which drastically affects the device performance [20]. This topic is still not understood clearly and is discussed at length among the scientific community. For example, the nature of contact between the layer and Au or Pd as metal contact is a topic of debate [21, 22]. Metals forming Schottky barrier can lead to Fermi-level pinning near the conduction band of MoS_2. Thus, a clear knowledge of the metal contacts is vital in constructing any successful devices with MoS_2. Various theoretical studies have been previously carried out to find out the behavior of metal–MoS_2 interface and how to maximize the carrier injection [23, 24]. It has been observed that metals with lower work function can help in increasing the carrier injection and thus leads to a lower contact resistance. Possibility of utilizing graphene as a contact material for construction of flexible MoS_2 transistors was investigated by Yoon et al. [25]. This configuration provided lowest contact resistance with high on–off ratio and charge mobility. Doping of graphene layer is also an effective method to modulate the properties of heterojunction FETs. For example, introduction of nitrogen-doped graphene [26] as electrode is reported to drastically the drain current as well as the

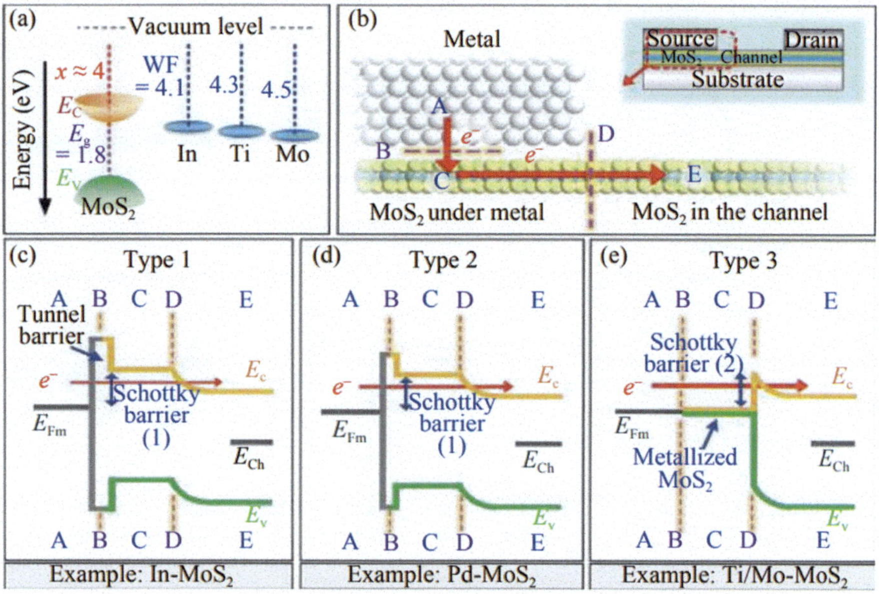

Fig. 1 Band alignment and band diagram of MoS2 with different metal contacts

mobility of MoS_2 FETs. The metal Mo was found to form flawless contact with MoS_2 and is shown to improve the device performance [27]. For back-gated MoS_2 monolayer FET, Bi was seen to provide lowest contact resistance to the group Ni, Ti, and Bi [28]. Metal carbides, nitrides, etc. are also potential candidates for electrode materials for MoS_2 FETs [29–31]. To conclude, choice of a suitable electrode is very crucial to achieve a high-performance low-power FET using MoS_2 (Fig. 2).

Impurity doping is another way to reduce the contact resistance between metal and MoS_2. Doping of MoS_2 with an external impurity atom reduces the Schottky barrier thereby enhancing the passage of current through the junction. There is still a long way of journey left to deeply understand the doping behavior of 2D materials in general. However, it was observed that polyethyleneimine (PEI)-doped FET gives 1.2 times reduction in contact resistance and there is also a huge improvement in charge mobility as well as the electrical nature of the device. It was concluded that molecular doping will indeed help in the reduction of contact barrier [32]. Chloride doping of MoS_2 and WS_2 is found to decrease the contact [33]. It was observed that the environmental factors hugely depend on the doping efficiency of 2D materials. For example, doping of MoS_2 with Mn is highly dependent on the substrate on which MoS_2 is grown as it may interact with the incoming Mn flux [34]. Thus, traditional doping techniques are usually not employed in the case of 2D materials. Surface adsorption is a method used for the doping of 2D layers [35, 36]. Impurity doping not only affects the contact resistance, but also modifies the band structure of the channel thereby changing the charge mobility as well as the on–off ratios [18, 37–39].

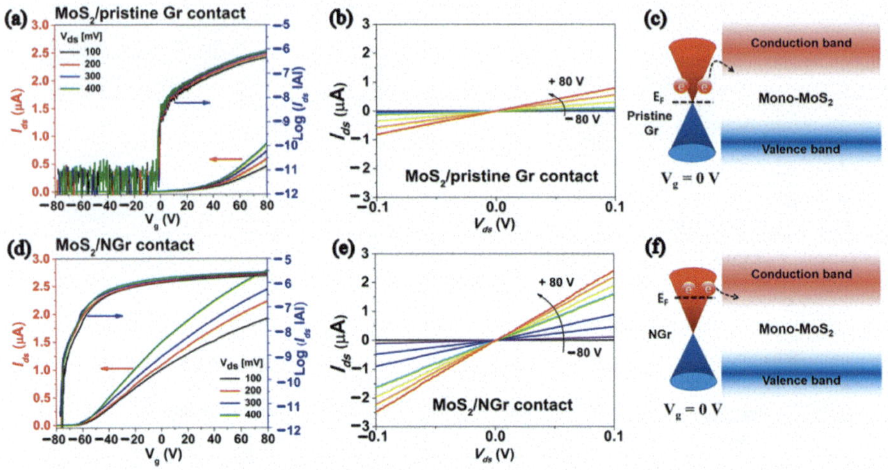

Fig. 2 Electrical characteristics of MoS_2/pristine Gr contact device: **a** transfer characteristic (left axis) and semilogarithm (right axis). **b** Output characteristic. **c** Schematic diagram of band alignment. Electrical characteristics of MoS_2/NGr contact device: **d** transfer characteristic (left axis) and semilogarithm (right axis). **e** Output characteristic. **f** Schematic diagram of band alignment

Engineering the dielectric materials can also significantly alter MoS_2 FET behaviors. Many theoretical studies have been carried out in this direction [21, 40, 41]. It is observed that charged impurity scattering is shielded by high K dielectric layers. It provides increased gate control and thus effectively improves the performance of MoS_2 FET [42, 43]. However, high k dielectrics are prone to generate trap charges and surface optical phonons which degrades the device performance further [43]. The dielectric materials can be modified using impurity metal atoms, polymers, etc., and this technique could eliminate the above-mentioned issues. Zhang et al. [19] recently suggested using high k and low k dielectric combinations for MoS_2 FET which provided low operating voltages and high stability. Even though the scattering and other issues can be suppressed by the high k dielectric, the interface problems multiply as the channel layer thickness reduces. Thus, optimization of the suitable dielectric thickness, addition of buffer layers, etc. is very crucial in the betterment of performance of MoS_2 FETs.

So far we have discussed the factors that can affect the performance of MoS_2-based FETs and how we can optimize the device performance using a suitable contact material, dielectric, etc.

MoS_2 with a suitable bandgap is overpowering than graphene in logic devices. Simulations show that monolayer MoS_2 FETs are superior than Si FETs in terms of size, reduced barrier height, and 13% lower subthreshold swing (SS) value [44]. ML MoS_2 grown on sapphire substrates by CVD method and HfO_2 as dielectric was used in logic circuits to obtain a high voltage gain [45]. Double-gated MoS_2 FETs is found to regulate the threshold voltage and SS in comparison with single-gated devices [46]. In the 5G communication period, RF devices are growing in

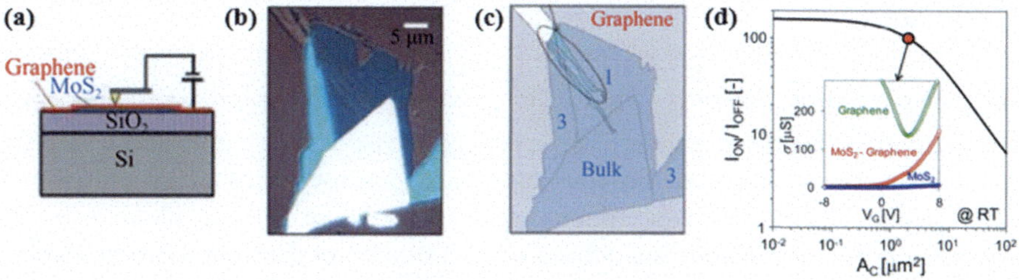

Fig. 3 **a** Cross-sectional view of graphene/MoS_2 heterostructure. **b** Optical microscope image. **c** Topographical diagram for a CVD graphene-coated MoS_2 single crystal **d** engineering on–off ratio in MoS_2-graphene FET

demand and popularity. To improve the bandwidth and operating frequency, newer materials are being studied for RF applications and MoS_2RF transistors are suitable candidates for that. Flexible MoS_2 RF device was fabricated by Cheng et al. which showed maximum frequency up to 50 GHz and cut off at 42 GHz [47]. There has been further studies on MoS_2 for RF transistor applications which mostly incorporated bilayer MoS_2 with varying substrates and dielectrics [48–50] (Fig. 3).

Heterojunction-based FETs are also employed for further improvement of MoS_2-based FETs. Graphene/MoS_2 heterostructures are reported to be used to tune on–off ratio and the mobility of electrons. The electron transport, charge transfer, and even photoluminescence can be controlled using the gate electrode and these values were investigated in the absence of any transverse electric field [51]. An n-type multilayer MoS_2 FET was constructed using Ti/Graphene as contact electrode which produced drain current of 160 mA/mm with gate length 1um. A single layer of graphene between metal contact and MoS_2 has improved the resistance value and the I–V characteristics [52]. Thin layers of metal oxides like MgO, Y_2O_3, etc. can also act as a buffer layer between metal and MoS_2 and thus enhances the device performance [53]. Combination of different 2D TMDs to form heterostructures is also investigated in depth. FETs formed using WSe_2/MoS_2 heterojunction were constructed by Kim et al. and it was further coated with PMMA to improve the device charactersistics [54]. These structures were used in logic devices. Various other heterojunction-based FETs have been studied and are still going on [55]. MoS_2-based FETs can completely outgrow the traditional Si-based devices. However, in terms of logic circuit applications, only simple logic can be handled by MoS_2-based FETs. Furthermore, construction of p-type MoS_2 FETs has performance issues and requires complex processing. Production of large area ML MoS_2 is also a technological hurdle now. Optimization of device structure and manufacturing processes is a crucial step at the moment. With impressive mechanical properties, MoS_2 as well as other layered materials can revolutionize the technological front by providing a platform to construct flexible electronic systems such as flexible wearable sensors and medical equipment. The application of MoS2 in the development of sensors and their applications in medical field is described in the next section.

3 Sensors

The van der Waal's bond between the MoS_2 layers and the bonding between Mo and S in the individual layers make them a perfect candidate for sensing applications. MoS_2-based sensing can be classified into four categories, biosensing, gas sensing, refractive index sensing, and photosensing (photodetectors) [56]. Out of these, photodetectors are generally FET-based sensors, which will be studied in detail in the next section (Fig. 4).

MoS_2 shows important applications in the field of biosensing. Biosensing usually involves electrochemical sensors [57]. The bioapplications include DNA studies, cancer, and corona virus detection [9, 10, 58]. It was observed that MoS_2 has no interaction with biological matter and hence is safe for human injection. A recent study

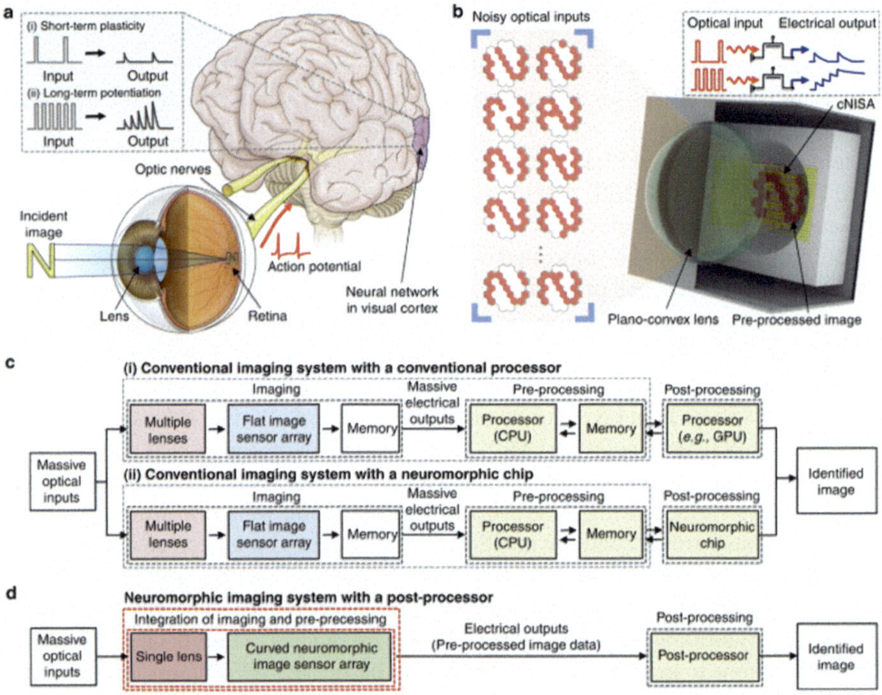

Fig. 4 **a** Schematic illustration of the human visual recognition system comprised of a single human-eye lens, a hemispherical retina, optic nerves, and a neural network in visual cortex. The inset schematic shows the synaptic plasticity (i.e., STP and LTP) of the neural network. **b** Schematic illustration of the curved neuromorphic imaging device. The inset in the dashed box shows the concept of photon-triggered synaptic plasticity that derives a weighted electrical output from massive optical inputs. **c** Block diagram showing the sequence of the image recognition using the conventional imaging and data processing system (e.g., conventional imaging system with a conventional processor (top) or with a neuromorphic chip (bottom)). **d** Block diagram showing the sequence of the image recognition using cNISA and a post-processor (e.g., GPU or neuromorphic chip)

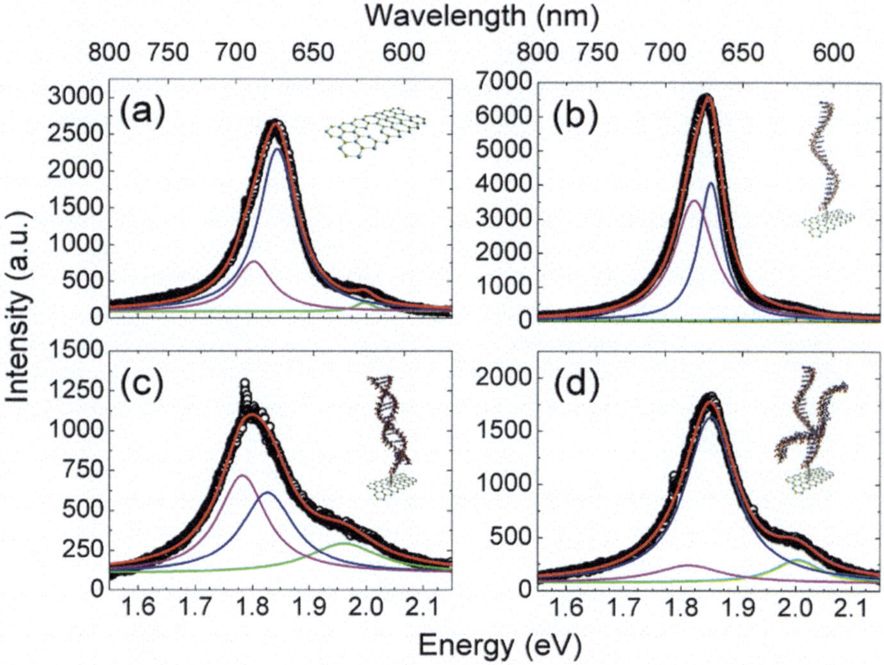

Fig. 5 PL spectra for samples in a different stages: **a** pristine as-grown MoS_2, **b** flake after ss-DNAp-SH functionalization, **c** after hybridization with the complementary sequence miRNA21c, and **d** with the non-complementary sequence miRNA21nc. The variation in the PL spectra with each step of functionalization is shown in the spectra

conducted by Sobanska [11] showed that MoS_2 may be used to cure Alzheimer's disease and cancer. A neuromorphic image sensor made with MoS_2 was observed to act similar to human vision [59]. It was thus used in making a biocompatible device to restore some vision in the visually impaired (Fig. 5).

A microfluidic device using MoS_2 is proposed to detect a COVID variant [9]. H_2O_2, a marker for Alzheimer's disease, can be detected using MoS_2/graphene oxide FET-based sensor [60]. It was also observed that MoS_2 nanosheets can detect prostrate antigens [61]. Based on the PL properties of MoS_2 flakes, they can be made use as biomarkers for breast cancer [10]. There are a wide range of applications of MoS_2 in microfluidic devices and immunosensors. They make use of the chemical and PL properties of MoS_2, [57, 58, 62]. The detection of biomolecules is easy using MoS_2 as they tend to stick on to the layers quite easily. Based on the PL properties, detection of amino acids using MoS_2 is quite investigated [63]. Machine learning combined with these biosensing behavior helps to detect various amino acids with accuracy up to 99.6% in a short time span [64]. Electrochemical lactate biosensor was developed using MoS_2 sheets and lactose oxidase [65]. Optimal 90 nm nanosheet of MoS_2 was identified and used for this purpose. Fluorescent biosensors based on MoS_2 were also studied in detail [66, 67] which utilizes the optical properties of

MoS_2 layers. Encapsulating MoS_2 in graphene can achieve increased biocompatibility and can be used for biomedical applications. GO/MoS_2 nanocomposites were used for sensing H_2O_2 through electrochemical reaction [68]. Various other electrochemical [69, 70, fluorescent [71, 72] and biosensors have been developed with graphene MoS_2 hybrids.

SPR sensors work on the principle of emission of electromagnetic waves or plasmons. The propagation of plasmons on a surface is related to the refractive index of the medium. Thus, SPR biosensors and refractive index biosensors are all related. MoS_2 can be combined with other materials to enhance the SPR sensitivity [73]. For example, graphene MoS_2 nanocomosite decorated with Al layers is used to achieve high sensitivity [74]. Flexible biosensors are also constructed using MoS_2–graphene hybrids [75, 76] and are expected to play a key role in the further development of biosensors. In the next section, we will discuss about the gas sensors based on MoS_2 and their application.

Gas-sensing nature of a layered material depends on the amount of charge transferred which in turn changes the electrical nature of the 2D layers. The gas species reorganizes the electron distribution on the surface of the material. For example, reducing gases like NH_3 on n-type MoS_2 donates charge to the layer whereas oxidizing gases like CO_2, NO, etc. accept charges from the layer [77]. The subtle changes in conductance or resistance values can be then used to sense the presence of a particular type of gas. In general, gas sensors are based on FET, but other configurations like chemiresisitive sensors, optical sensors, conductometric sensors, impedance sensors, etc. are studied [78, 79].

Chemiresistive sensors work based on the measurement of change in resistance of the sensing layer due to the presence of gas molecule [80]. These are cost-effective and consume low power and can also be reused. In general, FET-based sensors are used for toxic gases. The sensing layers act as a channel between source and drain electrodes whose conductance is varied by the applied voltage. The measurement of drain–source voltage can be used to predict the presence of gas molecule [81]. In optical sensors, the variation in optical properties, such as absorbance, reflectance, and fluorescence, is studied in the presence of gas [82]. Under atmospheric condition, conductometric gas sensors are in popular use because of their low power consumption, high sensitivity, and ease of use [83–85]. The working principle is the modification of electric current or resistivity linked to the gas adsorption and its concentration [86]. In impedance sensors, application of sinusoidal signal to obtain the impedance spectrum and the subsequent modification to that by gas adsorption is used as a base for sensing [87].

Lee et al. suggested an FET-based MoS_2-sensing device for NH_3 detection with a detection limit as low as 300 ppb [88]. This was followed by further investigation on monolayer MoS_2-based NH_3 and NO_2 sensors [89]. There was a 20 and 40% variation in conductance for NH_3 (1 ppm) and NO_2(20 ppb), respectively. The charge transfer between gas molecule and MoS_2 as well as modulation in the Schottky barrier height was considered here. Mechanically exfoliated single- and few-layer MoS_2 was used for NO gas sensing [90] which displayed higher sensitivity with 0.8 ppm. It was concluded that single-layer MoS_2 sensor showed rapid response

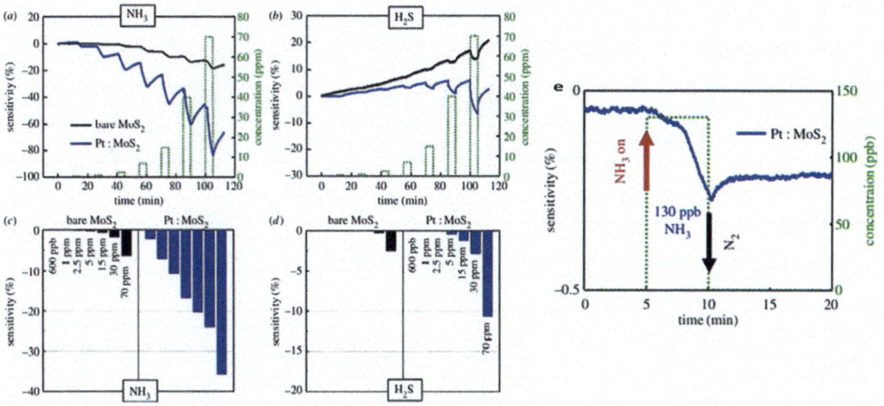

Fig. 6 **a–b** Gas-sensing characteristics of NH_3 and H_2S, **c–d** sensitivity based on bare MoS_2- and Pt-decorated MoS_2 for NH_3 and H_2S at different concentrations, **e** Pt-decorated MoS_2 gas sensor sensitivity on exposure of NH_3 to 130 ppb

to the target gas molecule. Further, NO_2 gas sensor formed based on MoS_2 thin film demonstrated detection in the range 1–256 ppm [91]. A recent study by Yan et al. showed a successful integration of flexible gas sensor based on MoS_2 and graphene as electrodes. It was used for the detection of NO_2 at room temperature [92]. In these cases which require electrode printing, various methods like sputtering, CVD, solution-based techniques, etc. can be employed, but post-processing works are complex and costly [93, 94]. Laser irradiation is one of the techniques used for printing on flexible electrodes. Such devices are reported to show 100 ppm NO_2 sensing and quick response (Fig. 6).

Noble metals like Au, Pt, Ag, etc. are used to enhance the gas-sensing properties of MoS_2. This method not only enhances the sensing ability but also makes them selective toward certain gas molecules. Pd is one of the most used materials for H_2 sensing. MoS_2 nanorods decorated with Pd particles can sense even 1% H_2 gas at room temperature [95, 96]. Pt-absorbed MoS_2 is used for NH_3 and H_2S sensing at room temperature [97]. The catalytic activity of Pt enhanced the sensing behavior with the limit of detectivity (LOD) of 130 ppb (NH_3) and 5 ppm (H_2S).

H_2 leakage studies involving MoS_2-Pd composites have been carried out by Kuru et al. [98] The variation in work function of Pd is detected which gives a direct implication of the H_2 exposure. A chemiresistor made of Pt-functionalized MoS_2 was used to study the presence of 1% H_2 at room temperature with enhanced response and recovery time [99]. CVD-grown MoS_2 was functionalized and used for sensing investigations (NO_2) which resulted in an optimized Au concentration of 12 μl to obtain maximum sensitivity.

A heterostructure of $MoS_2/SiO_2/Si$ was synthesized by magnetron sputtering technique and decorated with Pd particles [100]. This arrangement was used for H_2 sensing at RT and showed an excellent sensitivity of $9.2 \times 10^3\%$ with faster response and recovery time. It was observed that the thickness of Pd layers hugely impacted

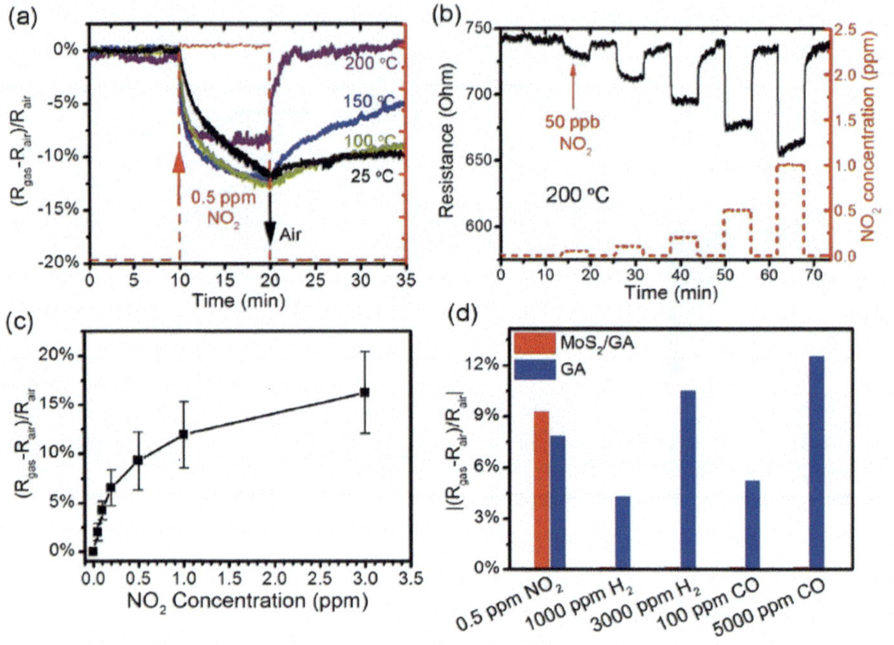

Fig. 7 **a** Sensing response of the device toward 0.5 ppm of NO_2 at different temperatures, **b** dynamic resistance change of device at 200 °C, **c** sensing response of device based on prepared sample at various concentrations of NO_2, **d** histogram representation of sensing response obtained by MoS_2/graphene aerogel and graphene aerogel at 200 °C

the sensing behavior of the structure. Thus, decoration of metal nanoparticles can be beneficial for the gas-sensing behavior of MoS_2 [101, 102] (Fig. 7).

Combining MoS_2 with graphene is another technique to enhance the gas-sensing nature. This is found to improve the catalytic and electrochemical properties. For a MoS_2/graphene aerogel, a detection limit of 50 ppb was observed for MoS_2 gas at RT [103]. The recovery and response of these structures was found to enhance at an elevated temperature of 200 °C. On comparison to CO, these aerogels showed higher selectivity toward H2 gas. rGO/MoS_2 composites are used for NO gas sensing [104] which was enhanced up to 200% compared to the pristine MoS_2 due to the presence of rGO. However, rGO/MoS_2 composite fiber demonstrated a higher sensitivity and selectivity for NO_2 and NH_3 [105]. For NH_3, an increase in resistivity is observed, whereas NO_2 presence caused a decrease in resistivity. MoS_2/rGO-based nanocomposites are fabricated using solvothermal methods in Ref. [76]. Stretchable NO_2 gas sensors with various specific areas and temperatures are studied in this report.

Even though MoS_2 has exciting wide range of applications, controlling of the impurities, lattice imperfections, and lack of cost-effective high yield synthesis techniques, etc. acts as challenged to full-scale technical revolution using these materials. In this section, we have given an outline of MoS_2-based sensors and its combinations to give better sensing behaviors. The next section will review the photosensing or photodetector applications of MoS_2.

4 Photodetector

Photodetectors (PD) are highly used in sensing and detecting, communications, lasers etc. The basic working principle of semiconductor-based PDs lies in the formation of electron–hole pairs due to the absorption of photons with energy larger than the bandgap of the materials. The e–h pair thus formed is separated by the external voltage, collected at the respective electrodes thus generating photocurrent [106]. In order to keep up in Moore's world, newer materials are always studied to meet the technological requirement. 2D materials [107–109] like graphene have already been studied for their photodetection application. However, the zero-bandgap nature of these materials causes high noise levels which makes them severely compromised for their candidature in photodetection applications.

2D TMDs with high mobility [110, 111], high optical absorption [112, 113] and tunable bandgaps [3] are now studied widely to overcome this limitation posed by graphene. The 2D nature offers high flexibility and the absence of dangling bonds makes the assembly of these materials into heterostructures quite easy. Broadband light detection is also possible owing to the layer-dependent bandgap tuning [114–116]. MoS_2, among the family of 2D TMDs, has shown remarkable photoelectric performance with high responsivity and light detection [112, 117, 118]. The tunable bandgap changes from 1.9 eV to 1.3 eV for monolayer to bulk [119]. A giant photoluminescence is shown by MoS_2 as we reach the monolayer regime which enhances the application in optoelectronic devices. [120, 121]. Furthermore, weak light absorption [122], short carrier lifetime along with persistent light conductance [123] in general limit the MoS_2-based optoelectronic devices. Bandgap modulation, chemical doping, construction of heterostructures, etc. can reasonably improve these issues [124–127]. Various unique properties of MoS_2 PD have been studied profoundly over the past few years.

The ample photocurrent and reduced noise levels are the main necessities for an efficient PD. The trap states at the MoS_2 surface are a main factor that reduces the charge carrier and slow response. The fast charge carrier recombination and lower light absorption coefficient significantly reduce the efficiency and lead to poor photodetection. Various methods have been devised over time to overcome these difficulties. Formation of impurity levels and shifting the fermi level significantly enhance the light absorption and broadening the absorption spectra [128–130]. Formation of heterostructures can modify the electric fields thus effectively separating the photogenerated carriers [131, 132]. Inclusion of transport channels with varying dimensions on the surface of the channel also improves the light absorption [133–135]. In addition, the choice of contact metal also largely influences the PD gain as the band alignment determines the contact resistance, which further regulates the photocurrent generation (Fig. 8).

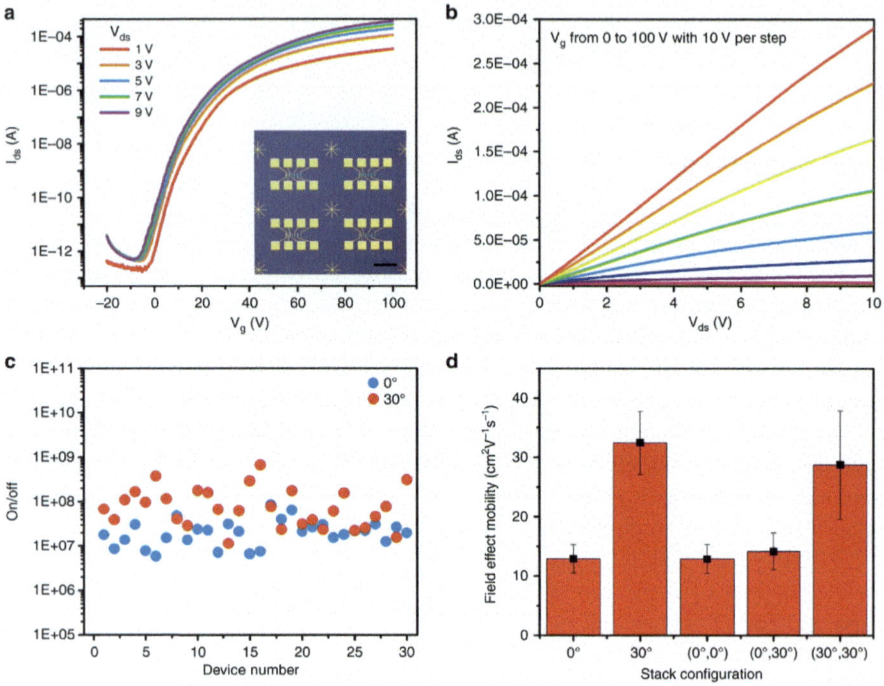

Fig. 8 **a** Transfer properties of MoS_2 bilayers with twist angle 30°. Inset shows the optical images of devices (scale bar 400 nm) **b** Electrical output curves of a 30° twisted bilayer MoS_2 FET. **c** On/off ratio of 0° and 30° devices. d Mobility of twisted multilayer MoS_2 films. Standard deviation shown in error bars

Modified designs including interface engineering, bandgap modulations, multi-functional integration, etc. are used to increase the device performances [136, 137]. With 2D layers, the presence of van der Waal's forces introduces a new degree of freedom, the twist angle. The twist angle between layers in bilayers MoS_2 is shown to highly tune the optical absorption [138] by modifying the interlayer coupling and bandgap. However, the most used method for performance upgrade of MoS_2-based PD is the layer number control and doping [118, 124] which highly increases their responsivity and gain [139, 140].

Optical absorption and ballistic transport in MoS_2 depend on the layer number as it modulates the photoelectric coupling and photon–phonon coupling [141, 142]. By tweaking the layer number or thickness, the bandgap of MoS_2 can be modified which in turn allows a photoresponse over a broad frequency range. The photoluminescence from MoS_2 has a negligible yield [82], which can be adjusted to a broader spectral range [82, 143]. The absence of interlayer coupling produces a strong PL in ML MoS_2 [144] while the quantum yield reduces as we move away from the ML regime [145]. Thus, photocurrent in MoS_2-based PD can be modulated just by adjusting the layer numbers. As the layer number increases, there is a reduction in the interface trap states. But this occurs at the cost of charge mobility. With an increase in layer thickness, we can see a subsequent reduction in the charge mobility from 10^4 to 10^2 [146].

A commonly used method for making MoS_2 layers is the mechanical exfoliation which provides films with fewer defects and good crystal quality. A phototransistor based on monolayer MoS_2 was fabricated on Si/SiO_2 which produced large photogain in the visible wavelengths [114, 147]. Lopez -Sanchez reported a very low dark current values (2pA) although a strong bias of −70 V was used to achieve this [147]. A.R. Klots fabricated a similar device with faster response but with lower light response [148]. Strain can also be a parameter to tune optical response as demonstrated by Gant and coworkers [149]. They showed that the ML MoS_2-based PD showed 2–3 orders of magnitude variation in photoresponse in the visible region in the presence of strain [149]. The formation of trap states is considered to be the reason for this behavior.

A few-layered MoS_2 obtained through wet synthesis technique was deposited on to SiO_2/Si substrate with Au electrodes to obtain a responsivity of 0.57 AW^{-1} [150]. Yang et al. suggested a modified CVD technique wherein they introduced an NaCl-coated Mo foil to guarantee an even release of Mo ions which gives a control on the thickness [151]. They have observed the highest responsivity of 10^4 AW^{-1} for ML at 660 nm.

The optoelectronic properties of multilayer MoS_2 on Si were optimized w.r.t layer numbers by Shin et al. [124]. A maximum photoresponse and detectivity was shown by 48-nm-thick MoS_2 film at 12 nW and 2 V. Yang and his coworkers observed that MoS_2 photodetector with thickness of 9 nm showed a dark current of 8 nA at 3 V bias voltage. It was observed that the on–off ratio increased to 10^4 as thickness was increased to 9 nm but further shows a dip as the thickness grows. A subsequent investigation by Ling et al. [152] also suggests the broadband photodetection displayed by multilayer MoS_2 due to its small indirect bandgap. This essentially spreads the optical range in comparison with the ML MoS_2. However, there is still room for improvement in the responsivity [153].

Most of the MoS_2 photodetectors are used in the visible region of the spectrum. Lesser number of devices are fabricated with a response in the UV and near-IR spectrum as the absorption edge of MoS_2 lies closer to 600 nm and cuts off any longer wavelength. The constraint in the photoresponse of layered MoS_2 comes from the limitation in bandgap as well as lesser light absorption. Various strategies like control of layer numbers, formation of heterostructure, doping, defect engineering, etc. are adopted to overcome the limitations in the optoelectronic properties of MoS_2 layers. Devices thus formed show high responsivity even in the regions outside the visible spectra but has slow response due to the trap states which are introduced by the external materials [154–156]. For UV–vis–IR applications, graphene is frequently used as a contact material which utilizes the high mobility of graphene along with the versatile nature of MoS_2 [157].

Composite formed by reduced graphene oxide (rGO) and MoS_2 is reported to show significant light absorption and subsequently high responsivity [158]. These could be performed in the NIR and above range of wavelengths due to the defect-assisted light absorption from MoS_2 layers which are then transferred to the RGO layer thereby increasing the photocurrent. Chemical doping using rhodamine [159], zinc phthalocyanine [127], Al_2O_3 [160], etc. were reported to display enhanced photodetection capabilities. It is even observed that the n-type nature of MoS_2 can be changed to p-type by doping with Nb [128]. This doped structure shows a photoresponsivity of 2×10^5 AW^{-1} which is almost 10^6 times greater than the pristine MoS_2 layers measured under identical conditions. The photocurrent values also increased to 10^5 times owing the better charge transfer processes. Hybridization with organic molecules is also a strategy used to improve the properties of MoS_2 photodetectors [161] (Fig. 9).

Plasmonic nanostructures like Au nanoparticles are observed to increase the responsivity of ML MoS_2 by 10-fold. Similarly, the surface plasmons from Ag nanowire/ML MoS_2 were observed to hugely enhance the photoresponse [162]. It was seen that there is almost 10^4 order difference in the on–off ratio, a 560- and 250-fold increase in PL and photocurrent is observed for this structure in comparison with the pristine ML MoS_2. There is a 470% increase in photoresponsivity reported for Ag nanoparticles (25 nm)/ML MoS_2 structures as per a study conducted by Jing and

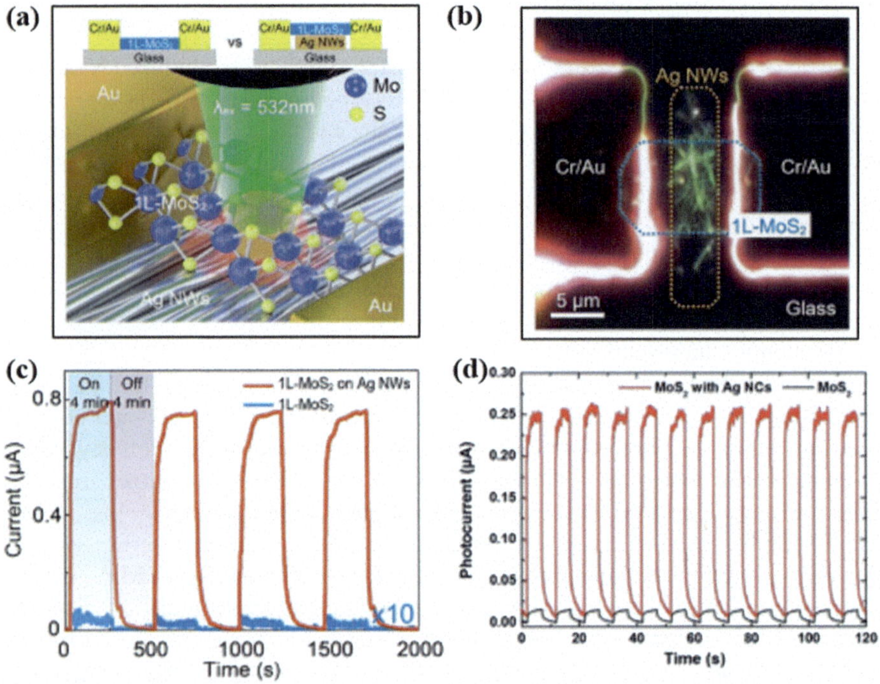

Fig. 9 **a** Schematic diagram of the 1L-MoS_2 and Ag nanowire (NW)-based photodetectors **b** dark field image of the 1L-MoS_2/AgNWs hybrid photodetector (60% density of AgNWs). **c** Time-dependent photocurrent of the pristine 1L-MoS_2 (blue) and 1L-MoS_2/AgNWs hybrid photodetectors (red) recorded with an Ion/Ioff time of 4 min **d** a comparison of the transient photocurrent of MoS_2 and MoS_2/Ag nanocubes (NCs) photodetectors under periodical Ion/Ioff illumination at 1 V

coworkers [163]. For all these structures, the light–matter interaction in ML MoS_2 is enhanced by the presence of plasmonic nanostructures (Fig. 10).

It is also possible to combine different TMD materials with MoS_2 to form heterostructures, the built-in potential of which causes faster photogenerated charge separation. $MoS_2/MoTe_2$ stacking formed by mechanical exfoliation produced a response time of 60 μs with responsivity 0.046 AW^{-1} to a 637 nm light in the absence of bias voltage [164]. The MoS_2/WS_2 heterostructure formed by CVD was investigated by Tan et al. [165] and they reported an outstanding photoresponsivity at 532 nm. For heterostructures, in general, it was observed that the responsivity was at least one order more than what is observed for a single TMD device. Triple layer

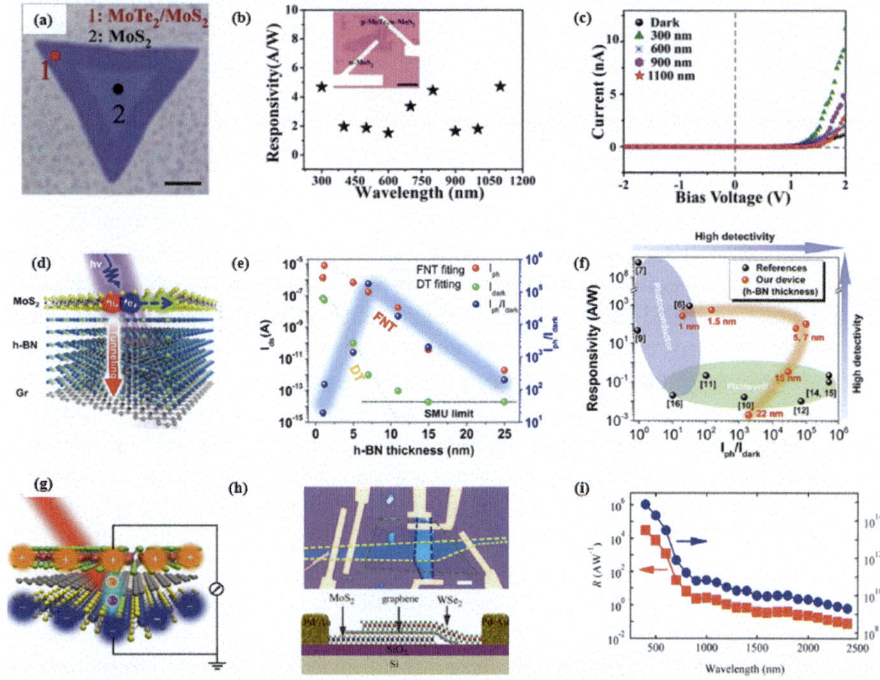

Fig. 10 **a** Optical image of the $MoTe_2/MoS_2$ flake **b** photoresponsivity of $MoTe_2/MoS_2$ photodetectors as a function of wavelength between 300 and 1100 nm region. **c** I–V current characteristics of the $MoTe_2/MoS_2$ photodetectors at different applied bias voltage. **d** Schematic of **e–h** pair generation with laser illumination and selective hole carrier tunneling through h-BN layer. **e** The photocurrent (I ph), dark current (I dark), and their I ph/dark ratio concerning h-BN thickness. **f** The photoresponsivity and I ph /I dark ratio of the photodetector with various h-BN thicknesses (red dots) and other 2D material-based PDs (black dots). Red numbers indicate the thickness of h-BN in the MoS_2 /h-BN/graphene PDs. **g** Schematic drawing of the p–g–n heterostructure for photodetection. **h** Top: the optical image of a fabricated device based on vdW-assembled MoS_2–graphene–WSe_2 heterostructure. Scale bar, 5 μm. Bottom: schematic side view of the heterostructure. **i** Photoresponsivity R (left) and specific detectivity D ∗ (right) of a typical device at V ds = 1 V and V g = 0 V

stacked heterostructures are also studied which incorporate MoS_2 with two other layered materials and it is shown to display broad spectrum responsivity [165, 166].

h-BN insulating layer can be sandwiched between MoS_2 and the graphene contact layer, whose thickness variation controls the photocarrier tunneling which was investigated by Vu et al. [132]. For 7-nm-thick h-BN layers, the heterostructure exhibited a photoresponsivity of 180 AW^{-1} and detectivity of 2.6X10 [13] Jones. Graphene sandwiched between p–n junction formed between MoS_2 and WSe_2 displays a broader photodetection from 400 to 2400 nm with a maximum photoresponsivity of the order of 10^4 AW^{-1} which decreased to a few mAW^{-1} in the visible and at 2400 nm wavelengths. It was observed that the overlapping region of the three individual layers is responsible for the photocurrent generation, rather than the electrode region. These heterostructures show faster photoresponse in a broader wavelength range at room temperature. Various highly sensitive photodetectors have been developed based on MoS_2 van der Waal's heterostructures [167–169]. The $MoTe_2$/MoS_2 bilayer photodetectors showed a responsivity of 4.17 AW^{-1} (4.67AW^{-1}) and quantum efficiency of 533% (1935%) at 1100 nm (300 nm) [170]. The phototdetectivity can be further enhanced by utilizing interface coupling effect (ICE) as demonstrated in Ref [171, 172]. Flexible photodetectors are another field which has been utilizing the unique properties of 2D layers including MoS_2. The substrate-dependent properties of MoS_2 photodetectors were investigated by Zhang et al. [173]. The commonly used substrates for wearable application include SiC, Kapton polymer like PET, polyacrylate, etc. [173–175] and have applications in surveillance purposes, biomedical purposes, etc. Heterostructures of MoS_2 with graphene, CNT, TMD layers, ZnO, and even surface functionalization have been investigated earlier.

All these investigations point toward a successful integration of MoS_2 layers into developing high-performance photodetectors. Generally, the working environment of photodetectors is harsh, which has a negative impact on their lifetime. For MoS_2–hybrid heterostructure-based photodetectors, an extended environmental life is observed [176–178]. The applications of such devices are endless starting from optical communication, remote sensing, lasers, data storage, military, etc., to name some. With all these developments, there is still room for development in the process of integration of MoS_2 with existent flexible electronic, optoelectronic devices. With the growing family of 2D materials, there are still materials that need to be investigated coupled with MoS_2 for their photodetection abilities. Hybrid heterostructures based of layered materials could be utilized in flexible photodetectors with immense uses in biomedical applications such as imaging and sensing monitoring. However, production of low-cost, high-performance devices in large scale for commercial applications is still a challenge.

5 Conclusions

2D MoS_2 has attracted wide attention as promising materials to construct devices like FETs, sensors, optoelectronic devices, etc. A major advantage of such materials is their natural atomic thickness and intrinsic bandgap values that can be tuned according to the layer number. A brief insight into the application of MoS_2 has been discussed in this chapter. We have discussed the formation of MoS_2-based FETs, sensors, and photodetectors. Various combinations and techniques to improve the performance of such devices have also been discussed in brief. Functional 2D MoS_2 FETs find applications in logic and RF devices, sensors, and even optoelectronic devices like photodetectors. The important challenge, however, lies in the formation of uniform large flakes of MoS_2. Another problem lies in the balance of high photoresponsivity and fast response speed of the PDs. Even though MoS_2 provides broadband detection, the responsivity and detectivity are rather poor. However, assembly of MoS_2 with other materials like graphene, h-BN, etc. has the potential to overpower Si in integrated circuits.

References

1. Zhao G-Y, Deng H, Tyree N, Guy M, Lisfi A, Peng Q, Yan J-A, Wang C, Lan Y (2019) Appl Sci 9:678
2. del Alamo JA (2011) Nature 479:317
3. Geim AK, Grigorieva IV (2013) Nature 499:419
4. Butler SZ, Hollen SM, Cao L, Cui Y, Gupta JA, Gutiérrez HR, Heinz TF, Hong SS, Huang J, Ismach AF, Johnston-Halperin E, Kuno M, Plashnitsa VV, Robinson RD, Ruoff RS, Salahuddin S, Shan J, Shi L, Spencer MG, Terrones M, Windl W, Goldberger JE (2013) ACS Nano 7:2898
5. Bhat N (2018) 2018 IEEE international conference on semiconductor electronics. IEEE, pp. C1–C1
6. Novoselov K (2009) Phys World 22:27
7. Yoo G, Lee S, Yoo B, Han C, Kim S, Oh MS (2015) IEEE Electron Device Lett 36:1215
8. Iwai H, Kakushima K, Wong H (2006) Int J High Speed Electron Syst 16:43
9. Weng X, Neethirajan S (2018) IEEE Sens J 18:4358
10. Catalán-Gómez S, Briones M, Cortijo-Campos S, García-Mendiola T, de Andrés A, Garg S, Kung P, Lorenzo E, Pau JL, Redondo-Cubero A (2020) Sci Rep 10:16039
11. Sobanska Z, Zapor L, Szparaga M, Stepnik M (2020) Int J Occup Med Environ Health 33:1
12. Liu L, Liu Z, Huang P, Wu Z, Jiang S (2016) RSC Adv 6:113315
13. Flores-Silva PA, Borja-Hernández C, Magaña C, Acosta DR, Botello-Méndez AR, Serkovic-Loli LN (2020) Phys. E Low-Dimens Syst Nanostructures 124:114282
14. Feng X, Huang X, Chen L, Tan WC, Wang L, Ang K-W (2018) Adv Funct Mater 28:1801524
15. Liu B, Ma Y, Zhang A, Chen L, Abbas AN, Liu Y, Shen C, Wan H, Zhou C (2016) ACS Nano 10:5153
16. Larentis S, Fallahazad B, Tutuc E (2012) Appl Phys Lett 101:223104
17. Han B, Zhao Y, Ma C, Wang C, Tian X, Wang Y, Hu W, Samorì P (2022) Adv Mater 34:2109445
18. Li H, Cheng M, Wang P, Du R, Song L, He J, Shi J (2022) Adv Mater 34:2200885
19. Zhang Z, Su M, Li G, Wang J, Zhang X, Ho JC, Wang C, Wan D, Liu X, Liao L (2020) IEEE Electron Device Lett 41:1036

20. Zhao P, Jin H, Lv X, Huang B, Ma Y, Dai Y (2018) Phys Chem Chem Phys 20:16551
21. Radisavljevic B, Radenovic A, Brivio J, Giacometti V, Kis A (2011) Nat Nanotechnol 6:147
22. Fontana M, Deppe T, Boyd AK, Rinzan M, Liu AY, Paranjape M, Barbara P (2013) Sci Rep 3:1634
23. Li Z, Li X, Yang J, Appl ACS (2015) Mater Interfaces 7:12981
24. Chuang H-J, Chamlagain B, Koehler M, Perera MM, Yan J, Mandrus D, Tománek D, Zhou Z (2016) Nano Lett 16:1896
25. Yoon J, Park W, Bae G-Y, Kim Y, Jang HS, Hyun Y, Lim SK, Kahng YH, Hong W-K, Lee BH, Ko HC (2013) Small 9:3295
26. Seo D, Lee DY, Kwon J, Lee JJ, Taniguchi T, Watanabe K, Lee G-H, Kim KS, Hone J, Kim YD, Choi H-J (2019) Appl Phys Lett 115:012104
27. Kang J, Liu W, Banerjee K (2014) Appl Phys Lett 104:093106
28. Shen P-C, Su C, Lin Y, Chou A-S, Cheng C-C, Park J-H, Chiu M-H, Lu A-Y, Tang H-L, Tavakoli MM, Pitner G, Ji X, Cai Z, Mao N, Wang J, Tung V, Li J, Bokor J, Zettl A, Wu C-I, Palacios T, Li L-J, Kong J (2021) Nature 593:211
29. (John) Zhang C (2021) J Energy Chem 60:417
30. VahidMohammadi A, Rosen J, Gogotsi Y (2021) Science 80:372
31. Zhou Y, Zhuge X, An P, Du S (2020) Nanotechnology 31:395203
32. Du Y, Liu H, Neal AT, Si M, Ye PD (2013) IEEE Electron Dev Lett 34:1328
33. Yang L, Majumdar K, Liu H, Du Y, Wu H, Hatzistergos M, Hung PY, Tieckelmann R, Tsai W, Hobbs C, Ye PD (2014) Nano Lett 14:6275
34. Zhang K, Feng S, Wang J, Azcatl A, Lu N, Addou R, Wang N, Zhou C, Lerach J, Bojan V, Kim MJ, Chen LQ, Wallace RM, Terrones M, Zhu J, Robinson JA (2015) Nano Lett 15:6586
35. Fang H, Tosun M, Seol G, Chang TC, Takei K, Guo J, Javey A (2013) Nano Lett 13:1991
36. Rastogi P, Kumar S, Bhowmick S, Agarwal A, Chauhan YS (2014) J Phys Chem C 118:30309
37. Kim E, Lee Y, Ko C, Park Y, Yeo J, Chen Y, Choe HS, Allen FI, Rho J, Tongay S, Wu J, Kim K, Grigoropoulos CP (2018) Appl Phys Lett 113:013105
38. Wei Z, Tang J, Li X, Chi Z, Wang Y, Wang Q, Han B, Li N, Huang B, Li J, Yu H, Yuan J, Chen H, Sun J, Chen L, Wu K, Gao P, He C, Yang W, Shi D, Yang R, Zhang G (2021) Small Methods 5:2100091
39. Li M, Wu X, Guo W, Liu Y, Xiao C, Ou T, Zheng Y, Wang Y (2022) J Mater Chem C 10:7662
40. Yoon Y, Ganapathi K, Salahuddin S (2011) Nano Lett 11:3768
41. Liu H, Ye PD (2012) IEEE Electron Dev Lett 33:546
42. Yu Z, Ong Z-Y, Li S, Xu J-B, Zhang G, Zhang Y-W, Shi Y, Wang X (2017) Adv Funct Mater 27:1604093
43. Song X, Xu J, Liu L, Lai P-T (2021) Appl Surf Sci 542:148437
44. Liu L, Lu Y, Guo J (2013) IEEE Trans Electron Dev 60:4133
45. Wang L, Chen L, Wong SL, Huang X, Liao W, Zhu C, Lim Y, Li D, Liu X, Chi D, Ang K (2019) Adv Electron Mater 5:1900393
46. Yi J, Sun X, Zhu C, Li S, Liu Y, Zhu X, You W, Liang D, Shuai Q, Wu Y, Li D, Pan A (2021) Adv Mater 33:2101036
47. Cheng R, Jiang S, Chen Y, Liu Y, Weiss N, Cheng H-C, Wu H, Huang Y, Duan X (2014) Nat Commun 5:5143
48. Lai R, Mei XB, Deal WR, Yoshida W, Kim YM, Liu PH, Lee J, Uyeda J, Radisic V, Lange M, Gaier T, Samoska L, Fung A (2007) 2007 IEEE international electron devices meeting. IEEE, pp. 609–611
49. Gao Q, Zhang Z, Xu X, Song J, Li X, Wu Y (2018) Nat Commun 9:4778
50. Dragoman M, Aldrigo M, Connolly J, Povey IM, Iordanescu S, Dinescu A, Vasilache D, Modreanu M (2020) Nanotechnology 31:06LT01
51. Shih C-J, Wang QH, Son Y, Jin Z, Blankschtein D, Strano MS (2014) ACS Nano 8:5790
52. Du Y, Yang L, Zhang J, Liu H, Majumdar K, Kirsch PD, Ye PD (2014) IEEE Electron Dev Lett 35:599
53. Zou X, Wang J, Chiu C-H, Wu Y, Xiao X, Jiang C, Wu W-W, Mai L, Chen T, Li J, Ho JC, Liao L (2014) Adv Mater 26:6255

54. Kim JY, Park HJ, Lee S, Seo C, Kim J, Joo J, Appl ACS (2020) Mater Interfaces 12:36530
55. Marin EG, Marian D, Perucchini M, Fiori G, Iannaccone G (2020) ACS Nano 14:1982
56. Zhang W, Zhang P, Su Z, Wei G (2015) Nanoscale 7:18364
57. Singh C, Ali MA, Kumar V, Ahmad R, Sumana G (2018) Sens Actuators B Chem 259:1090
58. Liu Y, Zhang J, Shen Y, Yan J, Hou Z, Mao C, Zhao W (2017) RSC Adv 7:54638
59. Choi C, Leem J, Kim M, Taqieddin A, Cho C, Cho KW, Lee GJ, Seung H, Bae HJ, Song YM, Hyeon T, Aluru NR, Nam S, Kim D-H (2020) Nat Commun 11:5934
60. Zheng C, Jin X, Li Y, Mei J, Sun Y, Xiao M, Zhang H, Zhang Z, Zhang G-J (2019) Sci Rep 9:759
61. Kong R-M, Ding L, Wang Z, You J, Qu F (2015) Anal Bioanal Chem 407:369
62. Faramarzi V, Ahmadi V, Fotouhi B, Abasifard M (2019) Sci Rep 9:6230
63. Huang Y, Shi Y, Yang HY, Ai Y (2015) Nanoscale 7:2245
64. Barati Farimani A, Heiranian M, Aluru NR (2018) NPJ 2D Mater Appl 2:14
65. Parra-Alfambra AM, Casero E, Vázquez L, Quintana C, del Pozo M, Petit-Domínguez MD (2018) Sens Actuators B Chem 274:310
66. Xiao M, Chandrasekaran AR, Ji W, Li F, Man T, Zhu C, Shen X, Pei H, Li Q, Li L, Appl ACS (2018) Mater Interfaces 10:35794
67. Roush S, Slack FJ (2008) Trends Cell Biol 18:505
68. Yoon J, Lee T, Bapurao BG, Jo J, Oh B-K, Choi J-W (2017) Biosens Bioelectron 93:14
69. Jeong J-M, Yang M, Kim DS, Lee TJ, Choi BG, Kim DH (2017) J Colloid Interface Sci 506:379
70. Bhardwaj H, Marquette CA, Dutta P, Rajesh, Sumana G (2020) Anal Bioanal Chem 412:7029
71. Schmetzer O, Moldenhauer G, Nicolaou A, Schlag P, Riesenberg R, Pezzutto A (2012) Immunol Lett 143:184
72. Jung YK, Woo M-A, Soh HT, Park HG (2014) Chem Commun 50:12329
73. Rahman MM, Rahman MS, Rikta KA, Rana MM, Anower MS, Paul AK (2020) 2020 IEEE regional 10 symposium. IEEE, pp 1197–1200
74. Wu L, Jia Y, Jiang L, Guo J, Dai X, Xiang Y, Fan D (2017) J. Light. Technol. 35:82
75. Wang Z, Dong S, Gui M, Asif M, Wang W, Wang F, Liu H (2018) Anal Biochem 543:82
76. Yi N, Cheng Z, Li H, Yang L, Zhu J, Zheng X, Chen Y, Liu Z, Zhu H, Cheng H (2020) Mater. Today Phys. 15:100265
77. Yue Q, Shao Z, Chang S, Li J (2013) Nanoscale Res Lett 8:425
78. Yuan W, Shi G (2013) J. Mater. Chem. A 1:10078
79. Donarelli M, Ottaviano L (2018) Sensors 18:3638
80. Cho B, Kim AR, Park Y, Yoon J, Lee Y-J, Lee S, Yoo TJ, Kang CG, Lee BH, Ko HC, Kim D-H, Hahm MG, Appl ACS (2015) Mater Interfaces 7:2952
81. Late DJ, Huang Y-K, Liu B, Acharya J, Shirodkar SN, Luo J, Yan A, Charles D, Waghmare UV, Dravid VP, Rao CNR (2013) ACS Nano 7:4879
82. Splendiani A, Sun L, Zhang Y, Li T, Kim J, Chim CY, Galli G, Wang F (2010) Nano Lett 10:1271
83. Yang S, Jiang C, Wei S (2017) Appl Phys Rev 4:021304
84. Hashtroudi H, Mackinnon IDR, Shafiei M (2020) J Mater Chem C 8:13108
85. Rajkumar K, Kumar RTR (2019) Fundamemtals and sensing applications of 2D materials, pp 205–258. Elsevier
86. Perkins FK, Friedman AL, Cobas E, Campbell PM, Jernigan GG, Jonker BT (2013) Nano Lett 13:668
87. Li N, Chen X-D, Chen X-P, Ding X, Zhao X (2017) IEEE Electron Dev Lett 38:806
88. Lee K, Gatensby R, McEvoy N, Hallam T, Duesberg GS (2013) Adv Mater 25:6699
89. Liu B, Chen L, Liu G, Abbas AN, Fathi M, Zhou C (2014) ACS Nano 8:5304
90. Li H, Yin Z, He Q, Li H, Huang X, Lu G, Fam DWH, Tok AIY, Zhang Q, Zhang H (2012) Small 8:63
91. Kim S, Park H, Choo S, Baek S, Kwon Y, Liu N, Yang JY, Yang C-W, Yoo G, Kim S (2020) Commun Mater 1:86
92. Yan W, Yan W, Chen T, Xu J, Tian Q, Ho D, Appl ACS (2020) Nano Mater 3:2545

93. Cho B, Yoon J, Lim SK, Kim AR, Kim D-H, Park S-G, Kwon J-D, Lee Y-J, Lee K-H, Lee BH, Ko HC, Hahm MG, Appl ACS (2015) Mater Interfaces 7:16775
94. Tian B, Yao W, Zeng P, Li X, Wang H, Liu L, Feng Y, Luo C, Wu W (2019) J. Mater. Chem. C 7:809
95. Kumar R, Varandani D, Mehta BR, Singh VN, Wen Z, Feng X, Müllen K (2011) Nanotechnology 22:275719
96. Baek D-H, Kim J (2017) Sens Actuators B Chem 250:686
97. Park J, Mun J, Shin J-S, Kang S-W, Soc R (2018) Open Sci 5:181462
98. Kuru C, Choi C, Kargar A, Choi D, Kim YJ, Liu CH, Yavuz S, Jin S (2015) Adv Sci 2:1
99. Park CH, Koo W-T, Lee YJ, Kim YH, Lee J, Jang J-S, Yun H, Kim I-D, Kim BJ (2020) ACS Nano 14:9652
100. Hao L, Liu Y, Du Y, Chen Z, Han Z, Xu Z, Zhu J (2017) Nanoscale Res Lett 12:567
101. Sarkar D, Xie X, Kang J, Zhang H, Liu W, Navarrete J, Moskovits M, Banerjee K (2015) Nano Lett 15:2852
102. Zhang Z, Chen K, Zhao Q, Huang M, Ouyang X (2020) Mater Res Express 7:015501
103. Long H, Harley-Trochimczyk A, Pham T, Tang Z, Shi T, Zettl A, Carraro C, Worsley MA, Maboudian R (2016) Adv Funct Mater 26:5158
104. Zhou Y, Liu G, Zhu X, Guo Y (2017) Sens Actuators B Chem 251:280
105. Niu Y, Wang R, Jiao W, Ding G, Hao L, Yang F, He X (2015) Carbon N Y 95:34
106. Taffelli A, Dirè S, Quaranta A, Pancheri L (2021) Sensors 21:2758
107. Chen X, Shehzad K, Gao L, Long M, Guo H, Qin S, Wang X, Wang F, Shi Y, Hu W, Xu Y, Wang X (2019) Adv Mater 1902039
108. Deng S, Sumant AV, Berry V (2018) Nano Today 22:14
109. Vicarelli L, Vitiello MS, Coquillat D, Lombardo A, Ferrari AC, Knap W, Polini M, Pellegrini V, Tredicucci A (2012) Nat Mater 11:865
110. Liu H, Jiao L, Yang F, Cai Y, Wu X, Ho W, Gao C, Jia J, Wang N, Fan H, Yao W, Xie M (2014) Phys Rev Lett 113:066105
111. Zeng H, Dai J, Yao W, Xiao D, Cui X (2012) Nat Nanotechnol 7:490
112. Khan MF, Nazir G, Lermolenko VM, Eom J (2016) Sci Technol Adv Mater 17:166
113. Qin C, Gao Y, Qiao Z, Xiao L, Jia S (2016) Adv Opt Mater 4:1429
114. Xia F, Wang H, Xiao D, Dubey M, Ramasubramaniam A (2014) Nat Photonics 8:899
115. Wang F, Wang Z, Yin L, Cheng R, Wang J, Wen Y, Shifa TA, Wang F, Zhang Y, Zhan X, He J (2018) Chem Soc Rev 47:6296
116. Liu C, Ye H-F, Shi Y-L (2022) Chip 1:100005
117. Kim JS, Kim J, Zhao J, Kim S, Lee JH, Jin Y, Choi H, Moon BH, Bae JJ, Lee YH, Lim SC (2016) ACS Nano 10:7500
118. Dai Y, Wang Y, Das S, Xue H, Bai X, Hulkko E, Zhang G, Yang X, Dai Q, Sun Z (2020) ACS Nano 14:8442
119. Liu B, Zhao W, Ding Z, Verzhbitskiy I, Li L, Lu J, Chen J, Eda G, Loh KP (2016) Adv Mater 28:6457
120. Venkata Subbaiah YP, Saji KJ, Tiwari A (2016) Adv Funct Mater 26:2046
121. Tonndorf P, Schmidt R, Böttger P, Zhang X, Börner J, Liebig A, Albrecht M, Kloc C, Gordan O, Zahn DRT, Michaelis de Vasconcellos S, Bratschitsch R (2013) Opt Express 21:4908
122. Bernardi M, Palummo M, Grossman JC (2013) Nano Lett 13:3664
123. Lopez-Sanchez O, Lembke D, Kayci M, Radenovic A, Kis A (2013) Nat Nanotechnol 8:497
124. Shin GH, Park J, Lee KJ, Lee G-B, Jeon HB, Choi Y-K, Yu K, Choi S-Y, Appl ACS (2019) Mater Interfaces 11:7626
125. Yoo H, Hong S, On S, Ahn H, Lee H-K, Hong YK, Kim S, Kim J-J, Appl ACS (2018) Mater Interfaces 10:23270
126. Förster A, Gemming S, Seifert G, Tománek D (2017) ACS Nano 11:9989
127. Huang Y, Zhuge F, Hou J, Lv L, Luo P, Zhou N, Gan L, Zhai T (2018) ACS Nano 12:4062
128. Kim Y, Bark H, Kang B, Lee C, Appl ACS (2019) Mater Interfaces 11:12613
129. Liu Y, Wang G, Luo F, Li H, Zhu M, Liu X, Yang SA, Liu S (2021) Chem Asian J 16:1756

130. Pak S, Cho Y, Hong J, Lee J, Lee S, Hou B, An G-H, Lee Y-W, Jang JE, Im H, Morris SM, Sohn JI, Cha S, Kim JM, Appl ACS (2018) Mater Interfaces 10:38264
131. Xiong Y-F, Chen J-H, Lu Y-Q, Xu F (2019) Adv Electron Mater 5:1800562
132. Vu Q, Lee J, Nguyen O, Shin Y, Lim S, Lee O, Heo J, Park S, Kim K, Lee Y, Yu W (2016) Nano Lett 17:453–459
133. Pak Y, Mitra S, Alaal N, Xin B, Lopatin S, Almalawi D, Min J-W, Kim H, Kim W, Jung G-Y, Roqan IS (2020) Appl Phys Lett 116:112102
134. Ye K, Liu L, Liu Y, Nie A, Zhai K, Xiang J, Wang B, Wen F, Mu C, Zhao Z, Gong Y, Liu Z, Tian Y (2019) Adv Opt Mater 7:1900815
135. Jiang B, Yang Z, Liu X, Liu Y, Liao L (2019) Nano Today 25:122
136. Qiu F, Chu J, Liu Z, Xiang J, Yang J, Wang C (2020) Sol RRL 4:2000452
137. Tong L, Qiu F, Zeng T, Long J, Yang J, Wang R, Zhang J, Wang C, Sun T, Yang Y (2017) RSC Adv 7:47999
138. Liao M, Wei Z, Du L, Wang Q, Tang J, Yu H, Wu F, Zhao J, Xu X, Han B, Liu K, Gao P, Polcar T, Sun Z, Shi D, Yang R, Zhang G (2020) Nat Commun 11:2153
139. Li H, Tsai C, Koh AL, Cai L, Contryman AW, Fragapane AH, Zhao J, Han HS, Manoharan HC, Abild-Pedersen F, Nørskov JK, Zheng X (2016) Nat Mater 15:48
140. Lien D-H, Uddin SZ, Yeh M, Amani M, Kim H, Ager JW, Yablonovitch E, Javey A (2019) Science (80):364, 468
141. Kuc A, Zibouche N, Heine T (2011) Phys Rev B 83:245213
142. Yun WS, Han SW, Hong SC, Kim IG, Lee JD (2012) Phys Rev B 85:033305
143. Subbaiah YPV, Saji KJ, Tiwari A (2016) 2046
144. Islam KM, Ismael T, Luthy C, Kizilkaya O, Escarra MD, Appl ACS (2022) Mater Interfaces 14:24281
145. Mak KF, Lee C, Hone J, Shan J, Heinz TF (2010) Phys Rev Lett 105:2
146. Huang X, Shi J, Yao Y, Peng S, Zhang D, Jin Z (2021) Nanotechnology 32:135204
147. Yue Q, Kang J, Shao Z, Zhang X, Chang S, Wang G, Qin S, Li J (2012) Phys Lett A 376:1166
148. Klots AR, Newaz AKM, Wang B, Prasai D, Krzyzanowska H, Lin J, Caudel D, Ghimire NJ, Yan J, Ivanov BL, Velizhanin KA, Burger A, Mandrus DG, Tolk NH, Pantelides ST, Bolotin KI (2014) Sci Rep 4:6608
149. Gant P, Huang P, Pérez de Lara D, Guo D, Frisenda R, Castellanos-Gomez A (2019) Mater Today 27:8
150. Tsai D-S, Liu K-K, Lien D-H, Tsai M-L, Kang C-F, Lin C-A, Li L-J, He J-H (2013) ACS Nano 7:3905
151. Yang P, Zhang Z, Sun M, Lin F, Cheng T, Shi J, Xie C, Shi Y, Jiang S, Huan Y, Liu P, Ding F, Xiong C, Xie D, Zhang Y (2019) ACS Nano 13:3649
152. Ling ZP, Yang R, Chai JW, Wang SJ, Leong WS, Tong Y, Lei D, Zhou Q, Gong X, Chi DZ, Ang K-W (2015) Opt Express 23:13580
153. Nalwa HS (2020) RSC Adv 10:30529
154. Ra H-S, Kwak D-H, Lee J-S (2016) Nanoscale 8:17223
155. Sahatiya P, Jones SS, Badhulika S (2018) Appl Mater Today 10:106
156. Gomathi PT, Sahatiya P, Badhulika S (2017) Adv Funct Mater 27:1701611
157. Li X, Wu J, Mao N, Zhang J, Lei Z, Liu Z, Xu H (2015) Carbon N Y 92:126
158. Xiao P, Mao J, Ding K, Luo W, Hu W, Zhang X, Zhang X, Jie J (2018) Adv Mater 30:1801729
159. Yu SH, Lee Y, Jang SK, Kang J, Jeon J, Lee C, Lee JY, Kim H, Hwang E, Lee S, Cho JH (2014) ACS Nano 8:8285
160. Wu G, Wang X, Chen Y, Wang Z, Shen H, Lin T, Hu W, Wang J, Zhang S, Meng X, Chu J (2018) Nanotechnology 29:485204
161. Sun M, Yang P, Xie D, Sun Y, Xu J, Ren T, Zhang Y (2018) Adv Electron Mater 1800580
162. Bang S, Duong NT, Lee J, Cho YH, Oh HM, Kim H, Yun SJ, Park C, Kwon M-K, Kim J-Y, Kim J, Jeong MS (2018) Nano Lett 18:2316
163. Jing W, Ding N, Li L, Jiang F, Xiong X, Liu N, Zhai T, Gao Y (2017) Opt Express 25:14565
164. Chen Y, Wang X, Wu G, Wang Z, Fang H, Lin T, Sun S, Shen H, Hu W, Wang J, Sun J, Meng X, Chu J (2018) Small 14:1703293

165. Tan H, Xu W, Sheng Y, Lau CS, Fan Y, Chen Q, Tweedie M, Wang X, Zhou Y, Warner JH (2017) Adv Mater 29:1702917
166. Long M, Liu E, Wang P, Gao A, Xia H, Luo W, Wang B, Zeng J, Fu Y, Xu K, Zhou W, Lv Y, Yao S, Lu M, Chen Y, Ni Z, You Y, Zhang X, Qin S, Shi Y, Hu W, Xing D, Miao F (2016) Nano Lett 16:2254
167. Liu H, Li D, Ma C, Zhang X, Sun X, Zhu C, Zheng B, Zou Z, Luo Z, Zhu X, Wang X, Pan A (2019) Nano Energy 59:66
168. Yang S, Wu M, Wang B, Zhao L-D, Huang L, Jiang C, Wei S-H, Appl ACS (2017) Mater Interfaces 9:42149
169. Yang S, Wang C, Ataca C, Li Y, Chen H, Cai H, Suslu A, Grossman JC, Jiang C, Liu Q, Tongay S, Appl ACS (2016) Mater Interfaces 8:2533
170. Ding Y, Zhou N, Gan L, Yan X, Wu R, Abidi IH, Waleed A, Pan J, Ou X, Zhang Q, Zhuang M, Wang P, Pan X, Fan Z, Zhai T, Luo Z (2018) Nano Energy 49:200
171. Deng J, Shao J, Lu B, Chen Y, Zaslavsky A, Cristoloveanu S, Bawedin M, Wan J (2018) IEEE J Electron Dev Soc 6:557
172. Liao F, Deng J, Chen X, Wang Y, Zhang X, Liu J, Zhu H, Chen L, Sun Q, Hu W, Wang J, Zhou J, Zhou P, Zhang DW, Wan J, Bao W (2020) Small 16:1904369
173. Zhang K, Peng M, Yu A, Fan Y, Zhai J, Wang ZL (2019) Mater Horizons 6:826
174. Kim T-Y, Ha J, Cho K, Pak J, Seo J, Park J, Kim J-K, Chung S, Hong Y, Lee T (2017) ACS Nano 11:10273
175. Singh E, Meyyappan M, Nalwa HS, Appl ACS (2017) Mater Interfaces 9:34544
176. Wang L, Jie J, Shao Z, Zhang Q, Zhang X, Wang Y, Sun Z, Lee S-T (2015) Adv Funct Mater 25:2910
177. Heo K, Jo S-H, Shim J, Kang D-H, Kim J-H, Park J-H, Appl ACS (2018) Mater Interfaces 10:32765
178. Kang D-H, Pae SR, Shim J, Yoo G, Jeon J, Leem JW, Yu JS, Lee S, Shin B, Park J-H (2016) Adv Mater 28:7799

Chapter 10
MoS_2 for Optoelectronics and Photovoltaic Devices

Kasturi Kamat, Shweta Jagtap, Som Datta Kaushik, and Dattatray J. Late

1 Introduction

Over the years in order to achieve the desired applications at the nanoscale level, alteration in shape and structure of materials has been exercised in order to gain significant development in their evolution. For over a decade, carbon nanotubes have been seen as the promising material for applications in electronic as well as optoelectronic devices. Although, lately the scope of the use of carbon nanotubes has declined due to several reasons. Hence, development of 2D quasi-materials that entailing materials such as graphene and its derivatives, black phosphorus, earth-abundant transition metal dichalcogenides (2D-TMDs) like tungsten disulphide (WS_2), tungsten diselenide (WSe_2), titanium diselenide ($TiSe_2$), iron disulphide (FeS_2), molybdenum diselenide ($MoSe_2$) and molybdenum disulphide (MoS_2) has increased.2D TMDs exhibit a strong interaction with light, this can be attributed to their ability of possessing tunable bandgap, thus making them susceptible for developing various heterostructures with other materials including 0D–2D hybrid heterostructures, TMD-perovskite-based heterostructures, van der Waals heterostructures, etc. Amongst several 2D TMDs that can form stable two-dimensional crystal structures, molybdenum disulphide attracts the most attention. 2D MoS_2 exhibits a unique property of variation in its bandgap

K. Kamat · S. Jagtap (✉)
Department of Electronic and Instrumentation Science, Savitribai Phule Pune University, Pune 411007, India
e-mail: sdj_disc@unipune.ac.in

S. D. Kaushik
UGC-DAE Consortium for Scientific Research Mumbai Centre, R-5 Shed, BARC, Mumbai 400085, India

D. J. Late (✉)
Department of Physics, Federal University of Lavras, Campus Universitário, PO Box 3037, Lavras, Minas Gerais, Brazil
e-mail: datta099@gmail.com

D. J. Late and C. S. Rout (eds.), *Science and Technology of 2D MoS_2*,
Materials Horizons: From Nature to Nanomaterials,
https://doi.org/10.1007/978-981-97-7367-1_10

from 1.29 to 1.8 eV, depending upon the thickness of MoS_2 layers. Semiconductor material consisting of all MoS_2 films/blocks comprises of a natural indirect band gap of 1.2 eV, apart from a single-layered MoS_2 film structure that possesses a direct bandgap of 1.8 eV [1]. In this chapter, the current research advances for improving the photo detective as well as photovoltaic properties of atomic-layered MoS_2 have been discussed. The enhancement in its properties such as the charge carrier mobility, incident light absorption through methods such as chemical doping of MoS_2 with other TMDs, surface functionalization of MoS_2 atomic layers, MoS_2-based hybrid heterostructures with other materials to create halide-perovskites MoS_2 structures, 0D–2D QD (quantum dots) heterostructures, MoS_2-based van der Waals heterostructures has also been studied in detail. It was observed that by the application of the aforementioned methods, some of the obtained MoS_2-based heterostructures exhibited high responses including the large values of photoresponsivity, detectivity and striking decrement in response time value ranging from 10^{-3}A/W to 10^{10}A/W, 10^7 to 10^{15} Jones and sec to nano-sec, respectively. The natural abundance and attractive semiconductor properties of MoS_2 make it one of the most extensively studied 2D TMD compound. Bulk MoS_2 structure consists of a configuration wherein a stacking of a single molybdenum atom between 6 sulphur atoms takes place. The neighbouring layers in the bulk compound are connected via weak van der Waals forces which in turn enable the layers to slide over one another, this facilitates shear along the basal plane of the Crystal. Due to the presence of weak van der Walls forces within the layers, each plane interacts easily with the next plane and they are thus extensively used as lubricants in the mechanical industries. Further interest arises in MoS_2 due to its unique properties such as a good mobility value of 700 cm^2/ Vsec, a high current on/off ratio lying approximately in the range of 10^7–10^8, large visible range optical absorption of approximately 10^7 m^{-1} and high PL rise value for direct bandgap monolayer (1L) MoS_2. Like other TMD materials, MoS_2 displays rich polytyptism [2] having three polytypes format, which are 1 T (D_{6d}) [3, 4], 2H (D_{6h}) [5, 6] and 3R (C_{3v}) where T, H and R stand for tetragonal, hexagonal and rhombohedral crystal system, respectively.

In comparison with other optoelectronic materials, the characteristic that makes TMDs much suited for photodetector materials synthesis is its large electronic density of states, which results in eminent optical absorption and accelerated charge transfer. Bernardi et al. [7] observed 5%–10% absorption of incident sunlight by MoS_2, $MoSe_2$ and WS_2 TMD monolayers with the thickness under 1 nm. This value of incident sunlight absorption by the TMD materials was higher by one degree magnitude from that of value of conventional semiconductor thin film made of GaAs (gallium arsenide) and Si (silicon), and 2.3% higher than monolayer graphene, which are the preferred materials for solar energy applications. Figure 1a shows the comparison between the modulus of optical absorbance of TMD materials $MoSe_2$, MoS_2 and WS_2 in comparison with graphene. As stated, the TMD materials appear to show much more incident sunlight absorption than graphene. The flux value for absorbed photons in graphene was calculated as 2 mA/cm^2, while that for monolayer TMD materials MoS_2, $MoSe_2$ and WS_2 was 4.6, 3.9 and 2.3 mA/cm^2, respectively.[1] It was observed that the sunlight absorbed by a single layer TMD was equivalent to that

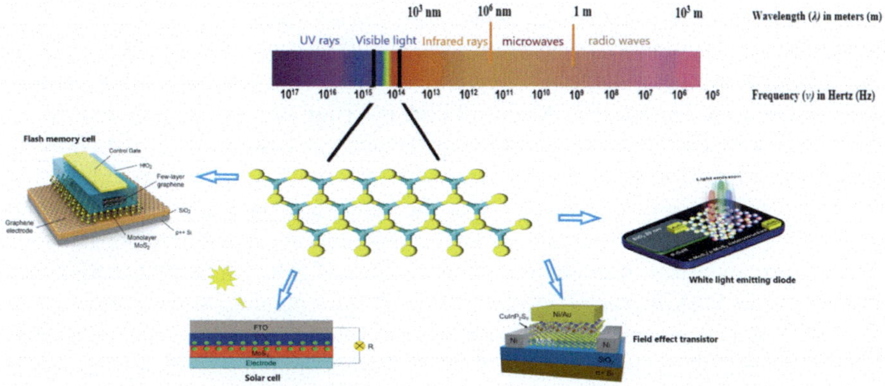

Fig. 1 Electromagnetic range of 2D MoS_2 from visible-NIR region and its applications

absorbed by a 50 nm thick Silicon film. A summarization of the band gaps of different 2D TMD materials along with corresponding photodetection range was done by Wang et al. [8], which included 2D materials ranging from MoS_2, to h-BN, having a photodetection range varying between near ultraviolet (NUV) and long infra-red region (LIR). A broad spectral range was observed for MoS_2, WSe_2, SnS_2 and black phosphorus in the spectral range wavelength of NUV to MIR (mid-infrared) this accredits that by the development of hybrid MoS_2-based heterostructures the optical wavelength value could be extended in MoS_2 thin films.

2 MoS_2-Based Hybrid Heterostructures for Photodetector Applications

As mentioned before, the band gap ofMoS_2 is directly dependent upon the thickness of MoS_2 layers. The bandgap in MoS_2 varies from monolayer (1.8 eV) to multilayer (ML) (1.2 eV). This property enables, in many TMDs including MoS_2, $MoSe_2$, WS_2, the ability to tune the band gap from direct (1L) to indirect (ML) while varying the thickness. This tunable band gap can then be used to improve the optical absorption of automatically thin TMDs (layered) by forming hybrid heterostructures. Different organic and 2D nanomaterials with MoS_2 hybrid heterostructures have been developed in order to improve the charge transfer process and extending the light absorption wavelength range. The following section highlights the performance of varied MoS_2-based heterostructures formed with graphene, perovskites, 0D–2D van der Waals heterostructures (vdWHs), etc.

Reduced graphene oxide (rGO)-MoS_2/pyramid Si heterostructure-based photodetectors were synthesized with three layered graphene and InGa (indium-gallium) alloy as top electrode and bottom electrode, respectively. Under the conditions of increasing photo voltage (180 mW to 276 mW) and power intensity (100 nW to 1 mW), evidence of a self-powering detector at zero bias voltage was observed for rGO-MoS_2/Si heterojunction [9]. Furthermore, the imperfect crystallinity of MoS_2 results in a narrowed band gap and due to the pyramid structure, high amount of light absorption is possible by the detector. Also, from the insertion of reduced graphene oxide, high charge transportation occurs in the as-formed heterostructure-based photodetector. Thus, large values of responsivity and detectivity of 21.8 A/Wand ~3.8 × 10^{15} Jones, respectively, are observed for the heterostructure, along with an ultra-broadband spectrum response from ultraviolet (350 nm) to mid-infrared (4.3 μm). On performing comparison between the photocurrents of rGO-MoS_2/planar Si device to that of rGO-MoS_2/pyramid Si device at 808 nm operating wavelength, it was observed that the former showed a photocurrent value of 1.1 μA, which is 5 times lesser than the latter having a value of 5.3 μA. Further applying an illumination over 1100 nm, the absorption of light in both devices drastically decreased, as a band gap of 1.1 eV was attained. Additionally, defects in MoS_2 crystal structure result in the absorption of light from NIR to MIR range and this results in a transfer of photoexcited carriers from MoS_2 to rGO layer. The transfer of photo excited carriers results in a rise in the photocurrent value. Also, sulphur vacancy defect in MoS_2 crystal results in a narrowing effect which enables pyramid heterojunction-based photodetector to operate within 350 nm to 4.3 μm spectral range. In yet another study [10], a hybrid photodetector was developed using an rGO conducting and perovskite/MoS_2 bulk heterojunction. The formed hybrid photodetector showed high responsivity, high detectivity, high EQE and fast photo response of 1.08 × 10^4 A/W, 4.28 × 10^{13} Jones, 2.0 × 10^6% and >45 ms, respectively. Selective electron trapping in MoS_2 material leads to a transfer of holes from perovskite to rGO layer, this in turn suppresses the combination of photo-generated carriers and higher carrier mobility magnitude in rGO channel.

Since a long time, layered MoS_2 has been used for optical absorption enhancement in semiconductors. In lieu of that MoS_2-based hybrid plasmonic heterostructures can be formed in order to obtain high-value photo-response detectors. In a study, through exploitation of surface of 1D silver nanowire plasmon network, the formation of 1L MoS_2 photodetector was done [11]. 1.39 × 10^4 current on/off ratio was observed for 1L MoS_2/AgNW hybrid photodetector while that for pristine 1L MoS_2 photodetector was observed to be 62.97. Similarly, comparative rise in photo response was also observed. The photosensitivity observed for 1L MoS_2/AgNW hybrid detector was 59.60 A/W compared to that of its pristine counterpart having a value of 0.05 A/W. A 1000 times enhancement is observed in detectivity of 1L MoS_2/AgNW hybrid photodetector (4.51 × 10^{10} Jones) to pristine 1L MoS_2 photodetector (4.11 × 10^7 Jones). Confocal Raman scattering was performed on heterostructure as illustrated in Fig. 2. A red shift from 386 to 383.5 cm^{-1} in $E^1{}_{2g}$ mode for 1L MoS_2 is observed, which can be attributed to strain relaxation. The highly localized electro-magnetic field due to AgNW may have resulted in strengthening oscillations of sulphur atoms in MoS_2 thus resulting in a blue shift from 403 to 404.5 cm^{-1} in $A^1{}_g$ mode for 1L MoS_2 in the hybrid region.

These results were attributed to the surface plasmon coupling of AgNW network, which improved the optoelectronic functioning of 1L MoS_2/AgNW hybrid

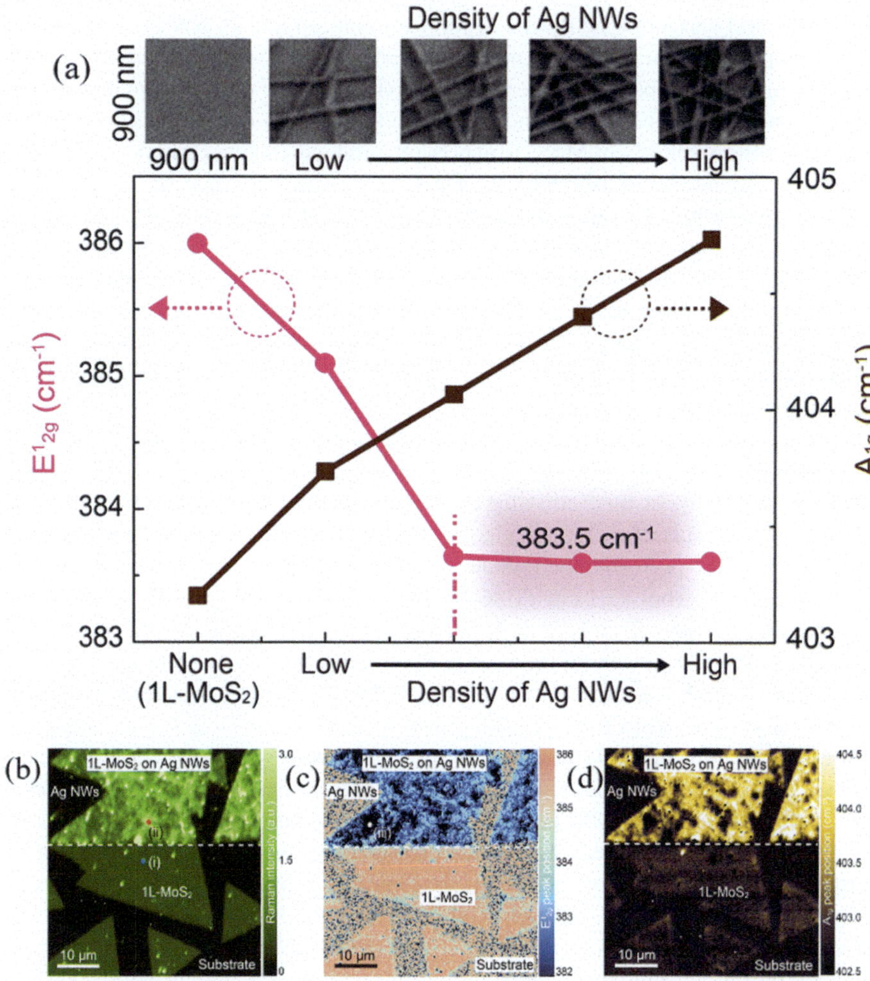

Fig. 2 **a** The tendencies of confocal Raman scattering for 1L MoS_2 with inset SEM images showing the morphology of heterostructure with increasing AgNW density. The peak shift of $E^1{}_{2g}$ and A_{1g} mode is shown by pink and brown colour, respectively; **b** Raman shift mapping of 1L MoS_2 and 1L MoS_2/AgNW heterostructure **c** Raman peak shift mapping for $E^1{}_{2g}$mode; **d** Raman peak shift mapping for A_{2g} mode. Reprinted with permission from Ref. [11], copyright © American Chemical Society

heterostructure. MoS_2/Ag nano-cube hybrids were prepared by Sun et al. [12] by coating silicon layer with silver nano-cubes (Ag-NCs) under optimized laser shock peening (LSP) functionalized in gap mode. A major difference in the photoluminescence spectrum (PL) was observed for MoS_2/Ag-NCs hybrid in comparison to MoS_2 films later modified by Ag-NC, with Ag placed beneath the films. This enhancement in PL spectrum could be attributed to the interaction between the excitons created by

MoS_2 layer and gap mope LSP. At 520 nm illumination and 46 nW incident power with the applied potential of 3 V, not much variation in the photocurrents of both photodetectors were observed. These values existed in the uniform bracket of 48–50 nA. Although under similar conditions, due to Ag nano-cubes, the photocurrent observed for the assembly of MoS_2 film present amid gap of silver and Ag nano-cube was at a heightened value of 0.91 μA. Thus, MoS_2 film modified by Ag-NCs had a photocurrent value, 19 times more than its primeval counterpart. Under similar lines, at a potential of 3 V with 2.2 pW incident power, the photo responsivity of MoS_2/AgNC flexible photodetector was 7940 A/W, 38-fold more than pristine MoS_2 film, whose decrease in photoresponsivity is the result of reduced scattering and recombination of charge carriers within material. Thus, through the inclusion of plasmonic nanostructures, the photo responsivity of MoS_2-based photodetectors could be increased.

The performance of MoS_2 layers can also be boosted via chemical doping. This was observed in a study wherein stable reversible n-type doping was performed on MoS_2flakes [13]. In this, utilizing the charge transfer from triphenylphosphine (PPh_3) to MoS_2 resulted in elevated electronic as well as optoelectronic performance of ML MoS_2. The performance enhancement is observed through the increase in mobility of charge from 12.1 cm^2/V sec to 241 cm^2/V sec, along with increased on/off current ratio values from 8.72×10^4 to 8.70×10^5. Additionally, on application of 5×10^{-12} W laser power, an increase in the photoresponsivity from 2.77 $X10^3$A/W to 3.92 $X10^5$A/W was observed in as-prepared photodetector. A 3-fold decrement of detectivity value from 6.82×10^{13} Jones to 2.36×10^{10}Jones, for increasing doping temperature (150 °C up to 350 °C) in ML MoS_2 photodetector was also observed. For durability check, the photodetector was exposed to air and, over a period of 14 days only 1.58% decrement in the photo responsivity of the detector was observed, which proved that it could offer long-term device stability under continuous usage. Transition of MoS_2 semiconductor from n-type to p-type was attained by Kim et al. [14], through Niobium (Nb) doping. The doping effect was confirmed using techniques of Raman, XPS, STEM, KPFM and electron transport characterization. Through these characterization techniques, it was also observed that on doping with Nb the obtained MoS_2 semiconductor consists more of n-type carriers, thus undergoing substitution method of doping. Figure label illustrates the detailed process used for the synthesis of Nb/Mn doped MoS_2, through which it was observed that each of the dopants, i.e., transition metals Nb and Mn were initially deposited individually on separate SiO_2/Si substrates, post which substitutional doping of MoS_2 on the deposited metal was done. Additionally, a steep contrast is observed between the synthesized Nb and Mn doped MoS_2 films, respectively, as illustrated from the optical microscopy images in Fig. 2b, it was proved that the doping of each MoS_2 film is continuous and uniform throughout. On application of illumination of 550 and 750 nm respectively, Nb-doped MoS_2 photodetector displayed photo responsivity of 2×10^5 A/W and 4.83 $X10^5$A/W. The highest photoresponsivity value attained by as-formed detector was observed to be 10^6 times greater than that of undoped MoS_2 detector (at 750 nm). Similar phenomena were observed with the EQE values of Nb-doped MoS_2 and undoped MoS_2 photodetector. On application of 450 nm illumination, the detectivity of the former was observed to be 5×10^{12} Jones, 10^5 times greater than the detectivity of the latter being 3.93×10^8 Jones. Also, at 550 nm illumination, Nb-doped MoS_2 device showed faster decay and rise time of 5.7 and 4.9 ms, respectively, as compared to the rise and decay time of undoped MoS_2 device being 0.63 and 1.63 s, respectively. The STEM image of Nb-doped MoS_2 is shown in Fig. 3f. It shows a hexagon structure as confirmed by fast Fourier transform pattern(inset) Fig. 3g. With similar atomic numbers, the atomic intensity for Nb-doped MoS_2structure was uniformly distributed in contrast to Mn-doped MoS_2 structure Fig. 3d. These results indicated effective substitution of Nb/Mn atoms with Mo atoms. Over a period of 30 days, under ambient conditions, Nb-doped MoS_2 device displayed good stability, which in turn accredited the method of substitutional doping for expansion of applications of 2D TMDS in electronic and optoelectronic fields.

Due to their tunable band gap, high efficiency of photo conversion, carrier charge mobility and easy solution processing of ultra-thin layers, perovskites have emerged well as an effective class of light-absorbing materials. For improving efficiency of MoS_2 photodetectors, Song et al. [15] reported successful combination of ML MoS_2 with $CsPbBr_3$ perovskite nano-sheets to fabricate high-efficiency photodetectors. These hybrid-perovskite-based photodetectors displayed values of photo responsivity, external quantum efficiency (EQE) and detectivity of 4.4 A/W, 302% and 2.5×10^{10} Jones, respectively. Such high values can be attributed to the separation of photoexcited carriers at the interface level, which is due to high absorption and carrier transportation of charges by the perovskite and MoS_2 layer respectively. In another study, Wu et al. [16] developed hybrid phototransistors of 2D 1L MoS_2 using QDs of 0D caesium lead halide perovskites ($CsPbBr_3$). The TEM image (Fig. 4a inset) of perovskite QD exhibits a typical cubic shape having a diameter of 15 nm. The UV spectroscopy characterization. In UV Spectroscopy Fig. 4c, two absorption peaks for primeval MoS_2 layer were observed between 600 to 700 nm. In lieu of that a significant increase in absorption was observed below 670 nm for MoS_2, which can be attributed to its optical band gap of 1.85 eV. Additionally, the bandgap of perovskite QD from Fig. 4c was determined to be 1.91 eV. Thus, the formed $CsPbI_{3-x}Br_x/MoS_2$ heterostructure, below 649 nm, showed enhanced absorption, due to the synergetic absorption effect of MoS_2 and perovskite QD layers. The mixed dimensional $CsPbBr_3$/1L MoS_2 van der Waals heterostructure photodetector displayed high values of photo responsivity 7.7×10^4 A/W, detectivity 5.6×10^{11} Jones and EQE of10^7%, at 1.5 μW optical power for 532 nm illumination. Such high values are obtained by the photo-gating effect caused in the hybrid material. For the same conditions, it was observed that MoS_2 hybrid detector combined $CsPbBr_3$ displayed 3-fold higher photoresponsivity than pristine MoS_2 photodetector. This indicated that 0D $CsPbI_{3-x}Br_x$ QDs exhibited high absorption and efficient separation of photo-generated carriers thus enhancing optoelectronic properties of MoS_2.

It is known in relation to its tunable bandgap, the rate of absorption of light in MoS_2 is directly proportional to its thickness. In this study, under optical illumination, the photo responsivity and detectivity of ZnCdSe/ZnS-based MoS_2 QD with varying thickness of MoS_2 (mono-layer, bi-layer, tri-layer and multi-layer) was compared

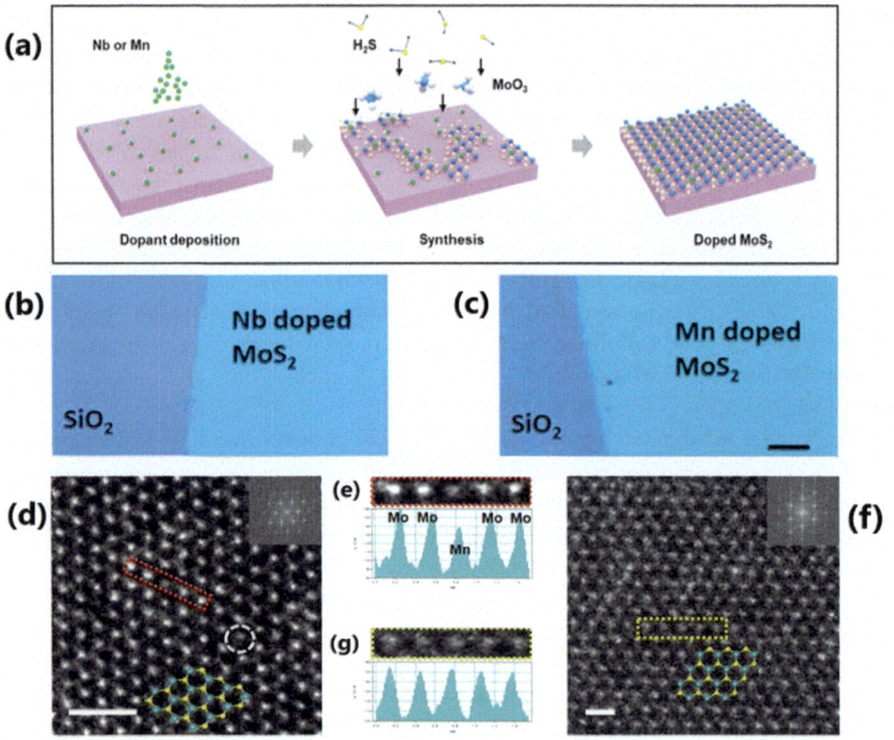

Fig. 3 **a** Schematic illustration of synthesis of doped MoS_2 and fundamental characterizations; **b** Optical microscopy image for Nb-doped MoS_2; **c** Optical microscopy image for Mn-doped MoS_2; **d** STEM image of Mn-doped MoS2 with overlayed atomic structure and FFT in inset. Mn atom is shown by white dash. The scale bar is observed at 1 nm; **e** the line intensity profile (experimental) for marked red rectangle in panel **f** STEM image of Nb-doped MoS2 with overlayed atomic structure and FFT in inset. The scale bar is observed at 0.5 nm; **g** the line intensity profile (experimental) for marked yellow rectangle in panel **f**. Reprinted with permission from Ref. [14], copyright © American Chemical Society

with primaeval MoS_2 photodetector. It was observed that with increasing thickness, the photocurrent value of as-formed hybrid detectors also increased. A thousand-fold increase in detectivity and a three time increase in photoresponsivity of hybrid ZnCdSe/ZnS-based MoS_2 QD was observed as compared to its pristine counterpart [17]. Also, the as-formed hybrid QDs displayed 50 times faster functionality than its pristine counterpart while displaying a reduction in response time from 15 to 0.3 s. The highest detectivity attained by hybrid MoS_2-QD was 1.0×10^{12} Jones, along with a five-order increase in gain equivalent to 1.08×10^5. This phenomenon can be due to the efficient energy transfer and increased absorption by the transfer of photoexcited charge carriers from QD to MoS_2 layer. These high gains and Ultra sensitivity are due to the quenching effect by MoS_2 layers, reducing the fluorescence offered by ZnCdSe/ZnS QDs while forming 2D-0D hybrid nanostructures. Liu

Fig. 4 Illustration of device model and optical characterizations of perovskite ($CsPbI_{3-x}Br_x$)/MoS_2 heterostructure. **a** Schematic model of device with typical TEM image of perovskite QDs in inset. The scale is observed to be 50 nm; **b** Optical microscopy image of $CsPbI_{3-x}Br_x$/MoS_2 heterostructure device. The scale is observed to be 10 μm; **c** UV–vis absorption spectrum for pristine MoS_2, primaeval perovskite QDs and $CsPbI_{3-x}Br_x$/MoS_2 heterostructure. Reprinted with permission from Ref. [16], copyright © Wiley

et al.[18] synthesized vertical stacked p-$SbTe_3$/n-MoS_2vdWH photodetector having high sensitivity, displaying photoresponsivity and conversion efficiency of 330 AW^{-1} and 4.5%, respectively. This assembly of heterostructures showed high response time, greater than 500 μs. In another study [19], SnSe/MoS_2 vdWHs showed a current on/off ratio of 1×10^5 s. The type-II SnSe/MoS_2 heterostructures displayed a photoresponsivity of 100 A/W, an EQE of 23.3×10^3% with a response time less than 10 ms. Thus, through fabrication of different 2D materials with vdWH-based FETs, better-functioning optoelectronic materials can be synthesized.

Flexible MoS_2 photodetectors are also important in the study of wearable applications. These can be developed with a variety of nanomaterials while controlling their performance through the process of material synthesis, fabrication of device, measurement of wavelength, illumination of incident laser, atomic layer processing, hybrid heterostructures, etc. [20–22]. The performance of an MoS_2-based heterostructure, like its photo conducting ability, photoresponsivity, detection limit, response time can be altered only when a thorough knowledge of the following parameters is known. These include the number of active layers in the structure, the nature of doping used and research on the properties of materials used in the process of heterostructure synthesis. Also, the information about the Vander wall interactions, bulk heterojunctions and the photo-generated carrier mobility of as synthesized material is also important including its spectral range and ability of charge transfer. Ferroelectric polymers including PVDF [poly (vinylidene fluoride)] have also been used in integrated devices for enhancing the mechanical strength, flexibility, piezoelectricity, pyroelectricity and ferroelectricity of sensors, transducers, actuators, etc. [23]. These polymers when used in the case of MoS_2 photodetectors resulted in improving the photo response by suppressing dark current in as-synthesized heterostructures. On the other hand, polymeric substrates like poly (ethylene terephthalate) PET, Kapton/polyamide/(PI), are also being used for developing wearable electronic and optoelectronic devices, as these devices provide natural mechanical strength and flexibility [20, 22]. In order to develop enhanced mechanical functioning, as well as flexibility in electric and optoelectronic devices, inherently flexible polymeric substrates can be used. Silicon carbide (SiC) with flexible Kapton as substrate was used for developing MoS_2 photodetectors. The as-formed photodetectors displayed a performance that was dependent on the substrate layer as reported by Zang et al. [24]. The enhancement in the photo response of the photodetector is attributed to high absorption efficiency of SiC as well as Kapton, which in turn induces dual photo-gating effect in MoS_2-SiC/Kapton interface. From observations, it was shown that at the interface, there is an energy level mismatch which could be due to high strength built-in electric field, resulting in photo-excited carriers from being separated in MoS_2 layers. The photo-response of MoS_2/SiC photodetector is observed to be more than MoS_2/SiO_2/Si photodetectors. The band gap and electron affinity of MoS_2 (2.7 and 4.2 eV) and SiC (3.2 and 1.8 eV) were also measured. The highest responsivity for MoS_2/SiC photodetector is observed to be approximately 10^4 AW^{-1} while the fastest current on/off ratio for MoS_2/Kapton photodetector is approximately 10^5 in magnitude (under illumination of 325 nm). Thus, significant improvement can be seen in the photo responsivity of MoS_2-SiC/Kapton substrate due inherent property of absorption of light. It is well known that graphene is an excellent photo-conducting material with high levels of charge mobility, extremely fast responsivity and ability to function in a broad spectrum of absorption. By combining the properties of graphene with MoS_2, excellent results for optoelectronic applications can be achieved. Ross et al. [25] constructed a hybrid structure by combining 1L graphene with nanosheets of ML MoS_2 placed on Si/SiO_2 substrate which yielded high mobility value of 1×10^4 cm^2/V sec. Also, high responsivity of 1×10^{10} A/W at 130 K (maintained at 1 fW μ/m^2 power) was attained as compared to responsivity value of 5×10^8A/W at standard temperature (maintained at 6.4 fW μ/m^2 power).

3 MoS_2 Based Heterostructures for Light-Emitting Diode Applications

In a study, MoS_2/GaN heterostructures were devised for functions in light-emitting diodes. Triangle-shaped 1L MoS_2 nanosheet (single-crystal) was synthesized on GaN substrates [26]. As shown in Fig. 5c, an arrow photoluminescence peak is observed at 1.84 eV under 532 nm laser radiation. This indicates high-quality MoS_2 growth on GaN. The significance of charge transfer from 2D 1L MoS_2 to GaN results in a redshift of 200 MeV for Ga(3d orbital), which suggested that further band bending could take place at the heterostructure interface. MoS_2/GaN heterojunction formation is owed to the surface potential value of 55 mV for GaN substrate, in turn having a work function value, larger than 55 meV than 1L MoS_2. This also results in the formation of a built-in potential in the heterostructure, indicating charge transfer taking place at the interface itself. At a bias voltage of 3 V, the EQE magnitude of MoS_2/GaN heterostructure is 6.6×10^5% and the photoresponsivity is 2200A/W, at 410 nm illuminating laser wavelength.

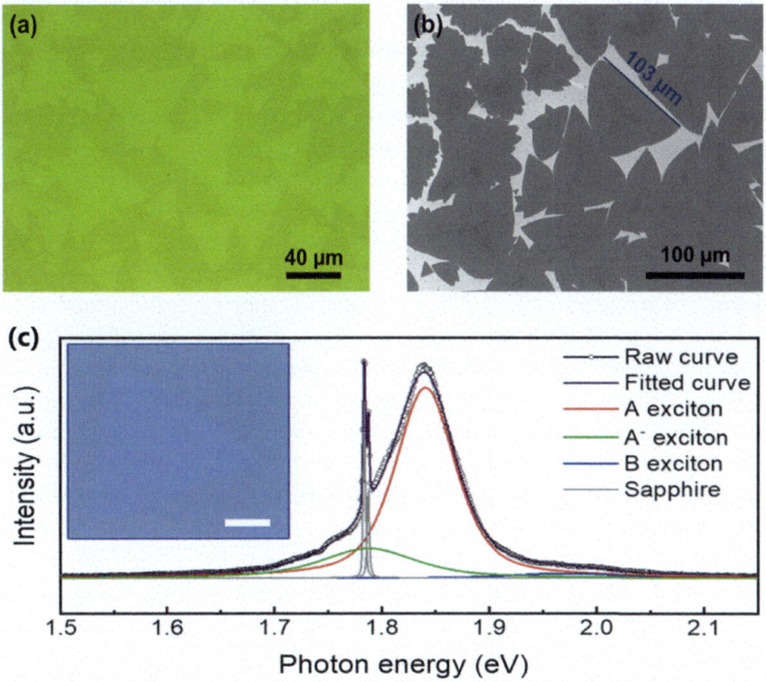

Fig. 5 Optical characterization observations for MoS_2/GaN heterostructures; **a** Optical microscopy image for MoS2 growth on GaN substrates; **b** optical microscopy image for MoS2 growth on sapphire substrates; **c** PL spectrum of 1L MoS_2 grown on GaN with optical image of MoS_2/GaN heterostructure in inset. Reprinted with permission from Ref. [26], copyright © American Chemical Society

Yin et al. [27] devised a single layer MoS_2-based top gated FET having super high channel mobility of ~200 cm^2V^{-1} s^{-1} along with an on/off current ratio of ~10^8. For proof of concept, light-emitting device with an active layer of n-SiC/p-MoS_2-MoO_3 heterojunction was fabricated (as seen in figure). The current–voltage characteristic curve observed for the as-prepared LED exhibited a turn on voltage of 4.5 V. The nature of light emission of the prepared device indicated that the MoS_2-MoO_3 hybrid layer worked actively as p-type hole injection layer resulting in enhanced device functioning. Choi et al. [28] demonstrated an ultrathin flexible organic light emitting diode (OLED) functioned by MoS_2-based thin-film transistors (TFTs). The device functioning was further improved by damping the Coulomb scattering, which was due to major dielectric difference at the Al_2O_3/MoS_2 interface. Also, reduction in trap charges and interface scattering was observed, through combining n-type doping of MoS_2 at the source, drain and channel regions leading to modifications in TFT structure. By having a channel sandwiched amongst two Al_2O_3 layers, the device functioning in turn was further improved. As the device electrodes were placed below the MoS_2 film, a subsequent deposition of Al_2O_3 (top dielectric) along with reduction in contact resistance of source and drain regions is observed. This leads to enhancement of charge carrier mobility facilitating OLED display, which needs high current value for operation. Thus, the study showed that an environment with high-k dielectric presence leads to high carrier mobility in MoS_2-based TFTs. In yet another study Nikam et al. [29] synthesized heterostructures with combining p-MoS_2/n-MoS_2 on p-GaN/sapphire substrates for LED device application. Due to the limitation of lateral heterostructures for forming alloys at the boundary regions, the formation of vertically directed 2D-TMDs heterostructures presented themselves as better alternatives [30]. This is attributed to their characteristics such as low value of contact resistance, high density of current offered for brighter LEDs and complete luminescence of device area [31, 32]. From the PL spectra as illustrated in Fig. 6a for p-type and n-type MoS_2 heterostructures, respectively, excited at an illumination of 488 nm, the signals for each respective heterostructures were observed at ~650 and ~520 nm. This showed imminent presence of direct excitonic transitions in the uppermost and lowermost layers of p-type and n-type MoS_2 heterostructures. Then, white light emitting diodes (WLEDs) were designed for study purposes by utilizing n-type MoS_2, p-type MoS_2 and p-type GaN heterostructures with emission colours orange, green & blue (Fig. 6c). A heterojunction of n-type MoS_2/p-type MoS_2/p-type GaN was formed, the heterojunction was then made to undergo electroluminescence, operating from 2 to 8 V. The application of different bias voltages on as-formed heterojunction resulted in displaying an EL spectrum functioning under forward bias. It was also observed that over a potential value of 4 V, heterojunction breakdown occurs and the EL spectrum is observed to decline. Additionally, from EL diagram and CIE (colour space), it was observed that, by combining blue emission of p-type GaN (~481 nm), green emission of p-type MoS_2 (~525 nm) and orange emission of n-type MoS_2 (~642 nm), successful fabrication of WLED can be achieved. The as-formed WLED device at a potential of 4 V displayed luminous efficiency of 29% and efficacy of 294 lm/W. These results show great potential of 2D materials, for forming cost-effective and colour-tunable light emitters with varying optical and lighting application skills.

4 MoS_2 Heterojunctions for Photovoltaic Applications

Lin et al. [33] designed 1L MoS_2/InP heterostructures for photovoltaic applications that can be electrically influenced. 1L MoS_2/InP-based type-I heterojunction was formed such that it consisted of MoS_2 as window layer, possessing a large bandgap. Upon application of positive gate voltage on the uppermost layer of MoS_2, an upward shift in the fermi level was observed. This causes further band bending in MoS_2/InP junction. The bending further results in creation of large barriers which in turn increases open circuit voltage (V_{oc}) of the solar cell. At an applied gate voltage of 3 V the V_{oc} of 27.2 mA/cm^2, short circuit density (J_{sc}) of 0.45 V and fill factor of 59% was observed, providing a power conversion efficiency (PCE) of 6%. At an application of higher gate voltage (6 V), the performance of solar cell was further improved. The characteristic reduction of series resistance in 1L MoS_2/InP heterojunction, under

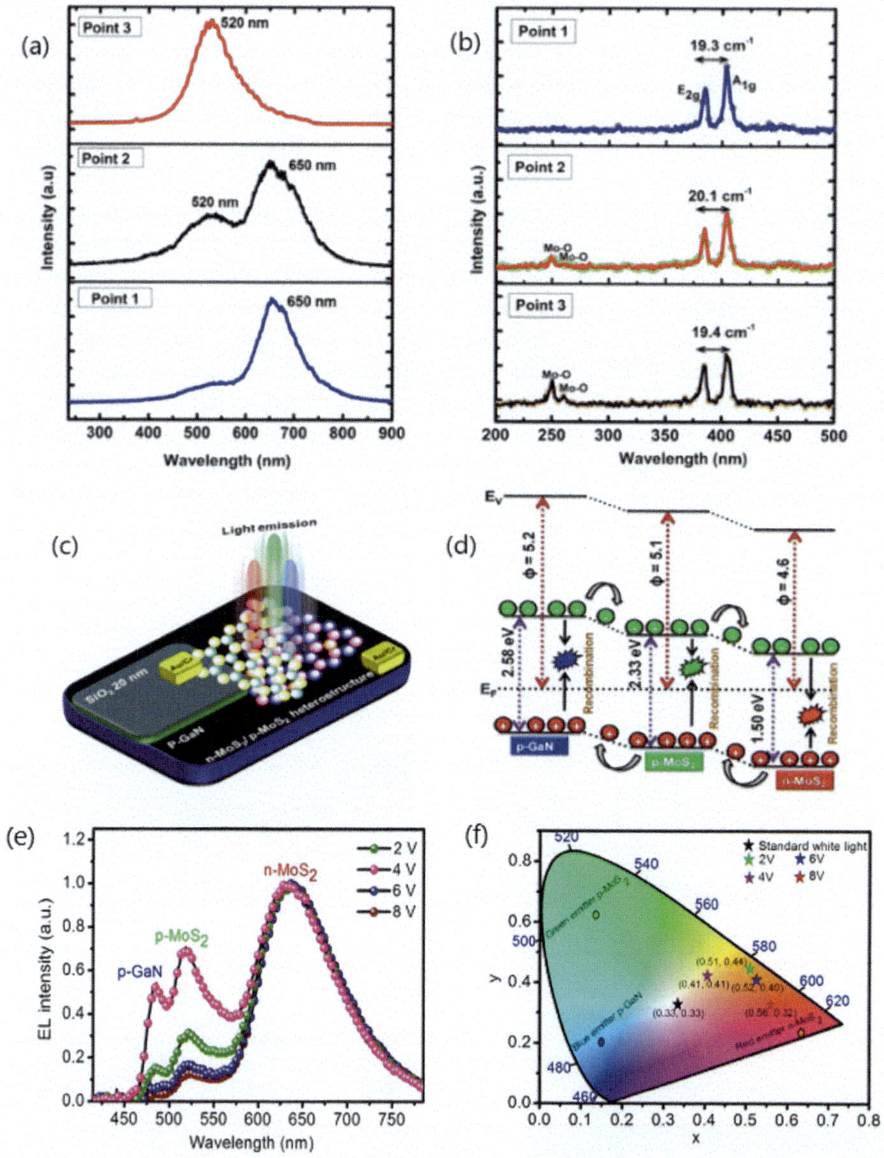

Fig. 6 The optical characterization for p-MoS2/n-MoS2 heterostructure **a** PL spectra at three chosen points (Points 1, 2 and 3), **b** Raman spectrum at three chosen points (Points 1, 2 and 3); **c** WLED device schematic structure fabricated through three emitters in the n-MoS_2/p-MoS_2/p-GaN heterostructure, **d** Band alignment of n-MoS_2/p-MoS_2/p-GaN heterostructure with respective band energies observed through UPS analysis. **e** EL spectra for WLED (four layers n-MoS_2/p-MoS_2/ p-GaN) device, functioning at applied forward bias from 2 to 8 V; **f** The CIE colour coordinates show purity of colour in emitted light, for as-fabricated device at respective bias voltage from 2 to 8 V. Reprinted with permission from Ref. [29], copyright © Elsevier Ltd. All rights reserved

the influence of electrical gating led to an improvement in the performance of the solar cell, resulting in higher illumination as the charge carriers were able to cross the junction and be collected effectively. This in turn provided better J_{sc} and PCE values for the solar cell. For enhanced photocatalytic applications of solar cells, Li et al. [34] designed two schemes comprising of MoS_2/TiO_2 and super-lattice heterostructures. Firstly, a MoS_2/TiO_2 heterostructure-based solar cell with an outer shell made of MoS_2 and TiO_2 layer was formed. This allowed quick transfer of the electrons and holes, thus enabling prompt catalytic recovery in photocatalytic process. In turn benefitting the cyclic utilization of photocatalysts and resulted in disabling the process of photocatalysis with core shell structures. In Fig. 6a, the band alignment of MoS_2/TiO_2 layers along with carrier transfer between the layers is illustrated. It was observed that, as per density functional theory, the value of valence band offset was 2.50 eV and conduction band offset (CBO) was 0.60 eV, between MoS_2/TiO_2 layers. Thus, it could be assumed that the overall bandgap of MoS_2/TiO_2 heterostructure was 1.3 eV. The CBO value of as-formed heterostructure is much larger than that for 1L MoS_2. Thus, leading to efficient exciton disassociation and recombination probability of photogenerated electrons and holes, improving the photoelectric conversion efficiency (Fig. 7).

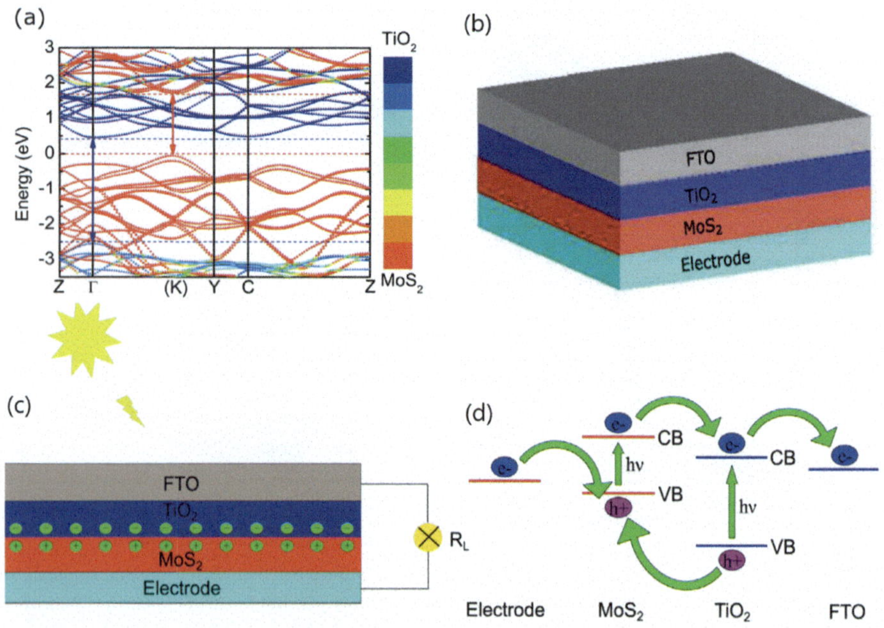

Fig. 7 Schematic illustrations of; **a** band structure observed for 2L MoS_2/TiO_2 heterostructure with respective contributions of MoS_2 (red curves) and TiO_2 (blue curves). The arrows indicate direct bandgap for MoS_2 (blue arrow) and TiO_2 (red arrow). (K) denotes the high symmetry point present in Brillouin zone of hexagonal MoS_2; **b** 3D design of solar cell; **c** 2D MoS_2/TiO_2 heterostructure solar cell; Reprinted with permission from Ref. [34], copyright © Elsevier Ltd. All rights reserved

In another study [35], perovskite solar cells fabricated on Indium tin oxide (ITO)-coated glass substrate, with MoS_2 as hole transport layer (HTL) was fabricated. Sivula et al. proposed the usage of surface ligand modified TMD sheets to increase dispersity, which would enable pin-hole free film fabrication. The device with MoS2|MAPBI3|PCBM p-i-n heterojunction (type-II band alignment) acted as solar cells upon illumination with certain conditions. The thickness of MoS_2 layers had a direct impact on J_{sc} of the cell as it attributed lesser carrier combination, while the FF of cell increased due to a better electron blocking by the TMD layer. Although, over 5 layers of thickness, an increase in device resistance is observed, this leads to subsequent decrease in both Jsc and FF values of the cell. Also, its role as HTL drastically affected the magnitude of V_{oc}. As an effective HTL, MoS_2 increased the electron blocking capability of the solar cell and lead to improved efficiency of the device by two-fold magnitude. The inferred hole-transport ability of MoS_2 causes an increase in shunt resistance of the cell and thereby lesser recombination takes place. Ma et al. [36] prepared high-efficiency graphene/MoS_2/Si Schottky barrier solar cells which showed an efficiency as great as 15.8% for AM 1.5 G radiation, depending upon the thickness of MoS_2. Inclusion of MoS_2 interlayers in as-formed solar cell caused lowering of large photo-generated electron charge transfer into graphene electrode. This causes recombination of holes at the graphene/Si interface and in turn lowers cell efficiency (Fig. 8).

The V_{oc}, J_{sc} and FF values of as prepared solar cell were observed to be 0.50 V, 26.6 mA/cm^2 and 62% respectively, exhibiting higher PV characteristics than graphene/Si solar cell. Also, the inclusion of 1L MoS_2 leads to enhancement in built-in electric field, this happens due to alignment of bands between MoS_2 and n-Si interface. For a MoS_2 interlayer thickness ~20 nm, degradation of PV performance takes place, which can be due to absorption of additional photons, reduction in band gap or poor electrical conductivity in a thick MoS_2 layer. Further TiO_2 coating resulted in better J-V characteristics as illustrated in Fig. 9.

5 MoS_2 Heterojunctions for Optical Memory Use

The formation of hybrid MoS_2-based devices with graphene/MoS_2 as top floating gates was demonstrated by Kim et al. [37] for non-volatile memory device applications. The top floating gates acted as charge trapping layers, in spite of their position as illustrated by EFM image and I_{DS}Vs V_G curves in Fig. 10a, b. These devices could be electrically read and programmed and erased optically.

The dependence of both devices on power of illumination was observed. On application of 532 nm illumination stepwise increase, I_{DS} for MoS_2-graphene device was observed, which for higher illumination showed lesser increment. A comparative study between MoS_2 with graphene as top-gate and MoS_2 with MoS_2 as top-gate devices was carried out in which a V_{DS} of 0.1 V at 532 nm illumination (P = 2 nW) was applied with varying pulse widths at each 30 s intervals. It was observed that both devices showed accelerated increase in I_{DS} with increase in pulse widths.

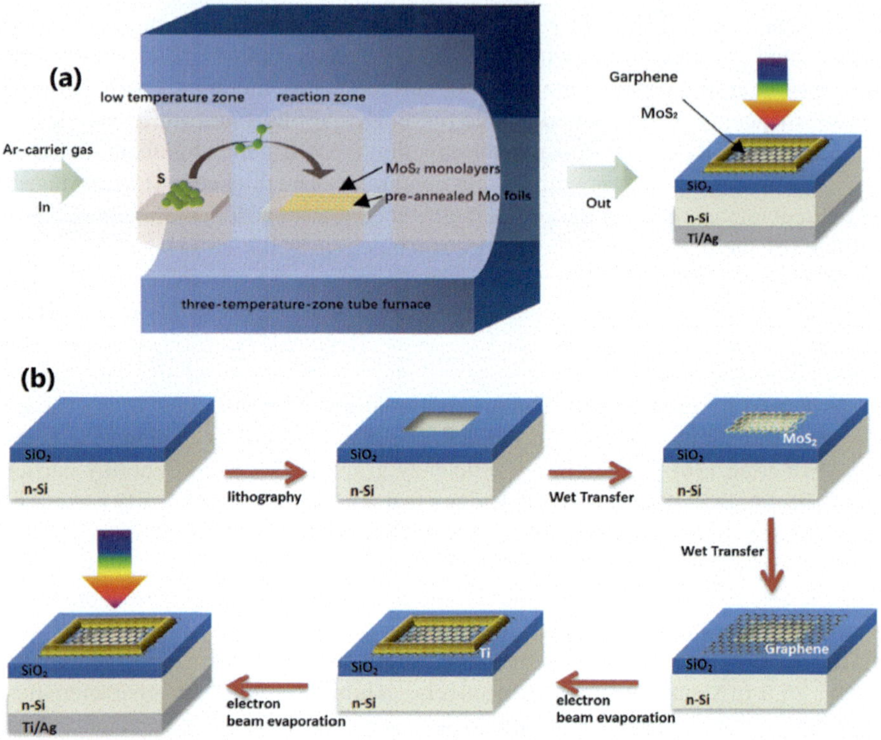

Fig. 8 Schematic illustration of **a** fabrication of MoS_2; **b** deposition process of as-fabricated solar cells; Reprinted with permission from Ref. [36], copyright © Elsevier Ltd. All rights reserved

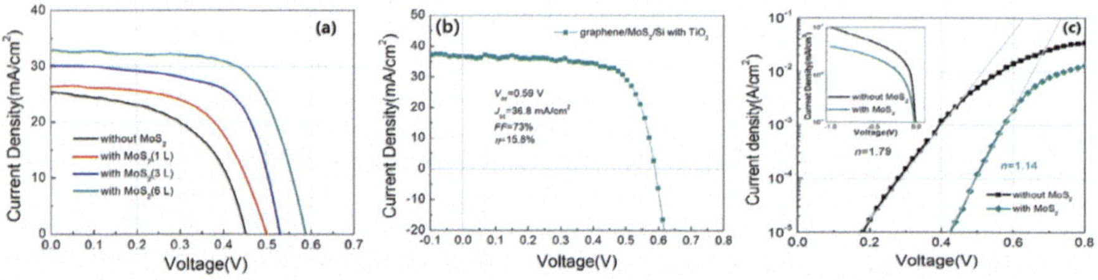

Fig. 9 J-V characteristics of graphene/MoS2/Si solar cell **a**, **b** Current density (J)-voltage (V) characteristics for as-fabricated solar cell for Initial condition and with anti-reflective TiO_2 coating, respectively; **c** dark J-V characteristics for as-fabricated solar cell with and without MoS_2 interlayers. Reprinted with permission from Ref. [36], copyright © Elsevier Ltd. All rights reserved

Although MoS_2 with MoS_2 as top-gate device exhibited more rapid increase in I_{DS} as compared to MoS_2 with graphene as top-gate device. Similar traits were shown by both devices on application of higher illumination intensity, thus indicating that the memory properties of multilevel optical device were dependent upon the characteristics of top floating gate material. Lee et al. [38] demonstrated 1L MoS_2-based

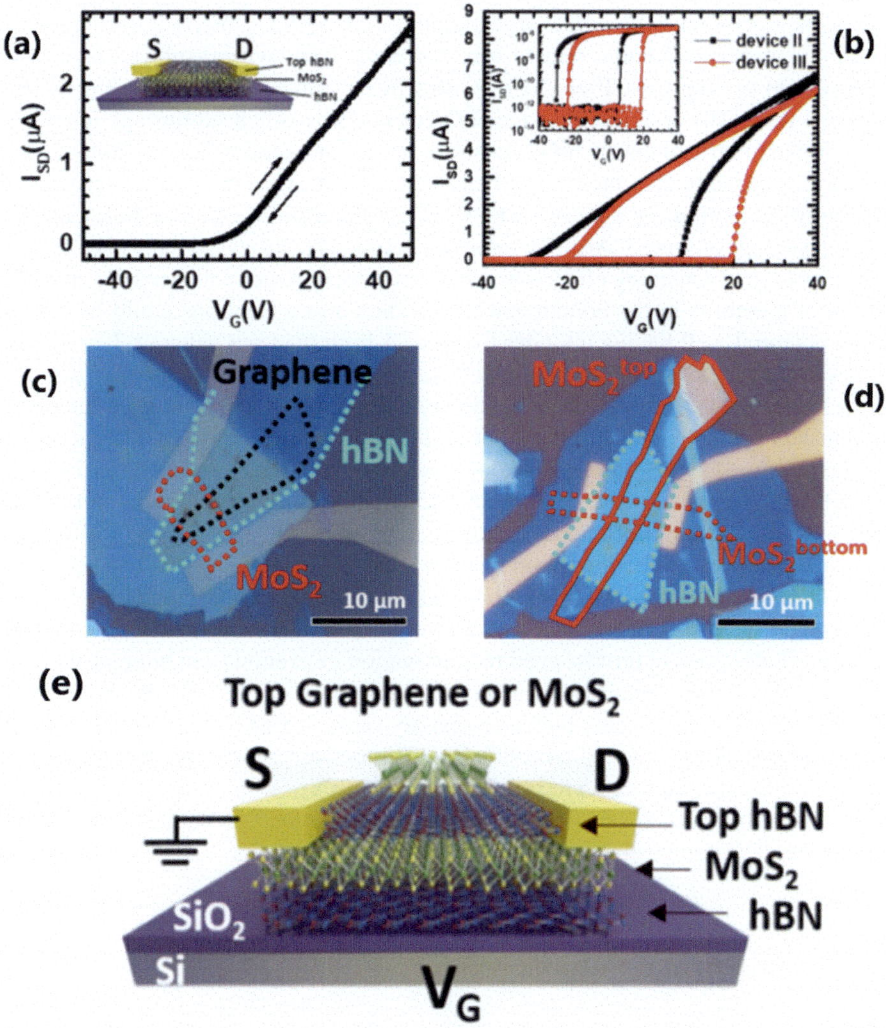

Fig. 10 **a** The characteristic I_{DS}-V_G transfer curve measured for MoS_2 with graphene as top gate (device II) and MoS_2 with MoS_2 as top gate (device III) sweeping from $V_G = -40$ V to +40 V, represented by colour black and red, respectively. Inset image represents the characteristic curves on logarithmic scale; **b** I_{DS}-V_G transfer curve measured for MoS_2-based FET, sweeping from $V_G = -50$ V to +50 V. The inset illustrates the structure of FET; **c** optical images of device II (MoS_2 with graphene as top gate); **d** optical image of device III (MoS_2 with MoS_2 as top gate); **e** schematic illustration of device structures for MoS_2 with graphene as top gate and MoS_2 with MoS_2 as top gate, respectively. Reprinted with permission from Ref. [37], copyright © American Chemical Society

optical memory device through charge trapping at MoS_2/SiO_2 interface, through illumination subsequent charge release was observed. The device displayed high ON/OFF current of 50 pA (at T = 333 K) with a linear optical dose response and long retention time. The as-formed device performance was enhanced through intensive oxygen plasma treatment of SiO_2 substrate before the application on MoS_2 layer, which in turn modulated the energy band structure of 1L MoS_2. On application of 450 nm illumination for a pulse width of 1 s, generation of electron–hole pairs in MoS_2 layer was observed. Also, by removing trapped electrons from the interface an elevation in gate screening effect was observed, which in turn enhanced the concentration of electrons in the conduction band. Thus, more electrons could be stored in the potential well having a longer lifetime. A high read-out value of 7.7 nA was observed for the as-formed device, which was also repeatable for number of devices under the same exposure conditions. The large dark current and prolonged retention time makes the device an attractive option for other optoelectronic applications. In another study, 1L-MoS_2 flakes were utilized as both channel and light absorption layers with Au nanoparticles employed as charge trapping layer. The as-fabricated device exhibited high performance memory operations. On application of illumination, an excitation of electrons from valence band (VB) to conduction band (CB) in MoS_2 layer was observed. Before recombination of excitons, a negative gate voltage was applied so as to induce charge transfer from Au nanoparticles to MoS_2 VB. The transferred electrons in turn stopped recombination of excitons, enabling prolonged storage of photonic signals. At an illumination power of 0.1 μW, at high V_G of 150 V ($V_D = 0$ V), extremely high responsivity value of 8000 A/W was observed as compared to 0.58 A/W at pristine conditions. This performance could be attributed to the origin of to the unprecedented application of high V_D, which was possible due to large area high-quality MoS_2 layers. Thus, by varying values of V_G and illumination power, the data storage ability of device could be configured.

6 Conclusion and Future Scope

MoS_2 has a wide spectrum of applications in electronic as well as optoelectronics. Its operating spectrum range from DUV to LWIR range makes it an excellent alternative for creating a wide range of products for industrial purposes in optical communications as chemical and biological sensors, lasers, optical memory cells, solar cells, LEDs and other generation optoelectronic-based devices. By possessing the ability to tailor photoresponsivity factor in 2D TMDs, especially atomic-layered MoS_2 with its variant heterostructures, development of high photo response optoelectronic devices can be achieved. The performance of 2D MoS_2 photodetectors can be improved by forming MoS_2 based heterostructures combined with halide-perovskites, 0D–2D QDs, other 2D materials, ferroelectric polymers. The semiconducting state of MoS_2 can also be altered via chemical doping. The doping of MoS_2 with Nb and Mn modulates it into p-type semiconductor. This makes it a versatile material for use in phototransistors, photodiodes and other photo-conducting applications. Depending upon the nature of doping agents, vdWHs, intensity of applied incident power and optical wavelengths, the MoS_2-based hybrid heterostructures display variation in photoresponsivity, detectivity and photo-responsive properties, thus offering a broader scope for future research and applications. The unique band structure and varying thickness led to efficient separation of electron–hole pairs and suppression of recombination of excitons as observed in graphene/MoS_2/Si Schottky barrier solar cells. The same properties enabled fabrication of MoS_2-based solar cell with other materials as well. MoS_2-based optical memory devices exhibited excellent memory, retentivity, effective ON/OFF current and erasing ratios, multi-level data storage capability and cyclic endurance. This highlights their capability for fabricating advanced optoelectronic devices. Many studies have been performed for creating efficient heterojunctions and heterostructures using MoS_2, for finer results further investigations and validations are required. The future of MoS_2 heterostructures seems brighter as compared to its other counterparts due to its earthly abundance, great production cost and discussed attributes for optoelectronic and photovoltaic applications.

Acknowledgements This work was financially supported by UGC DAE CSR under CRS project (CRS/2021-22/03/546).

References

1. Mak KF, Lee C, Hone J, Shan J, Heinz TF (2010) Atomically thin MoS_2: a new direct-gap semiconductor. Phys Rev Lett 105:136805
2. Wilson JA, Yoffe AD (1969) The transition metal dichalcogenides discussion and interpretation of the observed optical, electrical and structural properties. Adv Phys 18:193–335
3. Eda G, Yamaguchi H, Voiry D, Fujita T, Chen M, Chhowalla M, Eda M, Yamaguchi H, Voiry D, Fujita T, Chhowalla Chen M (2011) Photoluminescence from chemically exfoliated MoS_2. Nano Lett 11:5111–5116
4. Wypych F, Weber T, Prins R (1998) Scanning tunnelling microscopic investigation of 1T-MoS_2. Chem Mater 10:723–727
5. El-Mahalawy SH, Evans BL (1976) The thermal expansion of 2H- MoS_2, 2H-$MoSe_2$ and 2H-WSe_2 between 20 and 800 Deg C. J Appl Crystallogr 9:403–406
6. Beal AR, Hughes HP (1979) Kramers-Kronig analysis of the reflectivity spectra of 2H- MoS_2, 2H-$MoSe_2$ and 2H-$MoTe_2$. J Phys C Solid State Phys 12:881
7. Bernardi M, Palummo M, Grossman JC (2013) Extraordinary sunlight absorption and one nanometer thick photovoltaics using two-dimensional monolayer materials. Nano Lett 13:3664–3670
8. Wang J, Fang H, Wang X, Chen X, Lu W, Hu W (2017) Recent progress on localized field enhanced two-dimensional material photodetectors from ultraviolet—visible to infrared. Small 13:1700894
9. Xiao P, Mao J, Ding K, Luo W, Hu W, Zhang X, Zhang X, Jie J (2018) Solution-processed 3D RGO-MoS_2/pyramid Si heterojunction for ultrahigh detectivity and ultra-broadband photodetection. Adv Mater 30:1801729
10. Peng Z-Y, Xu J-L, Zhang J-Y, Gao X, Wang S-D (2018) Solution-processed high-performance hybrid photodetectors enhanced by perovskite/MoS_2 bulk heterojunction. Adv Mater Interfaces 5:1800505

11. Bang S, Duong NT, Lee J, Cho YH, Oh HM, Kim H, Yun SJ, Park C, Kwon M-K, Kim J-Y, Kim J, Jeong MS (2018) Augmented quantum yield of a 2D monolayer photodetector by surface plasmon coupling. Nano Lett 18:2316–2323
12. Sun B, Wang Z, Liu Z, Tan X, Liu X, Shi T, Zhou J, Liao G (2019) Tailoring of silver nanocubes with optimized localized surface plasmon in a gap mode for a flexible MoS2 photodetector. Adv Funct Mater 1900541
13. Heo K, Jo S-H, Shim J, Kang D-H, Kim J-H, Park J-H (2018) Stable and reversible triphenylphosphine-based n-type doping technique for molybdenum disulphide (MoS_2). ACS Appl Mater Interfaces 10:32765–32772
14. Kim Y, Bark H, Kang B, Lee C (2019) Wafer-scale substitutional doping of monolayer MoS_2 films for high-performance optoelectronic devices. ACS Appl Mater Interfaces 11:12613–12621
15. Song X, Liu X, Yu D, Huo C, Ji J, Li X, Zhang S, Zou Y, Zhu G, Wang Y, Wu M, Xie A, Zeng H (2018) Boosting two-dimensional $MoS_2/CsPbBr_3$ photodetectors via enhanced light absorbance and interfacial carrier separation. ACS Appl Mater Interfaces 10:2801–2809
16. Wu H, Si H, Zhang Z, Kang Z, Wu P, Zhou L, Zhang S, Zhang Z, Liao Q, Zhang Y (2018) All-inorganic perovskite quantum dot-monolayer MoS_2 mixed-dimensional van der Waals heterostructure for ultrasensitive photodetector. Adv Sci 5:1801219
17. Zhang S, Wang X, Chen Y, Wu G, Tang Y, Zhu L, Wang H, Jiang W, Sun L, Lin T, Shen H, Hu W, Ge J, Wang J, Meng X, Chu J (2019) Ultrasensitive hybrid MoS_2-ZnCdSe quantum dots photodetectors with high-gain. ACS Appl Mater Interfaces 11:23667–23672
18. Liu H, Li D, Ma C, Zhang X, Sun X, Zhu C, Zheng B, Zou Z, Luo Z, Zhu X, Wang X, Pan A (2019) Van der Waals epitaxial growth of vertically stacked Sb_2Te_3/MoS_2 p–n heterojunctions for high performance optoelectronics. Nano Energy 59:66–74
19. Yang S, Wu M, Wang B, Zhao L-D, Huang L, Jiang C, Wei S-H (2017) Enhanced electrical and optoelectronic characteristics of few-layer type-II $SnSe/MoS_2$ van der Waals heterojunctions. ACS Appl Mater Interfaces 9:42149–42155
20. Singh E, Nalwa HS (eds) (2019) Nanomaterial-based flexible and multifunctional sensors. American Scientific Publishers, Los Angeles, California
21. Singh E, Meyyappan M, Nalwa HS (2017) Flexible graphene-based wearable gas and chemical sensors. ACS Appl Mater Interfaces 9:34544–34586
22. Singh E, Meyyappan M, Nalwa HS (2019) Graphene-based flexible and stretchable strain sensors for wearable electronics (Chap. 1). In: Singh E, Nalwa HS (eds) Nanomaterial-based flexible and multifunctional sensors. American Scientific Publishers, Los Angeles, California, pp 1–50
23. Nalwa HS (ed) (1995) Ferroelectric polymers: chemistry: physics, and applications. CRC Press, Boca Raton, FL
24. Zhang K, Peng M, Yu A, Fan Y, Zhai J, Wang ZL (2019) A substrate-enhanced MoS_2 photodetector through a dual photogating effect. Mater Horiz 6:826–833
25. Ross JS, Klement P, Jones AM, Ghimire NJ, Yan J, Mandrus DG, Taniguchi T, Watanabe K, Kitamura K, Yao W, Cobden DH, Xu X (2014) Electrically tunable excitonic light-emitting diodes based on monolayer WSe_2 p–n junctions. Nat Nanotechnol 9:268–272
26. Yang P, Yang H, Wu Z, Liao F, Guo X, Deng J, Xu Q, Wang H, Sun J, Chen F, Bao W, Hu L, Liu Z, Chen Y, Qiu ZJ, Fang Z, Liu R, Cong C (2021) Large-area monolayer MoS_2 nanosheets on GaN substrates for light-emitting diodes and valley-spin electronic devices. ACS Appl Nano Mater 4(11):12127–12136
27. Yin Z, Zhang X, Cai Y, Chen J, Wong JI, Tay YY, Chai J, Wu J, Zeng Z, Zheng B, Yang HY, Zhang H (2014) Preparation of MoS_2-MoO_3 hybrid nanomaterials for light-emitting diodes. Angewandte Chemie (International ed. in English), 53(46):12560–12565
28. Choi M, Park YJ, Sharma BK, Bae S-R, Kim SY, Ahn J-H (2018) Flexible active-matrix organic light-emitting diode display enabled by MoS_2 thin-film transistor. Sci Adv 4(4):eaas8721
29. Nikam RD, Sonawane PA, Sankar R, Chen Y-T (2017) Epitaxial growth of vertically stacked p-MoS_2/n-MoS_2 heterostructures by chemical vapor deposition for light emitting devices. Nano Energy 32:454–462

30. Liu Y, Weiss NO, Duan X, Cheng H-C, Huang Y, Duan X (2016) Nat Rev Mater 16042
31. Choudhary N, Park J, Hwang JY, Chung H-S, Dumas KH, Khondaker SI, Choi W, Jung Y (2016) Sci Rep 6:25456
32. Withers F, Del Pozo-Zamudio O, Mishchenko A, Rooney AP, Gholinia A, Watanabe K, Taniguchi T, Haigh SJ, Geim AK, Tartakovskii AI, Novoselov KS (2015) Nat Mater 14:301–306
33. Lin S, Wang P, Li X, Wu Z, Xu Z, Zhang S, Xu W (2015) Gate tunable monolayer MoS_2/InP heterostructure solar cells. Appl Phys Lett 107(15):153904
34. Li Y, Cai C, Gu Y, Cheng W, Xiong W, Zhao C (2017) Novel electronic properties of a new MoS2/TiO2 heterostructure and potential applications in solar cells and photocatalysis. Appl Surf Sci 414:34–40
35. Dasgupta U, Chatterjee S, Pal AJ (2017) Thin-film formation of 2D MoS_2 and its application as a hole-transport layer in planar perovskite solar cells. Sol Energy Mater Sol Cells 172:353–360
36. Ma J, Bai H, Zhao W, Yuan Y, Zhang K (2018) High efficiency graphene/MoS_2/Si Schottky barrier solar cells using layer-controlled MoS_2 films. Sol Energy 160:76–84
37. Kim S, Yi SG, Park M, Lee C, Kim M, Yoo KH (2019) Multilevel MoS_2 optical memory with photoresponsive top floating gates. ACS Appl Mater Interfaces 11(28):25306–25312
38. Lee J, Pak S, Lee YW et al (2017) Monolayer optical memory cells based on artificial trap-mediated charge storage and release. Nat Commun 8:14734

Chapter 11
MoS_2 Based Field Emitters for Vacuum Nano Electronics Device Applications

Prashant K. Bankar, Mahendra A. More, and Dattatray J. Late

1 Introduction

The different ways electron emission takes place from the material surface are photoemission, optical emission, thermionic emission, electric field-assisted emission, field emission, etc. Field electron emission emits electrons from a metal or semiconductor by applying a strong electric field. The phenomenon was initially documented by R. W. Wood in 1897. Like other fundamental electron emission processes from solids, such as thermionic, photoelectric and secondary electrons, later field emission research has made a substantial contribution to basic science and technical advancements. A paper by Jenkins [1] provides a great overview of earlier work. The experimental data showing that field emission was temperature-independent in the zone much below the thermionic emission onset were not explained by the classical theories. The fundamental procedure was then explained by Fowler and Nordheim in 1928 [2] utilizing the wave mechanical formulation. The practical difficulty in producing the very high fields required for field emission (10^6–10^7 V/cm) severely limits obtaining reliable experimental results on this effect. E. W. Müller placed a cathode in the form of a sharply etched tungsten wire in front of a screen in a vacuum to see how the electrons were ejected from the tip. Field emission from a well-cleaned, rounded tip was obtained after heating the tip at high temperatures. Reproducible current–voltage (I-V) characteristics could be measured with this two-dimensional projection microscope, the now-called field emission microscope (FEM), by projecting the tip geometry of the surface onto the screen. Müller obtained

P. K. Bankar · M. A. More
Department of Physics, Savitribai Phule Pune University, Pune 411007, India

D. J. Late (✉)
Department of Physics, Federal University of Lavras, Campus Universitário, PO Box 3037, Lavras, Minas Gerais, Brazil
e-mail: datta099@gmail.com

D. J. Late and C. S. Rout (eds.), *Science and Technology of 2D MoS_2*,
Materials Horizons: From Nature to Nanomaterials,
https://doi.org/10.1007/978-981-97-7367-1_11

beautiful patterns of molybdenum (Mo), nickel (Ni), tungsten (W) and copper (Cu). These patterns demonstrated the several planes of the single crystal emitters' emission anisotropy. Müller measurement of the field emission, which occurred at a field strength of 10^7 V/cm, was in complete agreement with the Fowler–Nordheim (FN) theory.

Dyke and co-workers contributed substantially to developing field emission cathodes by employing continuous and pulsed emission modes [3]. The work showed the potential of the field emitters in obtaining high-current–density cathodes. There have been many theoretical contributions, together with the development of modern techniques, for measurements of energy distribution, work function, noise characteristics, and current stability [4].

2 Field Electron Emission Process

The Fowler and Nordheim proposed the idea of field emission on the base of Fermi–Dirac statistics and Sommerfeld's free electron model in 1928. Field emission is a quantum mechanical tunnelling, as this theory illustrates. The electron emission is independent of temperature when an electric field or bias is introduced. The F-N model, which is predicated on the application of a high electric field, is the standard theoretical approach to electron emission from metal into a vacuum.

1. The emitter's temperature is 0 °C.
2. The band structure of the emitter's electrons is unbound.
3. To provide a uniform application of the electric field, the emitter surface is assumed to be smooth and flat.
4. An image force potential and a potential resulting from the applied electric field make up the potential barrier at the surface in the vacuum area.
5. There is no effect of an external electric field on the local work function, which is constant throughout the emitting surface.

Figure 1 graphically depicts the potential energy E of an electron under the influence of an externally supplied electric field as a function of its distance x from the metal surface.

3 Fowler–Nordheim (F–N) Equation

The potential energy of the electrons inside the metal, Vp, is constant according to the Summerfield free electron model. Consequently,

$$V(x) = V_p \quad X\langle 0 \tag{1}$$

Outside near to the surface of the metal is,

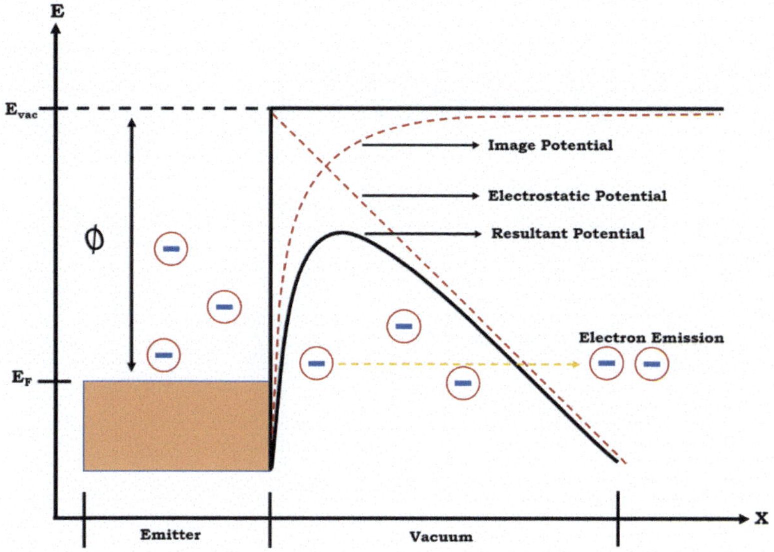

Fig. 1 Schematic representation of the field electron emission

$$\mathrm{V(x)} = -\frac{e^2}{4X} - \mathrm{e\ Fx\ X}\rangle 0 \tag{2}$$

where F is the intensity of the electric field and e is the charge of the electron. Equation (2) on the right-hand side has two terms: the contribution from the external electric field and the potential due to the image force on the electron as a result of the induced charge in the metal. According to the F-N electron emission model is based on tunnelling electrons through the one-dimensional barrier. If the E is the total energy and W is the x part of this energy, then

$$W = \frac{P_X^2}{2m} + V(x) \tag{3}$$

m is the mass of the electron, and PX is its momentum in the x direction. It is considered that the electron's chance of penetrating the potential barrier is solely determined by its average kinetic energy reaching the surface. The time-independent Schrödinger equation can be solved to determine the probability of the barrier penetration D (W). Utilizing the WKB approximation that is applicable for W Vmax(x), where Vmax(x) represents the greatest potential at the Schottky hump, and taking into account the fact that the electrons that penetrate the barrier begin their journey close to the Fermi level, the following results are obtained:

$$\phi D\ (W) = \exp\left[-c + \frac{W - \phi}{d}\right] \tag{4}$$

where

$$c = \frac{4\,(2\,m\,\phi^3)^{1/2}}{3\,\hbar\,e\,F} \cdot f\,(y) \tag{5}$$

$$d = \frac{\hbar\,e\,F}{2\,(2\,m\,\phi)^{1/2}\,t(y)} \tag{6}$$

And

$$t\,(y) = f\,(y) - \frac{2}{3}\,y\,\frac{df\,(y)}{dy} \tag{7}$$

where

$$y = \frac{\left(e^3\,F\right)^{1/2}}{\phi} \tag{8}$$

The functions of y are denoted by f (y) and t (y) [5]. Now, by multiplying D (W) by the supply function N(W) dW, which is the number of electrons with the X component of energy in the range W to W + dW incident standard to the surface per unit area per unit time, P (W) dW, the number of electrons in the range of X component of energy between W and W + dW tunnelling through the barrier, can be computed. Consequently, the usual energy distribution is provided by

$$\mathrm{P\,(W) = N\,(W)\,D\,(W)dW} \tag{9}$$

By multiplying P(W) dW by e, the electronic charge, and integrating across all values of W, one can find the current density J.

$$J = e\int_{V_P}^{\infty} P\,(W)\,dW \tag{10}$$

The Fermi–Dirac energy distribution obtains the supply function N(W) dW. While electrons escape from the metal surface the equilibrium distribution of electrons inside the metal is assumed,

$$N\,(W)\,d\,W = \frac{4\,\pi\,m\,k_B\,T}{h^3}\,\ln\left[1 + \exp\left(-\frac{W - \phi}{K_B\,T}\right)\right]dW \tag{11}$$

T is the absolute temperature, K_B is the Boltzmann constant and h is the Plank's constant. At low temperatures,

$$\left|\left(\frac{W - \phi}{k_B\,T}\right)\right| \rangle\rangle\ 1 \text{ and if } W \le \phi$$

$$k_B T \ \ln \left[1 + \exp \left(- \frac{W - \phi}{k_B T} \right) \right] = \phi - W \tag{12}$$

Combining (9), (11) and (12), and substituting in (10), integrating overall W, yields

$$J = \frac{e^3 F^2}{8 \pi \hbar \phi t^2(y)} \exp \left[- \frac{4 (2m)^{1/2}}{3 \hbar e} \frac{\phi^{3/2}}{F} f(y) \right] \tag{13}$$

The Fowler–Nordheim equation is obtained by putting J in A/cm^2, F in V/cm, ϕ in eV, and changing the numerical values of other constants. The resultant expression is

$$J = 1.54 \times 10^{-6} \frac{F^2}{\phi t^2(y)} \exp \left[-6.8 \times 10^7 \frac{\phi^{3/2}}{F} f(y) \right] \tag{14}$$

As discussed, this equation has been arrived at by following the normal energy distribution and assuming the free electron model inside the metal. The normal energy distribution is experimentally challenging to observe because of the stringent geometrical conditions of the emitter and the collector. However, the total current–voltage (I-V) measurements generally agree with the F-N equation.

4 Fowler–Nordheim (F-N) Plot

The electric voltage (V) applied between the cathode and anode and the field emission current (I) are the measurable quantities in a field emission experiment. These values are associated to J and E as follows:

$$\mathrm{I} = \mathrm{JA} \tag{15}$$

$$\mathrm{F} = \beta \mathrm{V} \tag{16}$$

where A is the emitter area and, β is the field enhancement factor. With the aforementioned replacements applied, the F-N equation becomes the following form.

$$J = \frac{1.54 \times 10^{-6} \ \beta^2 \ V^2 A}{\phi \ t^2(y)} \exp \left[\frac{-6.8 \times 10^7 \ \phi^{3/2} f(y)}{\beta V} \right] \tag{17}$$

A plot of ln (I/V^2) Vs (10^4/V) the Fowler–Nordheim plot, a straight line for metallic emitters, is the consequence of this equation. This straight line's slope is determined by

$$m = \frac{-6.8 \times 10^3 \,\phi^{3/2}\, s(y)}{\beta} \tag{18}$$

where

$$s(y) = f(y) - \frac{y}{2}\left(\frac{df(y)}{dy}\right) = 1 \text{ (unity)}$$

Straight line F-N plots emerge from the assumption that and A is independent of V and that the function s(y) has a value near to unity, causing m to be largely constant. A measure of the shift in the work function is the variation in the slope of the F-N plots following atom adsorption. It is anticipated that in this instance, the adsorption of atoms at low coverage will not change the β. To fully comprehend the work function modifications, the pre-exponent term will represent any changes in the emitting area A and the factor.

From the assumption of the F-N model, it is clear that the theory was developed for only metals, and it must be told that the field emission theory for semiconductors cannot be straightforward because band bending distorts the field penetration into the semiconductor [3].

5 Calculation of Effective Emission Area

The F-N plot's exceptional linearity indicates that the metallic emitter's field emission is consistent with the metal-vacuum field emission model given by the F-N equation. In addition, by knowing the work function of the emitter, the slope m of the F-N plot, can be used to estimate the lateral dimensions of the emitter with field enhancement factor β $\frac{1}{\alpha r}$, where α ~ 5 for the hemispherical geometry of the emitter. The effective emission area *A* is derived after substituting β and ϕ into the *y*-intercept of the F-N plot as

$$b = \ln\left(\frac{1.5 \times 10^{-6} A \beta^2}{\phi}\right) + \frac{10.4}{\phi^{1/2}} \tag{19}$$

The small emitting area is a desired attribute for high-resolution electron optical applications where point electron sources are required, even if the reason why only a small patch of the comparatively large area of the complete covered tip is triggered to emit electrons is yet unknown. By dividing the total emission current by the effective emission area, one can estimate the emission current density.

6 Field Emission Microscope (FEM)

A field emission microscope (FEM) allows the observation of a single crystal of material with a large magnification (10^5–10^6) and a high resolution (20–30 Å). It has the advantage of observing various crystallographic planes simultaneously in clean and UHV conditions. Figure 2 shows an optical photograph of a typical all-glass field emission microscope tube.

A field emission microscope is a projection microscope. It consists of a sharp needle of a refractory metal like tungsten or molybdenum of about 1–2 mm in length, spot welded onto a heating loop. The free end of the wire assumes a hemispherical shape of about 500–2000 Å in a radius after flashing at high temperatures in a vacuum. There is almost invariably a single crystal at this end of the needle, with a preferred orientation. At low applied voltages of a few kilovolts, emission occurs. The released electrons are directed onto the conducting screen coated with phosphor to create a pattern that may be seen that reflects the structural characteristics of the emitting region. The field at the surface of such an emitter is given by an expression.

$$F = \frac{V}{kr} \tag{20}$$

where r is the tip radius, and k is a geometrical factor arising because of the presence of a conical shank of the actual tip. In many cases, it has a value of about 5 near the tip apex. The magnification is given by

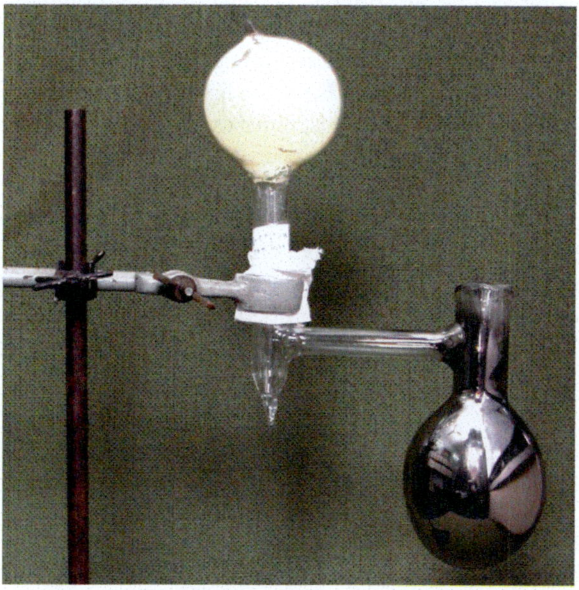

Fig. 2 Optical photograph of conventional field emission microscope tube (Reproduced with permission from [54])

$$M = \frac{X}{\gamma . r} \tag{21}$$

where X is the tip-to-screen distance (about 5–7 cm) and γ is a factor of value about 1.5 reducing the magnification. This results in an essentially evenly compressed image with axial symmetry over the exposed area of the emitter since the emitter shank reduces the field and compresses the lines of force towards the axis. Magnification of the order of 10^5–10^6 is usually obtainable.

The smallest distance on the tip surface between two objects that will result in distinct images on the screen is known as the resolution of a field emission microscope. The three factors governing the resolution are.

(1) The divergence of the electron beam.
(2) Statistical distribution of the transverse velocities.
(3) Diffraction of an electron wave passing through the emitting aperture.

Resolution is independent of tip-to-screen distance and varies with the square root of the tip radius. Ordinarily, the resolution of a field emission microscope ranges from 20 to 30 Å. The field emission microscope is an ideal tool for surface analysis; the crystal symmetry is reflected in the field emission pattern. The average work function determination is also possible. The characterization of the surface structure and electronic characteristics has made considerable use of field emission. It is possible to obtain quantitative information about the surface phenomena on the emitter surface through the field emission current fluctuation analysis.

An F-N plot of ln (I/V^2) versus 1/V yields a straight line due to the linear dependency of I on the voltage V, which is a useful tool for verifying the field emission mechanism. The work function is given by the slope, and the emission area is given by the pre-exponential, assuming that F can be calibrated with accuracy. A very strong vacuum is necessary for field emission; frequently, even at UHV, the clean surface is not the cause of emission. Typically, a field emitter needs to be "flashed" that is, cleaned by running current through the loop it is installed to. After flashing, the emission current is high but extremely erratic; with time, the current decreases and becomes more stable. This is due to contamination of the tip, either from the vacuum or more often from diffusion of adsorbed surface species to the tip. Further, the energy analysis of field-emitted electrons can be a useful probe in the band structure and other electronic properties studies of solid surfaces. More details regarding TED and FEM are elaborated in an excellent article by Gadzuk and Plummer [6].

7 Importance of Field Enhancement Factor in Nanometric Field Emitter:

The field enhancement factor determines the size of the emitter. The nanometric dimensions of materials are excellent field emitters due to their aspect ratio and size. An individual nanowire's field enhancement is roughly proportional to the aspect ratio, where l and r denote the nanowire's length and radius, respectively. The high current densities and low threshold field result from the high field enhancement factor.

The enhancement factor can also be estimated from the Filip model [7] for the planar geometry of the emitter using

$$F_{local} = \beta F = \frac{\beta V}{d} \tag{22}$$

The local field at the emitter tip is denoted by F local, the interelectrode spacing is represented by d, the applied voltage is represented by V, and the field enhancement factor, or, measures the emitter's capacity to magnify the electric field (F). The F-N law is stated in this framework as [8]

$$J = \eta a\left(\frac{\beta^2 F^2}{\phi}\right)\exp\left(\frac{-b\phi^{3/2}}{\beta F}\right) \tag{23}$$

where a = 1.54310^{-6} A. V^{-2} eV, b = 6.83 × 10^9 Vm^{-1} eV$^{-3/2}$, and ϕ is the work function. The geometrical efficiency of electron field emission is represented by the factor, which is the ratio of the real surface area that is emitted to the total surface area. It is simple to trace the F-N plot through ln (J/F2) Vs 1/F to find β because the slope of the plot is dependent on and β. The field enhancement factor β can be thus determined from the slope and the φ values.

8 Applications of Field Emission Microscope

8.1 Field Emission Cathode as a Current Source

Due to its two major characteristics—high electron optical brightness and a tiny virtual source size the field emission cathode might be a suitable point electron source for applications as compared to thermionic cathodes [5].

(1) The electron optical brightness "B" is given by

$$B\ \alpha\ \frac{J}{E_d}$$

Ed is the energy dispersion of the released electrons, and J is the current density. The field emission cathode has a comparable Ed (0.2–0.3 eV) to a thermionic cathode, but a J that can be up to 10^7 times higher.

(2) The optical brightness of the electron source is proportional to the maximal beam current "I," which can be concentrated into an incredibly small point.

For very small focal spots

$$I \alpha B d^{8/3},$$

where B is electron optical brightness and d is the diameter of the focal spot.

Low turn-on field, low beam energy spread, small virtual size ($<10^{-6}$ cm) and high emission current density are some of the characteristics of point electron sources, steady release of electrons, functioning under a moderate background pressure, increased lifespan of the emitter strong mechanical resistance to a strong electric field.

8.2 Field Emission Displays

A field emission display (FED) is a low-power flat cathode ray tube-type display that uses a matrix-addressed cold cathode to emit light from a screen coated with phosphor materials. When an electric field is applied to a field emission device, a field emission display normally has an electron emission tip set up to emit a stream of electrons. A visual display on a display panel can be formed by positioning a variety of miniature field emission devices on a plate. The emission tip is designed in a way that makes it easier for electrons to emit effectively. Its surface profile could be conical, pyramidal, or ridge-shaped, for instance. Alternatively, the tip could have a low work function material's flat emitter surface. The emitters of the first FEDs were made of metal tips known as Spindt-type (or microtips). Molybdenum serves as the primary component of the metal tips. On the back plate of a field emission display (FED), several hundred to thousands of carbon nanotubes (CNTs) or micro tips are used per pixel as an electron emission source. Figure 3a shows the typical small source size –fine beam small energy spread ~0.2 eV, and (b) shows the higher brightness and resolution. Figure 3c, d shows the typical CNT-Field Emitter Display marked by Samsung.

8.3 Field Emitter Arrays

In the early 1970s, field-emitter arrays (FEAs) were actively researched to produce vacuum microelectronic devices, such as flat panel displays. Historically, field emission technology has relied heavily on micro-structured tips made of materials such as molybdenum. Electron field emitters have several advantages over thermal and Schottky emitters, including great brightness, little energy spread, strong coherence, and the potential for miniaturization. The latter allows for reducing the operating voltage significantly and creating arrays of gated field emitters by micro and nanostructure. The most essential applications so far are flat panel displays and electron sources with high current densities for high-power microwave tubes. In this area, the

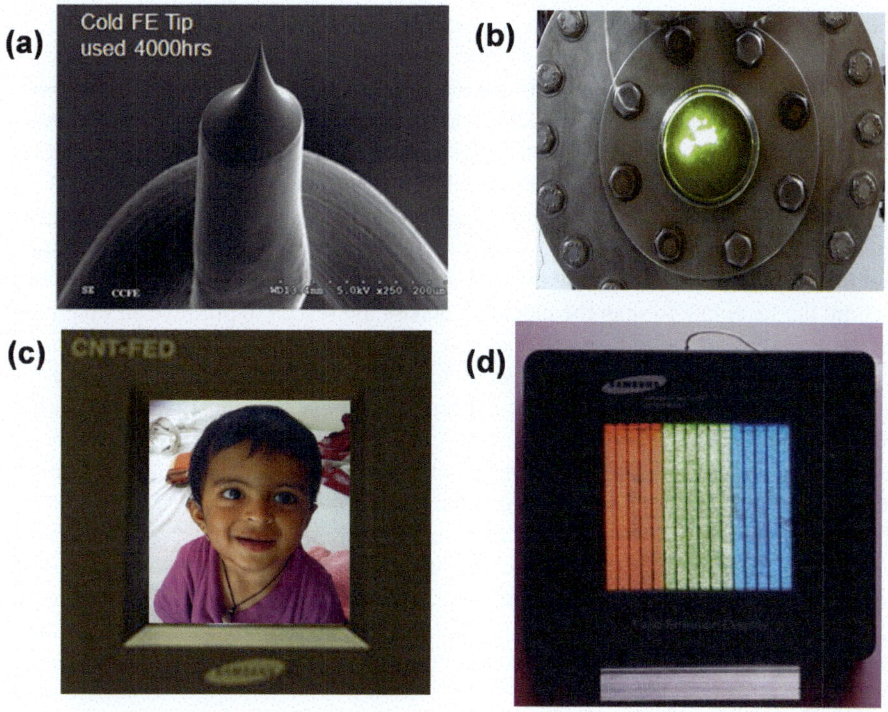

Fig. 3 **a** Small source size fine beam small energy spread: 0.2~0.3 eV, **b** field emission pattern with high brightness, **c** typical CNT-FED and **d** 30 inch CNT-based field emission display demonstrated by Samsung in 1999 [1]

emission characteristic of an individual emitter is not a critical issue since redundancy of 100 or even 1000 emitters per functional unit is high enough to compensate for instability and failure.

On the other hand, an electron source providing several independent beams may be applied to fast direct-write electron-beam lithography, electron-beam test systems or new types of memory devices. For this purpose, every single emitter has to meet much more severe demands on emission stability, lifetime, and uniformity. Thus, examining many emitters in parallel but independent from each other would be very helpful for the research and development of miniaturized field electron emitters. Moreover, an emitter tip with a relatively large and smooth surface could behave more stable due to a large emission area. Therefore, tips that are blunt compared to sharp silicon emitters or carbon nanotubes may be good candidates for individual stable field emitters. Therefore, this addressable field emitter array (FEA) is useful for experimentally designing field emitters with improved emission characteristics. The field emission tip is fabricated by electron beam-induced deposition and subsequent physical vapour deposition of a thin metal layer. Thus, it is possible to vary the geometrical setup of the tip and extraction electrode as well as the tip shape easily and also to define the exact material properties of the emitter, i.e. work function and sticking coefficient.

The market for relatively large home high-definition televisions and space-saving desktop monitors has grown rapidly, necessitating the development of flat panel displays (FPD) with image characteristics comparable to cathode ray tubes (CRTs). Regarding the physical design, the Field Emission Display (FED) and the CRT are the most similar types of FPDs FEDs use millions of micro-field emission electron guns directly behind the display screen to fluoresce the display, as opposed to three hot electron guns (as in the case of a colour CRT) that are positioned some distance behind the faceplate to scan the electrons across the screen to produce an image. As a result, FEDs can be reduced in size to a few millimetres [9]. Although a limited number of monochromatic FED devices have been on the market for the past few decades, there are still certain technological challenges that must be solved before the FED techniques (Fig. 3).

The anode front phosphor screen, which forms the image, the anode rear plane and a thin vacuum gap in the middle make up a typical FED construction [10, 11]. The most intricate structure among them is the cathode backplane, which is home to the micro-field emission cathodes. The Spindt type with molybdenum micro-cold field emitters is one effectively developed FED structure [9, 11–13]. According to

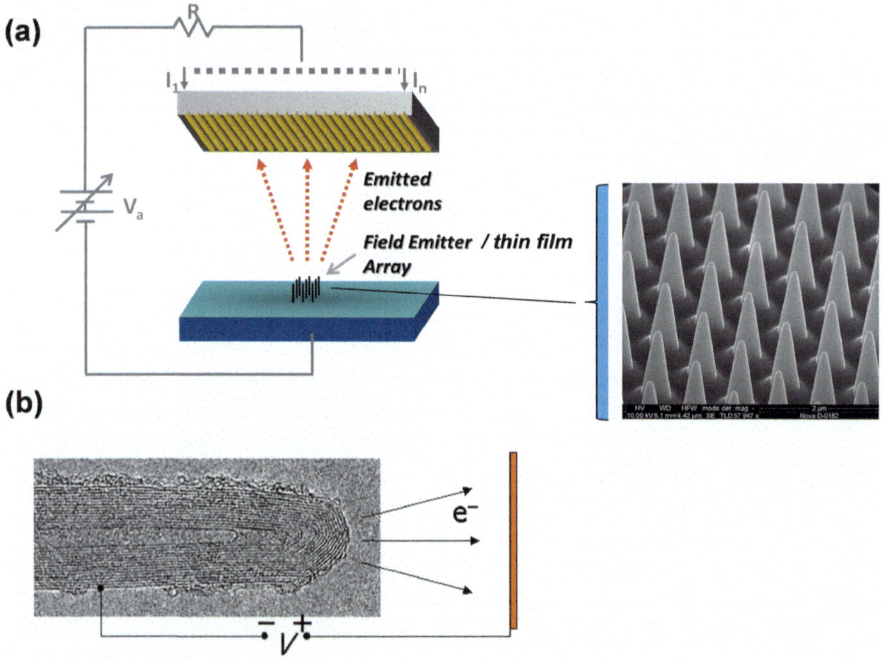

Fig. 4 **a** Typical close proximity/planer diode configuration for field emission (57a)). **b** A potential difference is applied between a nanotube (or an assembly of nanotubes) or arrays of emitter and a counter electrode (Reproduced with permission from [1])

field emission research, the micro-field emitter's electron emission current is unstable and has the potential to cause the device to irreversibly break down catastrophically [8, 14]. A ballast resistive layer was proposed to be included at the base of the micro-cone-shaped tips to solve this issue [15–18]. As more electron current exits the emitter, the ballast resistive layer can efficiently lessen the potential difference between the cathode tip and the gate electrode. By using a resistive layer, Meyer of LETI demonstrated the first FED ever observed and proved stable field emission from Spindt-type field emitters [19]. This encouraged other researchers to create field emitters as FED electron sources in the 1990s.

Now a days, soda-lime glass substrates are mostly used to create FEDs since they are compatible with LCD technology. Compared to other options, soda-lime glass has shown some advantages. Among its many advantages are its hoover compatibility, low cost, and ability to be laid extremely flat over a sizable surface. Huge-scale integrated (VLSI) thin film technology allows it to be coated with a variety of materials [20]. Nevertheless, it is challenging to fabricate a Spindt cathode that produces large-area field emitters with sufficient regularity when employing micro-fabrication techniques [10]. Thus far, only small and medium-sized structures have been used to manufacture FEDs with Spindt-type field emitters.

9 Field Emission from Nanomaterials

Field emission cathode development research has a long history, which starts with W. P. Dyke's field emission work in the 1940s. He devoted his research to developing field emission cathodes to achieve controlled, stable, electron emission, at high intensity levels, which have applications in various devices. He and his co-workers at Linfield College studied tungsten cathodes extensively. They might find use in flash X-ray systems, fine focus electron optical devices, and microwaves. However, some big challenges like field emission current stability and life of the field emission cathodes at industrial vacuum conditions remained unanswered. However, Dyke made comb-structure field emitter arrays and established a solid foundation for field emission cathode development research while working at Linfield College, USA. A good account of earlier research done by Dyke and his co-workers can be found in a review article by Charbonnier [5].

Compact multi-tip arrays were needed for microwave devices to have adequate conductance instead of a single tip emitter. In 1958, Shoulders [21] introduced the concept of field emitter arrays. After 10 years, in 1968, Spindt [12] described the creation and use of molybdenum tips and thin film technologies to create gated field emitter arrays. In 1972, Thomas and his co-workers [22] demonstrated silicon based ungated microfabricated field emitter arrays. Gray [23–26] introduced the concept of an integrated micro-fabricated vacuum transistor based on gated silicon, FEAs in 1980 and afterwards put the concept into practice. Till the end of last decade, there were several attempts reported by researchers to develop different kinds of field emission cathodes by cathode surface coating as well as without coating, which yield stable current at high-intensity levels and has a long life in the industrial vacuum conditions for making practical devices [26]. Dyke with his co-workers performed extensive research on tungsten field emitter cathodes [5]. He concluded that long-term field emission current stability is possible from a tungsten field emitter at the residual pressure well below 10^{-10} mbar during field emission cathode operation, which is not feasible to maintain in practical devices. To remove impurities and fix sputtering damage, the tungsten field emitter must be heated regularly, for example, once an hour at 10^{-9} mbar. Additionally, he concluded that all impurities on the tungsten surface must be desirable at a temperature of at least 1800 K to heal the sputtering damage. In this way, his conclusions about tungsten field emission cathodes were not very good to use in practical devices. The story of field emission from LaB_6-coated field emitters is not very old. The reports tell us that the low onset voltage, very high current density, and long-term current stability are obtained from coated LaB_6 field emitters [26].

Field emitter arrays (FEAs) are becoming more and more popular as micron-sized electron sources for a variety of cutting-edge, next-generation electronics, including ultrahigh-frequency devices and flat panel displays [13, 27]. However, the effectiveness and dependability of micro-fabricated FEAs are limited [28]. First, by sharpening the emitter tip; second, by precisely lowering the emitter-to-gate distance; third, by employing reduced work function emitter materials; and, finally, by enhancing uniformity and reproducibility, highly efficient and dependable FEAs can be achieved. These four prerequisites provide the low operating voltage that inhibits ionized residual gas bombardment and breakdown effects from causing FEA degradation while enabling extremely efficient field emission. Even though emitter materials have a significant impact on field emission characteristics due to their direct contribution to emission efficiency as per the F-N theory [2], they have not been well studied. One of the reasons is that the non-uniformity and non-reproducibility of emitter shapes make it extremely difficult to compare and evaluate emitter materials when using FEAs. For instance, the widely used method of coating low-work function emitter materials on conventional emitters, such as Mo and Si emitters, results in a decrease in emitter tip radii, irregular emitter forms, and challenges with the evaluation of emitter materials. Nakamota and Fukuda et al. employed the transfer mould approach to creating LaB_6 field emitter arrays [29–32] to address these issues and achieve LaB_6-based FEAs that are precise, consistent, and low operation voltage. In contrast to the 4.3–4.5 eV of traditional emitter material, transition metal borides and nitrides, such as titanium nitride (TiN) and LaB_6, are well recognised for having low work functions of 2.6 and 2.9 eV, respectively [32–41]. Using LaB_6 manufactured by transfer mould technology, Nakamota and Fukuda et al. [31] showed that it is possible to achieve highly sharp, uniform and low operation voltage FEAs by combining anisotropic Si wet etching, physical vapour deposition of emitter materials, and electrostatic bonding. Their electrical properties and FEA structures are contrasted with those of Mo FEAs. Table 1 illustrates their remarkable performance of low operation voltage at the low vacuum level of 4×10^{-7} mbar due to their emitter tip sharpness and low work function.

Table 1 Comparative performance of thermionic and field emission energy spread

Emitter	Thermionic emitter energy spread (eV)	Field emitter energy spread (eV)
Tungsten	3	0.3
LaB_6	1.5	0.7
Carbon fiber/CNT	–	≈0.3*

There have also been recent reports of low turn-on voltage, high current density, high field enhancement factor and persistent electron emission in nanostructures made of a variety of materials. These field-emission phenomena are believed to originate from the nanometric dimensions of the emitters such as CNTs [42, 43], ZnO [44], SnO_2 [45], GaN [46], MoO_3 [47], MoS_2 nanoflower [48], Mo nanowires [49], graphite nanofiber [50] and LaB_6 [51, 52]. However, these materials have a large work function, including carbon nanotubes, and require stringent working conditions for their utilization. The materials with good electrical, optical and mechanical properties were found to be the best emitter because a suitable contribution of mechanical and electronic properties makes the material a good field emitter [53, 54]. The atomically sharp edges create intense electric fields, which leads to excellent field emission properties. Further, metal emitters are known for their inherent potential for delivering high current density, whereas semiconducting emitters are more suitable for device applications.

10 Field Emission based MoS_2 Devices

It's well known that one-dimensional CNTs are the best field emitters owing to their high aspect ratio. Recently, it has been revealed that 2D graphene possesses better electrical and mechanical properties than CNTs. Hence, it is being widely explored for field emission properties. One of the essential reasons for exploring two-dimensional materials is their planar structure, which could be helpful for planar device technology. For example, layered MoS_2 [55] is an inorganic analogue of graphene. It possesses a direct band gap of 1.8 eV when thinned to a monolayer and has promising electronic properties. It also possesses sharp or nanometric edges like graphene, which can enhance the field emission behaviour. Because of the surface states and the finite amount of charge carriers in the conduction band, a field emitter is naturally semiconducting, which leads directly to nonlinearity. Hence, in a semiconducting emitter such as MoS_2, the observed nonlinearity is attributable to the semiconducting nature of the emitter. The nonlinearity of the F-N plot for semiconductors does indicate the involvement of surface states and band-bending effects [56]. The original Fowler–Nordheim hypothesis has been modified for several semiconducting materials. However, due to an inadequate understanding of the surface states, density, etc., these are challenging to apply to the real-world field emitter. [56].

Graphene analogues of inorganic layered MoS_2 layered materials have emerged as potential field emitter materials for the next generation of devices due to their useful and enhanced properties. Also, the observed turn-on values are comparable to the existing and known field emitters.

There has been less data on MoS_2's field emission. Few studies have been done on MoS_2 nanostructures such as edge-terminated vertically aligned films (ETVA), single- and multilayer flakes, and nanoflowers that have been subjected to field emission measurements [57]. Ultra-thin ETVA-MoS_2 films have also been reported to have field electron emission characteristics similar to those of carbon-based structures. [55, 58, 59].

For the first time, the enhanced and superior field emission performance of layered MoS_2 sheets has been investigated [55]. The field emission current density (J) versus applied electric field (E) plots for MoS_2 are shown in Fig. 5a). For a drawing current density of 10 $\mu A/cm^2$, the layered MoS_2 sheets have a low turn-on field value of 3.5 V/μm. At an applied field of 5.2 V/μm, the maximum current density of 844 $\mu A/cm^2$ is drawn. The related F-N figure is displayed in Fig. 5b. The F-N plot demonstrates a nonlinear character, with a propensity towards saturation in the high field region and nearly linear behaviour in the low field zone. Because of surface states and a finite amount of charge carriers in the conduction band, the MoS_2 field emitter is semiconducting, which directly results in nonlinearity. The inset of Fig. 5b displays many emission sites from the field emission pattern, which could be the atomically sharp edges of MoS_2 sheets extending from the Si substrate. Its field enhancement factor of 1138 is the cause of the enhanced field emission behavior. The sharply projecting MoS_2 nanoscale edge is responsible for the significant field enhancement factor that was found. Because of the high field enhancement factor in this instance, the low turn-on field is primarily the result of these sharp edges.

Dattatray Late et al. reported the field electron emission performance of pulsed laser-deposited (PLD) MoS_2 thin films on W-tip and Si substrates [60]. Figure 6

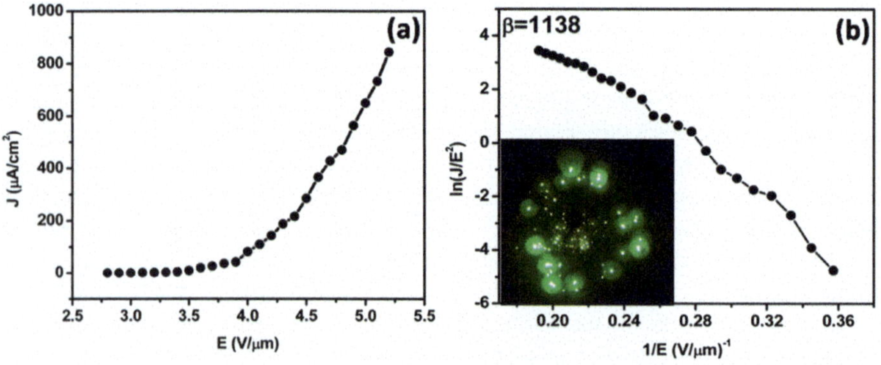

Fig. 5 Field emission from a few layers of MoS_2 sheet, **a** applied electrical field (J) as a function of current density (E), **b** F-N plot and inset shows field emission pattern of a few layered MoS_2 recorded at a current density of 50 $\mu A/cm^2$ (Reproduced with permission from [55])

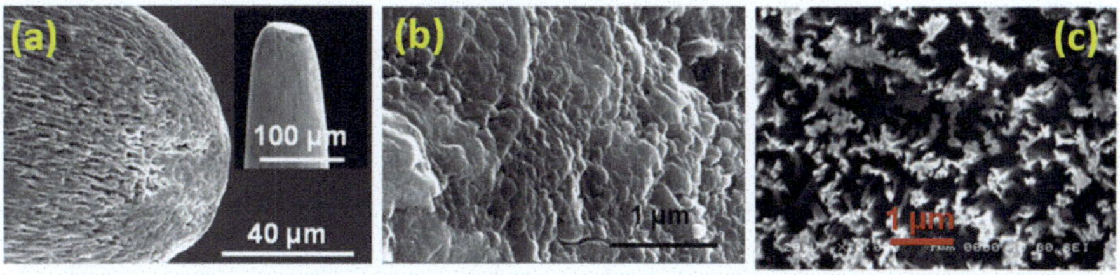

Fig. 6 Field emission scanning electron microscope images of pulsed laser deposited MoS_2 thin film; **a** and **b** on W-tip substrate, and **c** on Si substrate (deposited with Laser fluence 2 J/cm^2, Laser shots 900, thickness ~ 120 nm). Inset of **a** shows an image of bare W-tip (Reproduced with permission from [60])

shows the morphology of PLD MoS_2 films on Si and W-tip substrates detected using field emission scanning electron microscopy (FESEM). In the case of the MoS_2/W-tip sample, as can be seen in Fig. 6a and b, the MoS_2 film completely covers the W-tip substrate surface and is accompanied by an overgrowth of MoS_2 nanocrystals that range in size from 20 to 60 nm. This represents the film-substrate combination's Stransky–Krastinov growth mode, which is useful when using electron emitters. Figure 6a inset features a picture of a naked W-tip. The MoS_2 film equally covers the Si substrate surface, displaying nanometric protrusions that reveal the surface morphology of the MoS_2/Si sample (Fig. 6c). Both times, the dry PLD technique is shown to produce a densely nanostructured MoS_2 morphology under the selected growth conditions. This is crucial for local electric field amplification in field emission applications.

PLD MoS_2 thin film was subjected to FE measurements on planar Si and W-tip substrates. Using MoS_2 thin films for planner device technology, such as flat panel field emission displays and W-tips, offers a plethora of opportunities for basic research and real-world applications, including field emission displays and intense "point" electron sources. The MoS_2/W-tip field emitter's field emission current density (J) against the applied voltage (V) graph is shown in Fig. 7a. It's interesting to note that at a comparatively low applied voltage of around 3.8 kV, the MoS_2 film on the tungsten (W) tip emitter produces a significant emission current density of about 30 mA/cm^2 (Fig. 7). The emission grows with a steady increase in applied voltage, demonstrating that the emission is truly following the Fowler–Nordheim (F-N) theory. On the basis of the theory of field electron emission from semiconductors, the low electric field region corresponds to the electron emission from the conduction band. In contrast, the high electric field region corresponds to the electron emission from the conduction band and the valence band of MoS_2. The FE current stability characteristics of the PLD MoS_2/W-tip emitter studied at preset values of 1μA and 10μA for three hours, are shown in Fig. 7c. Moreover, a close examination shows that there are "spikes" superimposed on the excursions. Ion bombardment and the adsorption/desorption of leftover gas molecules are the primary sources of spike-like variations. Depending on the type of adsorbing molecule, the MoS_2 thin film emitter's local work function changes slightly during adsorption/

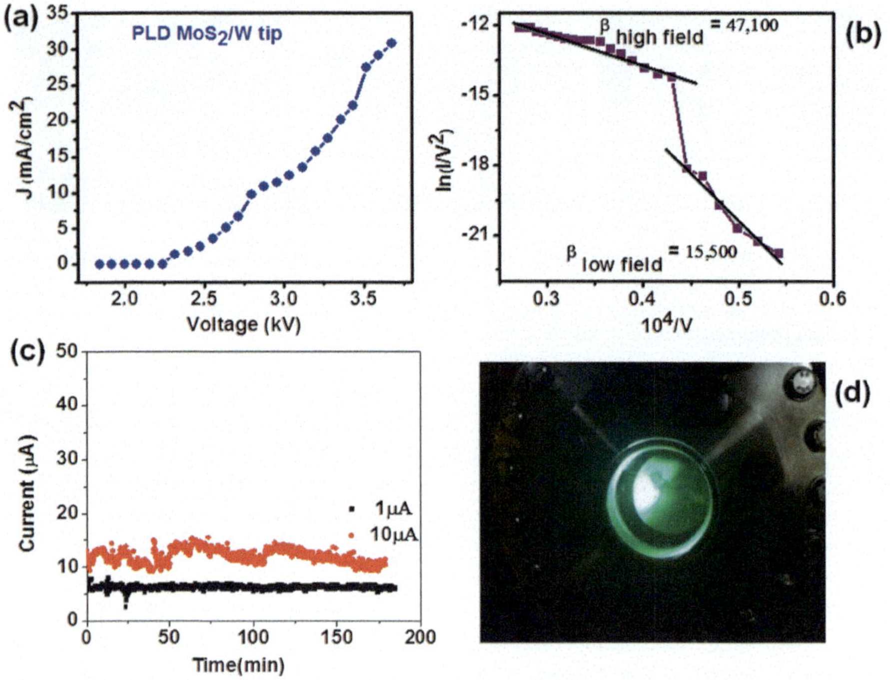

Fig. 7 Field emission of PLD MoS_2 films on W tip substrates (thickness ~ 120 nm), **a** J-V characteristics, **b** F-N plot showing non-linear behaviour indicating that the emission current from the semiconducting emitter, **c** Field emission current stability (I-t) plot for MoS_2/W tip emitter. **d** Field emission image recorded at 3.2 kV (Reproduced with permission from [61])

desorption events. The PLD MoS_2/W tip emitter's field emission image is shown in Fig. 7d. It displays the number of emission sites indicative of the existence of sharp, nanometric protrusions on the emitter surface.

The MoS_2 film on silicon (Si) requires a turn-on field of 2.8 V/μm to draw an emission current density of 10 μA/cm^2. The J-E plot for a PLD MoS_2/Si planar emitter is shown in Fig. 8a. It is discovered that to draw an emission current density of 1 μA/cm^2, a turn-on field of 2 V/μm is needed. A value of 2.8 V/μm is found for the electric field at 10 μA/cm^2, which is lower than the value of 3.5 V/m previously reported for an emitter consisting of several layers of MoS2 nanosheets [55]. The low turn-on could be explained by the sharp protrusions and enhanced electrical contact between the emitter (MoS_2) and substrate surface that are expected during PLD. Figure 8b shows the F-N plot for a PLD MoS_2/Si planar emitter. The emission current–time (I-t) behaviour observed at the preset current values of 1 μA and 10 μA is depicted in Fig. 8c. The observed emission current fluctuations are within 10% of the average value. The field emission image of the PLD MoS_2/Si planar emitter is seen in the inset of Fig. 8c. The typical experimental setup for PLD MoS_2/Si field emission studies is depicted in Fig. 8d.

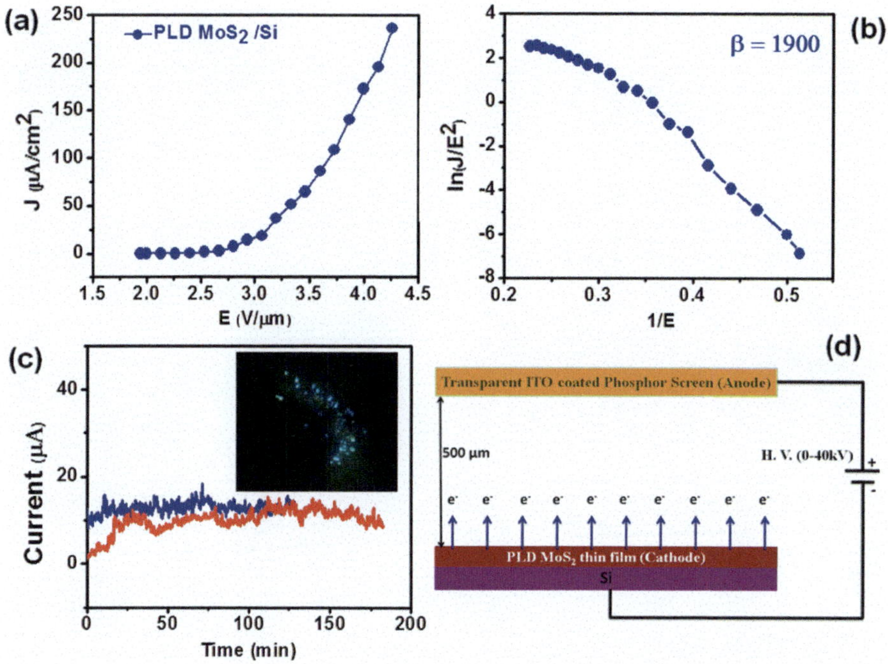

Fig. 8 Field emission of PLD MoS_2 thin film on Si substrates (thickness ~ 120 nm). **a** J-E characteristics **b** F-N plot ln (J/E^2) Vs 1/E, **c** Field emission current stability (I-t) plot and **d** Field emission set-up used for field emission investigations of PLD MoS_2/Si. Inset of **c** shows the field emission image recorded at an applied electric field of 3.5 V/μm (Reproduced with permission from [61])

Rupesh S. Devan et al. show core–shell nanostructure arrays, of 1D-TiO2 nanorods, and 2D MoS_2 layers as good field emitters [62]. Hot-filament metal vapour deposition (HF-MVD) method was used to create the large-area arrays of vertically aligned TiO_2 nanorods of the brookite phase. Furthermore, using the pulsed laser deposition (PLD) method, MoS_2 layers and shells of the necessary thicknesses were produced over TiO_2 nanorods. Researchers studied how the thickness of the MoS_2 layer affected the structural, chemical, and FE features. The surface morphology of MoS_2-loaded and pure TiO_2 nanorods synthesized on Si substrate shows a uniform distribution of TiO_2 nanorods over a large area in Fig. 9a. The surface morphologies of the MoS_2 loaded TiO_2 nanorods are depicted in Fig. 9b, c. A detailed examination of the top view of a section of the array reveals that MoS_2 layers are evenly distributed to the whole array of -TiO2 nanorods. The MoS_2 layers were created at optimum deposition rates, with average thicknesses of 40, 20 and 4 nm.

The FE measurements of MoS_2 coated β-TiO_2 nanorods were performed in a planar diode configuration, electron field emission study at various separations of 500, 1000, 1500 and 2000 μm. Initially, the pristine β-TiO_2 nanorod showed a larger emission current density of 470 μA/cm^2, lower threshold field of 4.8 V/μm, and lowest turn-on field of 3.9 V/μm were observed at 2000 μm. At an anode–cathode

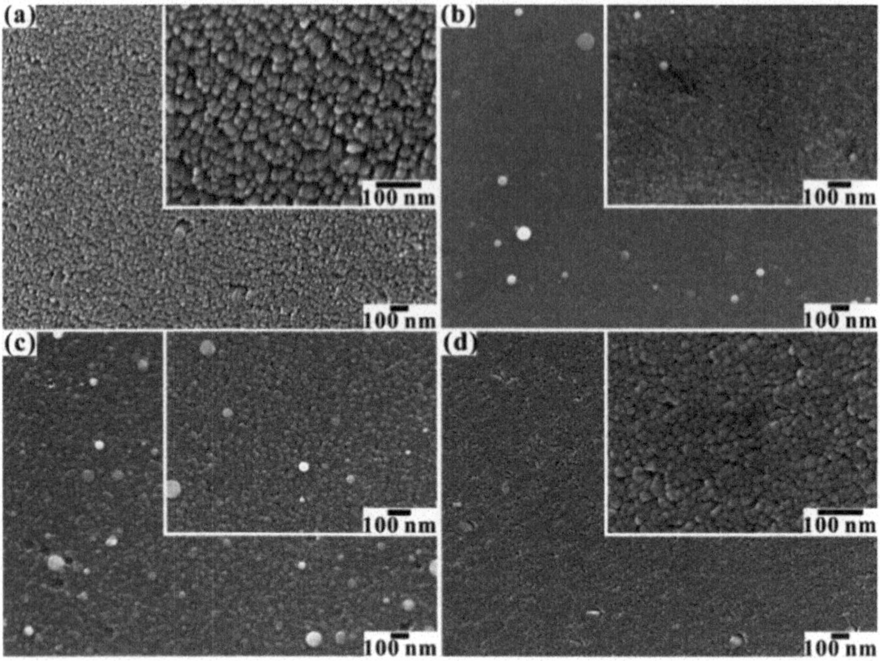

Fig. 9 FESEM images showing the top view of the large-area array of **a** vertically aligned pristine 1D β-TiO_2 nanorods on Si substrate, which were further decorated with **b** 40 nm, **c** 20 nm, and **d** 4 nm layer/shell of MoS_2. The inset shows their respective high-magnification images (Reproduced with permission from [62])

separation of 1000 μm, the pure MoS_2 more significant emission current density of 30.4 μA/cm^2 at 8.4 V/μm and turn-on field of 7.2 V/μm (at 10μA/cm^2) were obtained. The FE experiments of MoS_2 coated β-TiO_2 nanorods (MoS_2/β-TiO_2/Si) were completed at the same separation because of the lower turn-on for both 2D MoS_2/Si and β-TiO_2/Si, as well as the delivery of a bigger emission current density of β-TiO_2/Si emitters at 2000 μm separation. Figure 10 displays the FE characteristics of β-TiO_2 nanorods covered with MoS_2 at different thicknesses. Figure 10a displays the variation in the microscopic electron emission current density (J) of MoS_2/TiO_2/Si, TiO_2/Si and MoS_2/Si emitters as a function of the applied electric field (E) (i.e., J-E plot). Compared to both TiO_2/Si and MoS_2/Si emitters, the MoS_2/TiO_2/Si emitters notably produce a greater emission current density at relatively lower applied fields. Figure 10b displays the F-N graphs for MoS_2 controlled MoS_2/TiO_2/Si emitters. The F-N plots are well resolved into two distinct sections. The well-defined band alignment of MoS_2 and β-TiO_2 following their layer/shell development over one another is supported by the different separations of the F-N plots. In addition to the high aspect ratio of the vertically aligned TiO_2 nanorods, the modulation of the electrical characteristics due to heterostructure and interface effects can be attributed to the improved FE behaviour of the MoS_2@TiO_2/Si emitter. The additional boost

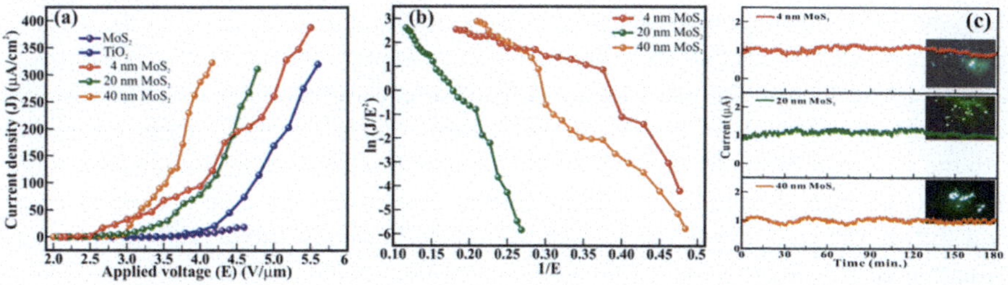

Fig. 10 Field emission **a** J-E curves of a large-area array of vertically aligned pristine 1D β-TiO_2 nanorods, 2D MoS_2 and MoS_2@ β-TiO_2 nanorods and MoS_2@β-TiO_2 nanorods decorated with ~40, 20 and 4 nm layer/shell of MoS_2; and their corresponding **b** F-N plots measured at vacuum separations of 2000 μm. **c** Field emission current stability (I-t) plot of 2D MoS_2@1D β-TiO_2 nanorods decorated with 40 nm (lower panel), 20 nm (middle panel) and 4 nm (upper panel) layer/shell of MoS_2 (Reproduced with permission from [62])

in FE behaviour is caused by the majority of the injected electrons moving easily towards emission sites due to the 4 nm overlayer of conducting MoS_2 along the β-TiO_2 nanorods. The emission current (I) obtained at a preset current value of 1 μA in Fig. 10c did not clearly exhibit any degradation for continuous emission for 180 min (t) [62].

One of the problems with MoS_2 nanostructures from an application standpoint is how easily they aggregate and rearrange themselves during synthesis It is challenging to exfoliate the stacked MoS_2 layers and bring out their sharp edges. Exfoliating the MoS_2 has always been easiest when done in a heterostructure. According to field emission (FE) studies by Sanjeewani R. Bansode et al. [63], the MoS_2-RGO nanocomposite performs better than a pure MoS_2 emitter. Exfoliating MoS_2 has always been made easier by heterostructures. Various MoS_2 heterostructures have been synthesised and characterized by research groups. The MoS_2-RGO nanocomposite emitter, which is made of MoS_2 nanosheets on RGO and was created using an easy, one-step hydrothermal process, is described in the following FE experiments. The influence of RGO concentration on the development, general morphology, and FE behaviour of the MoS_2-RGO nanocomposite has been attempted to be revealed. Intriguingly, as compared to the pristine MoS_2 emitter, the MoS_2-RGO nanocomposite emitters show improved FE behaviour in the form of lower turn-on and threshold field values as well as resilient long-term stability. The distinctive morphology and modulation of the electrical properties have been linked to a synergistic effect, which has improved the FE characteristics. Figure 11 shows the SEM images of the pure MoS_2 and MoS_2-RGO nanocomposites on a silicon substrate. Agglomerates of micron-sized MoS_2 sheets are seen in the pure sample under the current experimental circumstances (Fig. 11a). A close examination of the image captured at a greater magnification shows that the individual agglomerates are produced by the stacking of thin MoS_2 sheets, which result in structures that resemble "petals." The influence of the GO weight percentage on the final morphology is seen in the SEM pictures of MoS_2-RGO nanocomposites. Although the shape of the

MoS_2-RGO (1%) nanocomposite (Fig. 11b) is similar to that of a pure MoS_2 sample, close inspection reveals that there are very few GO sheets present in the SEM image. It's interesting to note that the surface shape changes significantly when GO weight percentage increases. Ultrafine MoS2 nanosheets and a few agglomerates, together with the presence of the MoS_2-RGO (3%) sample, define its morphology (Fig. 11c). The RGO surface is completely and uniformly covered by the MoS_2 sheets. Except for a shift in the areal density of the MoS_2 and RGO sheets, the overall morphology is unaffected by a further rise in the GO weight percentage.

All emitters' emission currents grew progressively up to the turn-on field with an applied voltage that was greater (by a step of 40 V). As predicted by the Fowler–Nordheim (FN) theory of cold emission, the emission current then increased dramatically with the same step increment in the applied field later on. The turn-on field values for MoS_2 nanosheets, MoS_2-RGO (1%), MoS_2-RGO (3%) and MoS_2-RGO (5%) are 4.7, 3.4, 2.8 and 2.6 μV/m, respectively. The turn-on field is defined as the field required by the material to draw an emission current density of 10 μA /cm^2. Table 2 contains a list of all emitters' observed FE characteristics. In contrast to other emitters, the MoS_2-RGO nanocomposite emitter has lower turn-on and threshold field values, and its overall FE characteristic has been improved. This is attributable to the combined influence of the composite's "morphology" and "modulation" of its electronic properties. Table 2. Field emission properties in present [64]. Figure 12c

Fig. 11 SEM images of hydrothermally synthesized **a** pristine MoS_2 **b** MoS_2-RGO (1%), **c** MoS_2-RGO (3%) and **d** MoS_2-RGO (5%) nanocomposites (Reproduced with permission from [64])

Table 2 Field emission properties of MoS_2 nanosheets, MoS_2-RGO nanocomposite

Emitter	Turn-on field (Vμm^{-2}) (for 10 μA cm^{-2})	Threshold field (V μm^{-2}) (for 100 μA cm^{-2})	Maximum current density (μA cm^{-2})	(β)
MoS_2 sheet	4.7	7	~300 at 8.8 V μm^{-1}	3042
MoS_2-RGO (1%)	3.4	4.2	~400 at 5.2 V μm^{-1}	3404
MoS_2-RGO (3%)	2.8	3.8	~540 at 5 V μm^{-1}	3538
MoS_2-RGO (5%)	2.6	3.1	~800 at 3.9 V μm^{-1}	4128

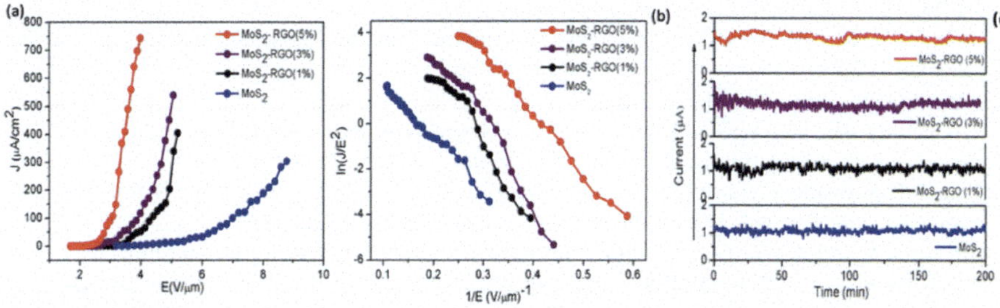

Fig. 12 Field emission characteristics of the pristine MoS_2, MoS_2-RGO (1%), MoS_2-RGO (3%) and MoS_2-RGO (5%) nanocomposite emitters **a** J-E curves, and **b** F-N plots, ln (J/E2) versus (1/E) **c** current stability (Reproduced with permission from [63])

shows the behaviour of the emission current stability of the pure MoS_2, MoS_2-RGO (1%), MoS_2-RGO (3%) and MoS_2-RGO (5%) nanocomposite emitters over 3 h.

11 Conclusion and Future Perspective

This chapter focus on field emission and its mechanism, developing several MoS_2field emitter materials. The field emission performance of several types of MoS_2 morphology has been examined. The field emission results' with high emission current density, low turn-on and threshold field, and extended emission stability make them a promising replacement for field emission-based existing devices. The turn-on and threshold voltages observed strongly suggest that vertically grown films and 1D nano forms of MoS_2 would be the best candidates for high current density electron beam applications. Large-area emitter applications, such as flat panel display displays, are better suited for large-area homogenous atomically thin nanosheet or thin films of MoS_2.

References

1. R. O. Jenkins, "Field emission of electrons", Report. Prog. Phys. Vol **9,** page 77, 1942.
2. Stern TE, Gossling BS, Fowler RH (1929) Further studies in the emission of electrons from cold metals. Proc R Soc (Lond) 124(795):699–723
3. Dyke WP, Dolan WW (1956) Field emission. Ad Electron Electron Phys 8:89–185
4. Swanson LW, Bell AE (1973) Adv Electron Electron Phys 32:194
5. Charbonnier F (1996) Developing and using the field emitter as a high intensity electron source. Appl Surf Sci 94/95:26–43
6. Gadzuk JW, Plummer EW (1973) Field emission energy distribution (FEED). Rev Mod Phys 45(3):487
7. Filip V, Nicolaescu D, Tanemura M, Okuyama F (2001) Modeling the electron field emission from carbon nanotube films. Ultramicroscopy. 89:39–49
8. Schwoebel PR, Brodie I (1995) Surface-science aspects of vacuum microelectronics. J Vac Sci Technol B 13(4):1391
9. Itoh S, Tanaka M, Tonegawa T (2004) Development of field emission displays. J Vac Sci Technol B 22(3):1362
10. Shah I (1997) Phys World Proc 45
11. Chalamala BR, Wei Y, Gnade BE (1998) FED up with fat tubes. IEEE Spectr 35(4):42–51
12. Spindt CA (1968) A thinfilm field emission cathode. J Appl Phys 39(7):3504
13. Spindt CA, Brodie I, Humphrey L, Westerberg ER (1976) Physical properties of thin-film field emission cathodes with molybdenum cones. J Appl Phys 47(12):5248
14. Brodie I, Spindt CA (1992) Vacuum microelectronics. Adv Electron Electron Phys 83:1–106
15. Py C, Baptist R (1994) Stability of the emission of a microtip. J Vac Sci Technol B 12(2):685
16. Ghis A, Mayer R, Ramboud P, Levy F, Leroux T (1991) Sealed vacuum devices: fluorescent microtip displays. IEEE Trans Electron Dev 38(10):2320–2322
17. Itoh S, Niiyama T, Taniguchi M, Watanabe T (1996) A new structure of field emitter arrays. J Vac Sci Technol B 14(3):1977
18. Baptist R, Bachelet F, Constancias C (1997) Microtips and resistive sheet: a theoretical description of the emissive properties of this system. J Vac Sci Technol B 15(2):385
19. Meyer R, Ghis A, Rambaud P, Muller F (1986) Techn Dig Jpn Disp 86:513
20. Ohsaki H, Kokubu Y (1999) Global market and technology trends on coated glass for architectural, automotive and display applications. Thin Solid Films 351(1–2):1–7
21. Shoulders KR (1961) Microelectronics using electron-beam-activated machining techniques. In: Advances in computers, 1 January 1961, vol 2, pp 135–293. Elsevier
22. Thomas RN, Nathanson HC (1972) Photosensitive field emission from silicon point arrays. Appl Phys Lett 21:384
23. Gray HF (1982) Proceedings of the colloquium on advances in electron emission, p 24
24. Gray HF (1982) Proceedings of the 29th international field emission symposium, p 111
25. Gray HF, Campasi GJ, Greene RF (1986) Technical digest of the international electron devices meeting IEDM, vol 86, p 776
26. Gray HF, Campasi GJ (1987) Mat Res Soc Symp Proc 76:25
27. Joag DS (1978) PhD thesis, Pune University, India, December 1978
28. Vaudaine P, Meyer R (1991) Technical digest of international electron devices meeting, p 197
29. Busuta HH, Pogemiller JE, Zimmerman BJ (1994) Scaling of emission currents and of current fluctuations of gated silicon emitter ensembles. J Vac Sci Technol B 12:689
30. Nakamoto M, Ono T, Nakamura Y (1993) Proceedings of the international vacuum microelectronics conference, Late News # 28
31. Nakamoto M, Ichimura K, Ono T, Nakamura Y (1995) Proceedings of the international vacuum microelectronics conference, p 186
32. Nakamoto M, Ono T, Sakai T, US Patent no. 5, no. 499, no. 938 (1996)
33. Nakamoto M, Fukuda K (2002) Field electron emission from LaB_6 and TiN emitter arrays fabricated by transfer mold technique. Appl Surf Sci 202:289–294

34. Lafferty JM (1951) Boride cathodes. J Appl Phys 22:299
35. Shimizu R, Kataoka Y, Tanaka T, Kawai S (1975) Field emission pattern of LaB6-single crystal tip. Jpn J Appl Phys 14:1089
36. Schmidt PH, Joy DC (1978) Low work function electron emitter hexaborides. J Vac Sci Technol 15:1809
37. Swanson LW, McNeely DR (1979) Work functions of the (001) face of the hexaborides of Ba, La, Ce and Sm. Surf Sci 83(1):11–28
38. Futamoto M, Nakazawa M, Usami K, Hosoki S, Kawabe U (1980) Thermionic emission properties of a single-crystal LaB_6 cathode. J Appl Phys 51(7):3869
39. Futamoto M, Nakazawa M, Kawabe U (1980) Thermion ice mission properties of hexaborides. Surf Sci 100:470–480
40. Futamoto M, Nakazawa M, Kawabe U (1983) High temperature surface composition of hexaboride thermionic electron emitters. Vacuum 33:727–732
41. Goebel DM, Hirooka Y, Sketchley TA (1985) Large-area lanthanum hexaboride electron emitter. Rev Sci Instrum 56:1717
42. Davis PR, Swanson LW, Hutta JJ, Jones DL (1986) Fabrication and characterization of rare earth hexaboride single-crystal materials. J Mater Sci 21:825–836
43. Toth LT (1971) Transition metal carbides and nitrides. Academic, New York
44. Rivory J, Behaghel JM, Berthier S, Lafait J (1981) Optical properties of substoichiometric TiNx. Thin Solid Films 78:161–165
45. Bonard JM, Kind H, Stöckli T, Nilsson LO (2001) Field emission from carbon nanotubes: the first five years. Solid-State Electron 45:893–914
46. de Jonge N, Lamy Y, Schoots K, Oosterkamp TH (2002) High brightness electron beam from a multi-walled carbon nanotube. Nature. 420:393
47. Ramgir NS, Mulla IS, Vijayamohanan K, Late DJ, Bhise AB, More MA, Joag DS (2006) Ultralow threshold field emission from a single multipod structure of ZnO. Appl Phys Lett 88:42107
48. Bhise AB, Late DJ, Ramgir NS, More MA, Mulla IS, Vijaymohanan K, Joag DS (2007) Field emission investigations of RuO_2-doped SnO_2 wires. Appl Surf Sci 253:9159–9163
49. Kim HM, Kang TW, Chung KS, Hong JP, Choi WB (2003) Field emission displays of wide-bandgap gallium nitride nanorod arrays grown by hydride vapor phase epitaxy. Chem Phys Lett 377:491–494
50. Li YB, Bando Y, Golberg D, Kurashima K (2002) Field emission from MoO3MoO3 nanobelts. Appl Phys Lett 81:5048
51. Li YB, Bando Y, Golberg D (2003) MoS_2 nanoflowers and their field-emission properties. Appl Phys Lett 83:1962
52. Zhou J, Xu NS, Deng SZ, Chen J, She JC, Wang ZL (2003) Large-area nanowire arrays of molybdenum and molybdenum oxides: synthesis and field emission properties. Adv Mater 15(21):1835
53. Zhang WJ, Wu Y, Wong WK, Meng XM, Chan CY, Bello I, Lifshitz Y, Lee ST (2003) Structuring nanodiamond cone arrays for improved field emission. Appl Phys Lett 83:3365
54. Late DJ (2008) Synthesis and characterization of LaB_6 thin films on tungsten, rhenium andsilicon substrates and their investigations as field emitters. PhD thesis, submitted to Savitribai Phule Pune University, formally University of Pune, India, February 2008
55. Talin AA, Dean KA, Jaskie JE (2001) Field emission displays: a critical review. Solid-State Electron 45(6):963–976
56. https://www.nist.gov/image/fieldemissiontoprt1png(Reproduced with permission from NIST. Accessed 9 April 2020
57. Bonard JM, Kind H, Stöckli T, Nilsson LO (2001) Field emission from carbon nanotubes: the first five years. Solid-State Electron 45(6):893–914
58. Kashid RV, Late DJ, Chou SS, Huang Y-K, De M, Joag DS, More MA, Dravid VP (2013) Enhanced field-emission behavior of layered MoS_2 sheets. Small 9:2730–2734
59. Rout CS, Joshi PD, Kashid RV, Joag DS, More MA, Simbeck AJ, Washington M, Nayak SK, Late DJ (2013) Superior field emission properties of layered WS2-RGO nanocomposites. Sci Reports 3:3282

60. Kong D, Wang H, Cha JJ, Pasta M, Koski KJ, Yao J, Cui Y (2013) Synthesis of MoS_2 and $MoSe_2$ films with vertically aligned layers. Nano Lett 13:1341–1347
61. Late DJ, Shaikh PA, Khare R, Kashid RV, Chaudhary M, More MA, Ogale S (2014) ACS Appl Mater Interfaces 6:15881–15888
62. Devan RS, Thakare VP, Antad VV, Chikate PR, Khare RT, More MA, Dhayal RS, Patil SI, Ma Y-R, Schmidt-Mende L (2017) Nano-heteroarchitectures of two-dimensional MoS_2@ one-dimensional brookite TiO_2 nanorods: prominent electron emitters for displays. ACS Omega 2:2925–2934
63. Bansode SR, Harpale K, Khare RT, Walke PS, More MA (2016) Morphological, structural and field emission characterization of hydrothermally synthesized MoS_2-RGO nanocomposite. Mater Res Express 3(11):115023
64. Li YB, Bando Y, Golberg D (2003) MoS_2 nanoflowers and their field-emission properties. Appl Phys Lett 82:1962–1964. https://doi.org/10.1063/1.1563307
65. Gaur APS, Sahoo S, Mendoza F, Rivera AM, Kumar M, Dash SP, Morell G, Katiyar RS (2016) Cold cathode emission studies on topographically modified few layer and single layer MoS_2 films. Appl Phys Lett 108:043103
66. Gesley M, Swanson LW (1984) A determination of the low work function planes of LaB_6. Surf Sci 146:583–599

Chapter 12
MoS_2 for Battery and Supercapacitor Applications

Anjali Singh, Dona Susan Baji, Shantikumar Nair, and Dhamodaran Santhanagopalan

1 Introduction

Renewable energy transition is essential for achieving the democratization of energy. Energy storage technologies, such as electrochemical technology (i.e., metal-ion batteries), fuel cells, electrochemical capacitors (i.e., supercapacitors) and electrolyzers, will play an important role in achieving the net-zero goals across the world. In the past decades, various types of energy storage systems (ESSs), including lithium (Li)-ion batteries (LIBs), Li–sulfur batteries (LSBs), sodium-ion batteries (NIBs), and supercapacitors, have been explored [1–3], owing to their advantages such as high energy density, low self-discharge, and long lifecycle. Nevertheless, these ESSs have limitations in terms of capacity and cycling. Metal dichalcogenides have been widely explored for energy storage applications due to its structure, surface area, and unique properties as well as high intrinsic fast ionic conductivity.

The general chemical formula of transition metal dichalcogenides is MX_2(M and X represent transition metal atom and chalcogen atom, respectively). Each layer consists of an X-M-X sandwich-like structure. In the bulk structure, layers are bonded with each other by weak van der Waals forces, and in each layer, the M atom is sandwiched in a trigonal prismatic coordination with two S layers of pyramidal geometry [4]. Exfoliation of bulk materials in their few layers mainly conserves their properties and leads to additional characteristics due to reduced dimensions [2, 5, 6]. In the transition metal dichalcogenide family, molybdenum disulfide (MoS_2) has gained attention in multiple applications such as lubrication [7], LIBs [5], supercapacitors

A. Singh (✉)
Indian Council for Research on International Economic Research, New Delhi 110017, India
e-mail: anjalisinghromi@gmail.com; asingh@icrier.res.in

D. S. Baji · S. Nair · D. Santhanagopalan
Amrita School of Nanosciences and Molecular Medicine, Amrita Vishwa Vidyapeetham, Kochi 682041, India

D. J. Late and C. S. Rout (eds.), *Science and Technology of 2D MoS_2*,
Materials Horizons: From Nature to Nanomaterials,
https://doi.org/10.1007/978-981-97-7367-1_12

[8], and drug delivery [9] due to its extraordinary catalytic activities and structure. The experimental development of MoS_2 has been explored through various synthesis strategies including mechanical or liquid exfoliation, chemical vapor deposition, and solvothermal and hydrothermal syntheses [10].

This book chapter provides a brief overview of both experimental and computational aspects of MoS_2 for energy storage applications. In this chapter, we briefly reviewed the computational aspects based on the density functional theory (DFT) and experimental reports on MoS_2 for energy storage applications, starting from metal-ion batteries (i.e., monovalent- and multivalent-ion batteries) followed by supercapacitors.

2 MoS_2 for Monovalent-Ion Batteries

LIBs are widely used in various applications, including consumer electronics, electric vehicles, and grid applications. LIBs have mainly three components: cathode, anode, and electrolyte. Usually, in most LIBs, the cathode is made of metal oxides whereas, the anode is made of carbon materials. Ideally, a cathode requires high Li intercalation voltage, whereas an anode should have a low intercalation voltage; both anode and cathode require high Li atom mobility [11, 12]. The MoS_2 layered structure offers high reversible Li intercalation/deintercalation due to its unique structure, with interlayer spacing, and has been considered an ideal electrode for advanced LIBs [13–16].

2.1 MoS_2 Electrode for Li-Ion Storage

Computational studies

Li et al. [17] investigated Li-atom adsorption and diffusion in the bulk, bilayer, monolayer, and zigzag nanoribbons of MoS_2 using DFT simulations. Bulk 2H-MoS_2 is considered for computational simulations, as it is the most stable form of MoS_2 among 2H and 3R crystal structures [18] (Fig. 1a). The adsorption of Li atoms at the octahedral site (binding energy = 3.08 eV) is preferable over that at the tetrahedral site, forming new Li–S bonds (bong length = 2.45 Å) uniformly with all the S atoms. The estimated Li diffusion barrier between two octahedral sites through a tetrahedral site is 0.45 eV, indicating faster diffusion of Li in bulk MoS_2 [17] (Fig. 1b). Enyashin et al. [19] studied the stability of 1T and 2H MoS_2 with lithiation as a function of Li content (x) and intercalation sites. Both the structures prefer Li adsorption on the octahedral site. Further, the 2H to 1T transition takes place for $x = 0.4$ Li content. In addition, the $x < 0.4$, 2H phase structure remains stable, whereas the 1T structure changes from hexagonal to monoclinic, making the monoclinic phase more stable with higher Li content $0.4 < x < 1$. The 2H Li_xMoS_2

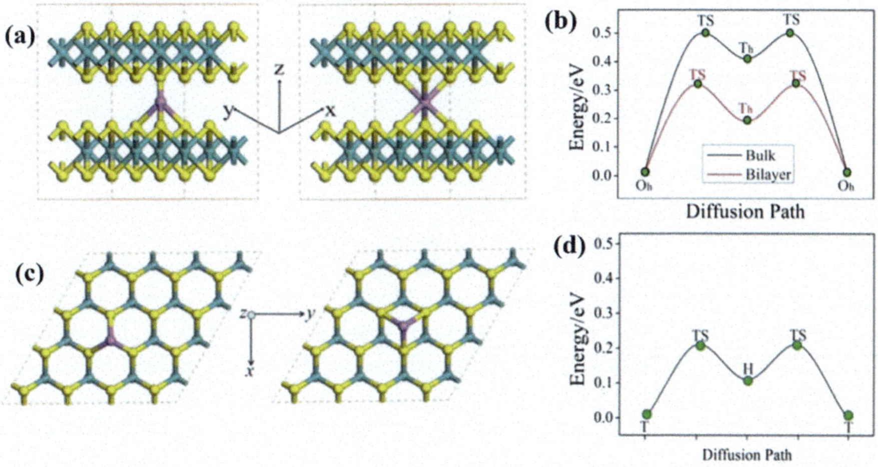

Fig. 1 Li adsorption on the tetrahedral site and octahedral site (**a**) and the diffusion barrier of Li migration on bulk and bilayer MoS_2 between two octahedral sites through a tetrahedral site (**b**). Li adsorption on the Mo site and hollow site on the monolayer MoS_2 (**c**) and the diffusion barrier of Li migration on the monolayer MoS_2 between two Mo sites through a hollow site (**d**). Reprinted with permission from [17] ©2012 American Chemical Society

structure shows an identical voltage for the entire composition (around 1.9–2.5 V). The volume change for both the phases (2H and 1T) is 15–20%, leading to maximum change in the c parameter [19].

In 2H MoS_2 bilayer, the Li atom prefers to be adsorbed at the octahedral site, similar to bulk 2H MoS_2 (Fig. 1b). The binding energy at this site is lower than that in the bulk by 0.37 eV. The interlayer distance is 0.2 Å, which is higher than that in the bulk [17]. This elongated interlayer distance facilitates faster diffusion, with a diffusion barrier of 0.17 eV, which is lower than that in bulk 2H MoS_2. In monolayer 2H MoS_2, the favorable site for Li adsorption is on top of the Mo atom, with a binding energy of 2.21 eV. The diffusion barrier for the monolayer is 0.21 eV, which is much lower than that in the bulk and bilayer (Figs. 1c–d). Reducing the dimensionality reduces the Li adsorption energy as well as the diffusion barrier for Li atoms in LIBs. Zigzag nanoribbons of MoS_2 increase the Li adsorption energy and preserve the diffusion barrier, similar to that in monolayer MoS_2, thereby making it suitable as an electrode material for battery applications [17]. The pre-lithiated MoS_2 offers a much lower diffusion barrier (0.007 eV) than clean MoS_2 (0.22 eV), leading to faster charging/discharging properties and shows a theoretical capacity of 308 mAh g^{-1} [20]. Kabiraj et al. [21] found that for $x_{max} = 1.625$, the reversible specific capacity is 272 mAh/g and the average open circuit voltage (OCV) is 0.74 V, which is much closer to experimentally reported results [22]. Pereira et al. [23] investigated MoS_2 nanotubes as an LIB electrode material. A comparative study of TiS_2 and MoS_2 nanostructures showed that the Li mobility is four times faster in MoS_2 than that in TiS_2. With an increase in the diameter of the nanotube, the voltage decreases, and the activation barrier increases for Li. The voltage, binding energy, and energy barrier for MoS_2 nanotubes of 10 Å are −0.10 V, 1.64 eV, and 198 meV, respectively [23].

Sukhanova et al. [24] investigated a new nanomaterial termed as A'-MoS_2(MOF) for LIB and NIB applications using DFT. In the A' phase, the MoS_2 monolayer crystallizes in rhombic symmetry with space group Pmn_21. The A' phase MoS_2 has a porous structure, which offers a lower area density (2.54×10^{-7} g cm^{-2}) than 1H MoS_2 (3.03×10^{-7} g cm^{-2}) and 1T MoS_2 (3.04×10^{-7} g cm^{-2}) phases. The porous structure of A'-MoS_2 facilitates higher diffusion, as the size of the pores is greater than the covalent diameter of Na- and Li-ions. A'-MoS_2 is a dynamically stable structure, with an indirect bandgap of 0.74 eV. The large surface area and porous structure of A'MoS_2 offer a high capacity of ~391 mAhg^{-1} (17% higher than 1H and 1T MoS_2 [both 334 mAhg^{-1}]), with maximum coverage of adsorption sites by Li- and Na-ions. The OCV of this monolayer is 1.2 V, which falls in the same range as that of 1H MoS_2 with Mo vacancies. The high adsorption energy of Li/Na atoms on the A'MoS_2 surface and small diffusion barriers (0.46 eV for Li and 1.05 eV for Na) make it suitable as an anode material for LIB and NIB applications.

Experimental investigations

Carbon-based materials for anodes in LIBs have drawbacks such as low theoretical capacity, safety issues, and poor rate capability, motivating researchers to investigate possible alternatives for LIB applications. Among transition metal dichalcogenides, MoS_2 has gained attention due to its theoretical capacity of 670 mAh/g and low cost. The unique layer structure of MoS_2 is also an added advantage, along with tunable interlayer spacing and band gap [2, 25].

Bulk MoS_2 for LIBs

He et al. [26] fabricated high mechanically integrated 3D MoS_2 nanoflowers by multistep hydrothermal processes. A specific capacity of 901 mAh/g was achieved at 1000 mA/g after 700 cycles were obtained for this MoS_2/graphene nanoflower. In another study [27], a 3D aerogel was fabricated using a simple hydrothermal method to construct layered ultrathin MoS_2/graphene nanosheets. This method did not utilize any polymer additive, conductive additive, or any metal current collector. A capacity of 870 mAh/g was achieved at a high current rate of 1 A/g. This binder-free electrode delivered a capacity of 1298 mAh/g at a rate of 0.1C. This architecture provides active sites for Li-ions and allows robust electron/Li-ion transportation. Wei et al. [28] fabricated a 2D MoS_2 nanosheet onto a 3D structure with a structural hierarchy of almost seven orders of magnitude. Thus, a 3D nanofoam was constructed, which allowed excellent charge transport and was mechanically resilient and chemically stable with a rapid ion diffusion property. Figure 2a–c shows the schematic representation of the printing of MoS_2 3D foam through dewetting-induced manufacturing, showing the electrohydrodynamic (EHD) setup that indicates the EHD and structural formation of the MoS_2 nanofoam. The rate performance study showed that it delivers a capacity of 1111 mAh/g at a rate of 10 A/g and 1200 mA/g at a rate of 5 A/g, maintained for about 200 cycles.

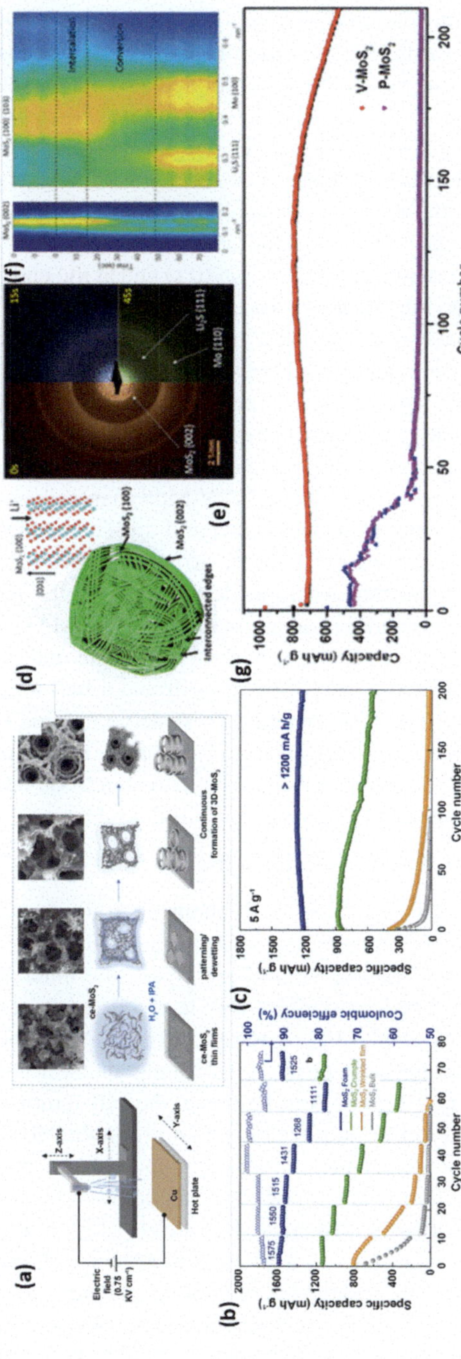

Fig. 2 **a–c** Schematic diagram of electrohydrodynamic (EHD) setup and MoS_2 foam structural evaluation, rate capability at different current densities and cycling comparison at 5A/g. Reproduced from [28] CC BY 4.0. **d–g** Schematic of Li insertion pathway in P-MoS_2. The selected area electron diffraction (SAED) analysis during the lithiation process with time sequential, SAED radial intensity profile with function of time and high-rate performance of MoS_2 at 2C. Reprinted with permission from [30] ©2019 American Chemical Society

Two-dimensional (2D) MoS_2 for LIBs

The potential of achieving strong electrode–electrolyte interaction and smaller diffusion pathways with high surface-to-volume ratio of 2D MoS_2 has become an interesting topic for researchers owing to the above-mentioned advantages [29]. Sun et al. [30] reported that 2D-layered structural materials are often known for their high capacity and faster transport within the layers. Figure 2d–g depicts the schematic representation of V-MoS_2 (MoS_2 nanocakes bonded by vertically stacked (002) planes with the {100} open surface), which is capable of tackling the volume expansion during the lithiation process. The selected area electron diffraction (SAED) analysis of V-MoS_2 was performed before lithiation (0 s) and at the end of intercalation (15 sand 45 s) after conversion. From the data, the diffraction spot of MoS_2 (002 plane) remains before the conversion, and after the reaction, it disappears to form Li_2S and Mo. The overall phase transition was observed through the radial intensity profile. The high-rate capability study and cycling comparison showed the stability of V-MoS_2. The cycling studies and rate capability tests of V-MoS_2 were compared with those of P-MoS_2 (MoS_2 nanoflowers comprising (002)-surfaced planar nanosheets) revealing the stability of V-MoS_2 at a high rate of 5C. V-MoS_2 nanocakes are interconnected and vertically stacked along the (100) plane, and this shows higher stability and higher capacity than (002)-terminated loosely arranged P-MoS_2 nanoflowers. Hao et al. [31] attempted to solve the structural degradation of 2D materials with a one-step metallothermic-synchronous strategy in constructing MoS_2 enclosed by few-layer graphene. It delivered a capacity of 550 mAh/g at the end of 500 cycles at a current rate of 5.0 A/g. Mateti et al. [32] demonstrated a ball milling process that resulted in high yield and low cost, providing a scalable method to develop a 2D MoS_2/Graphene hybrid electrode. This exhibited a capacity of 342 mAh/g at the end of 110 cycles at a rate of 250 mA/g, whereas the bare electrode only exhibited a capacity of 37 mAh/g at the end of 100 cycles at a similar rate.

Zero-dimensional/low-dimensional MoS_2 for LIBs

The conversion reaction with Li ions leads to mechanical strain; however, the 0D formation with carbon can effectively reduce this mechanical strain. As 0D can be formed by combining carbon spheres with MoS_2 layers, the inherent strain can be tackled by carbon spheres. This could effectively lead to the long-term stability of this material [1, 25]. Zhang et al. [10] reviewed different MoS_2 hollow structural materials with various dimensions through hard-template, soft-template, self-template, and template-free approaches. Among these, the self-template and hard-template approaches appeared to be complicated yet controllable. However, a scalable approach has been derived from the template-free and soft-template methods, but this method was not controllable. The researchers highlighted that the 0D hollow structure could enhance the diffusion kinetics. Jing et al. [33] formulated a few atomic-layered MoS_2 hollow nanospheres with an expanded interlayer and an N-doped monolayer carbon intercalation in the MoS_2 monolayer using the soft template approach. Figure 3a–d depicts the schematic representation of the synthesis method followed by the electrochemical analysis. A capacity of 1075 mAh/g remained even

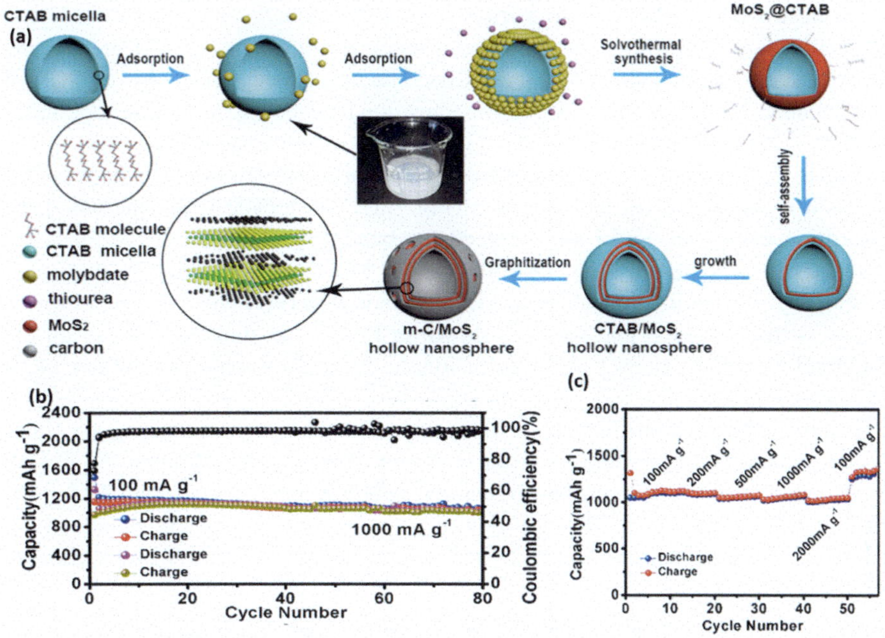

Fig. 3 **a** Schematic representation of fabrication of alternately intercalated m-C/MoS_2 hollow nanospheres with few atomic layers. **b** and **c** Cycling performance and rate capability of m-C/MoS_2. Reprinted with permission from [33] ©2018 Elsevier B.V.

after the 80th cycle. From the rate performance study, a capacity of 1035 mAh/g was obtained at 2000 mA/g. Wang et al. [34] utilized a solvothermal method to prepare hierarchical MoS_2 hollow nanospheres. A controlled solvothermal process followed by a sulfidation reaction was conducted to convert Mo-glycerate solid spheres to MoS_2. This structure delivered an impressive performance with a capacity of 1100 mAh/g at a rate of 0.5 A/g. The rate performance showed that a capacity of 576 mAh/g was delivered at a rate of 5A/g.

2.2 MoS_2 Electrode Applications Beyond LIBs (Na and K)

Computational studies

NIB research, driven by low cost, high abundance of Na, and safety aspect, has recently received great attention. Graphite is the most common material used as an anode in LIBs [35]. Li forms an intercalated compound, LiC_6, at a full-charge stage with a maximum capacity of 340 mAh/g. However, due to the large size of Na ions, intercalation of Na ions in graphite is inefficient, yielding a low capacity [35]. Unlike graphene, MoS_2 offers higher interlayer spacing to accommodate guest electronegative ions. As per a theoretical study [36], monolayer MoS_2 shows higher adsorption than bulk MoS_2. It exhibited a theoretical capacity of 335 mAh/g and could be used as an anode material in NIBs. The monolayer offers a lower OCV of 1.0 V than the bulk form (1.7–2.0 V). The transition from bulk to monolayer decreases the diffusion barrier from 0.7 eV to 0.11 eV, offering high Na-ion mobility.

Hao et al. [37] investigated the electrochemical properties of MoS_2 in the presence of strain using first-principles analysis (Fig. 4a). The tensile strain increased the adsorption energy of Na/Li and reduced the bandgap, which improved electrical conductivity with stability. The d-orbitals of the Mo atom were tuned by the strain, leading to s-d hybridization and hence Li/Na adsorption (Fig. 4b). The Li adsorption energy with Li content and strain is shown in Fig. 4e. In this case, the strain did not affect the energy barrier of ion diffusion, high storage capability, and low OCV (Fig. 1c–d).

K atom intercalation in graphite increases the interlayer distance by six times in comparison to Li, leading to faster decay in capacity [38]. Rehman et al. [39] investigated 2D MoS_2 for anodic application in K-ion batteries and showed that the loading of K atoms even at a low concentration yields an MoS_2 metallic conductivity behavior. The study reported an OCV and storage capacity of 0.24 V and 335 mAh/

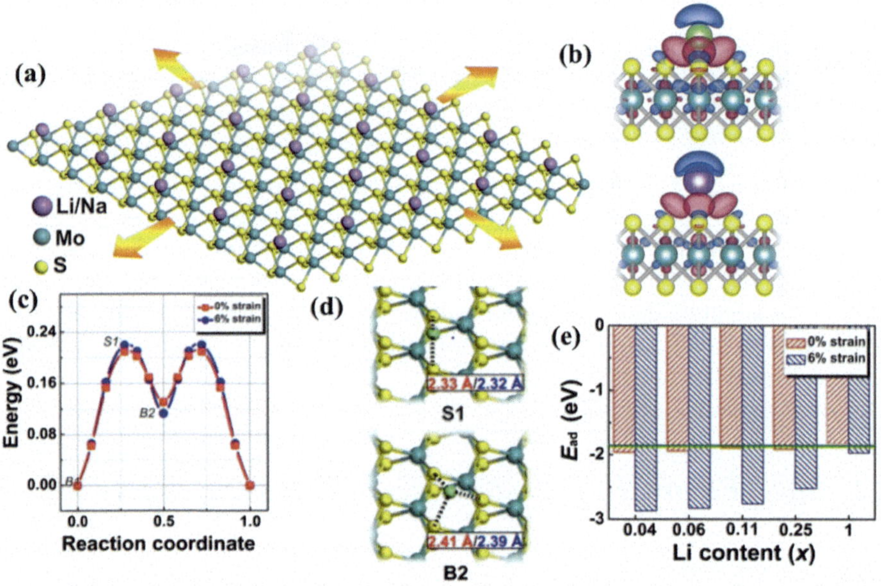

Fig. 4 Schematic of Li/Na adsorption on an MoS_2 sheet in the presence of tensile strain **a**. Iso surfaces of differential charge density of the most stable configuration of Li (top) and Na (bottom) on the MoS_2 monolayer **b**. Diffusion energy barrier for Li migration on the MoS_2 surface with strain **c**. Structure of adsorbed Li at S1 and B2 site **d**. Adsorption energy as a function of Li content with 0 and 6% strain on the MoS_2 monolayer **e**. Reprinted with permission from [37] ©2018 CC BY 4.0

g, respectively. The low energy barrier of 63 meV offered faster charge and discharge cycles.

Experimental investigations

Although portable electronic devices offer a wide operating window, good efficiency, high energy density, and lightweight, the uneven distribution of Li resources and its depletion necessitate a lookout for possible alternatives to LIBs, like NIBs and K-ion batteries (KIBs) [40]. Fagiolari et al. [41] explored MoS_2 as a potential candidate for KIB application and conducted preliminary studies. In a recent review paper in 2020, it was reported that the metal sulfide shows better reversibility and electronic conductivity than the counterpart metal oxide [42]. Wu et al. [43] developed dual-phase MoS_2 (DP-MoS_2) by mixing 1T and 2H phases to overcome the challenges of MoS_2. This DP-MoS_2 showed a capacity of 300 mAh/g at the end of 200 cycles at a current density of 0.5 A/g. In addition, a rate capability of 220 mAh/g at a rate of 2 A/g was obtained. A full cell was assembled with the DP-MoS_2 anode and $Na_3V_2(PO_4)_3$ cathode and delivered a capacity of 210 mAh/g at a rate of 0.5 A/g. Yao et al. [44] synthesized MoS_2 for NIB applications by developing a bundled defect-rich MoS_2 with plenty of vacancies. This architecture delivered a highly reversible capacity of 350 mAh/g at a rate of 2 A/g and a capacity of 272 mAh/g at a rate of 5 A/g at the end of 1000 cycles. Dong et al. [45] synthesized a 3D MoS_2/Nitrogen-doped graphene aerogel through a facile hydrothermal method. The 3D porous morphology tackled the volume expansion during charging and discharging. A capacity of 673 mAh/g at a rate of 100 mA/g and a specific capacity of 305 mAh/g at a rate of 2 A/g were reported for the NIB. This material was tested for KIB application and showed a capacity of 349 mAh/g at a rate of 100 mA/g. In another study [46], an MXene ($Ti_3C_2T_x$)/MoS_2 composite was designed for KIB application. This facilitated electron and ion transport and accommodated higher volume expansion. At 50 mA/g, a reversible capacity of 290 mAh/g was obtained. Figure 5a–e shows the electrochemical performance of spherical MoS_2@polydopamine to form an MoS_2@NC core shell structure used for NIB applications [47]. The MoS_2@NC shows high-rate capability, and even at a rate of 30 A/g, it delivered a capacity of 467 mAh/g. Cycling test at 20 A/g delivered 85% retention calculated for the 500–10,000th cycle. The inset in Fig. 5a shows that the MoS_2@NC layer remains intact with the current collector after cycling, while MoS_2 peeled off from the current collector. This is supported by the TEM images taken after cycling, as represented in the figure. The schematic representation is provided in Fig. 5e [47]. Figure 5f–j shows the utilization of MoS_2@C for KIB applications. The schematic diagram represents the synthesis procedure for MoS_2@C using the hydrothermal as well as three-roll milling process. Further, the cycling data at a rate of 0.1 A/g were compared between MoS_2 and MoS_2@C, and the cycling data at a rate of 0.5 A/g for MoS_2@C were reported. The rate performance of MoS_2 nanosheets and cycling at a high rate of 2 A/g for 150 cycles established its applicability for KIB applications [47].

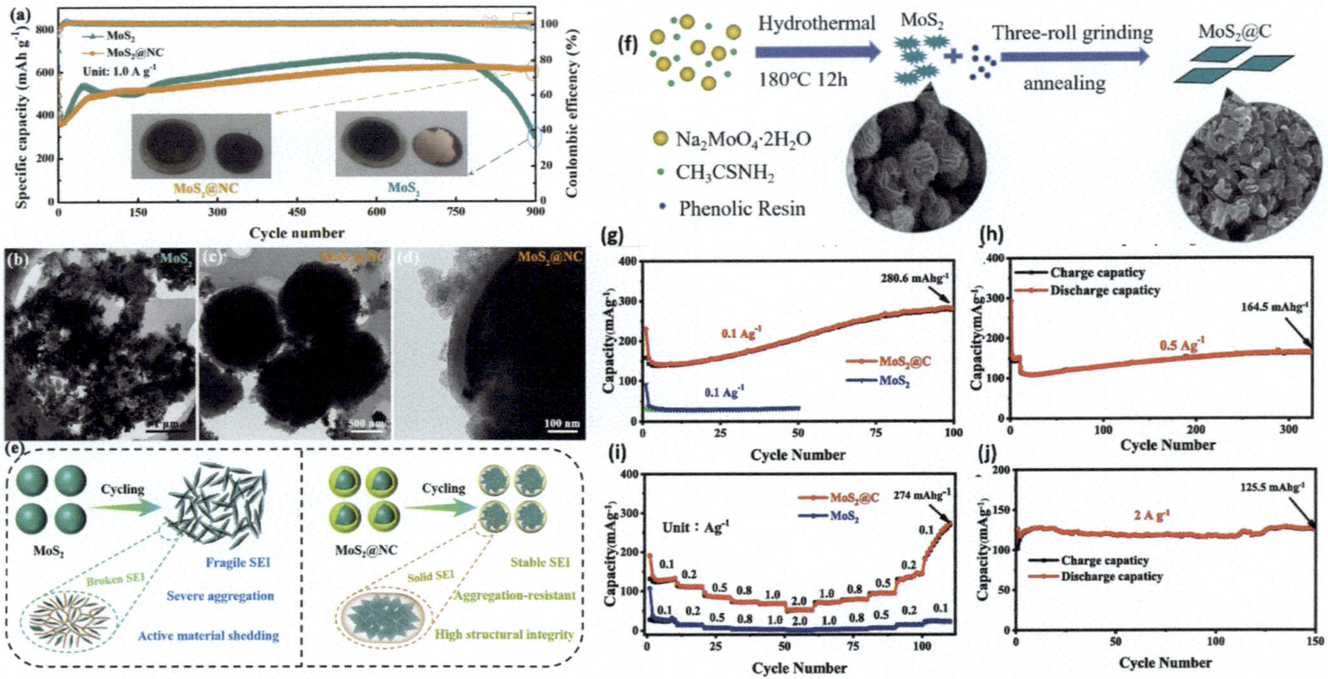

Fig. 5 **a** High-rate performance of MoS_2@NC for NIB, comparison of cycling of MoS_2 and MoS_2-Coreshell@NC at rate of 1.0A/g, TEM images of **b** spherical MoS_2@NC, **c** and **d** core–shell MoS_2 after 900 cycles, **e** schematic representation of the electrode during cycling. **f–j** shows the K-ion battery work, **f** represents schematically the synthesis procedure for MoS_2@C **g** the cycling comparison of MoS_2 and MoS_2@C at a rate of 0.1 A/g, **h** MoS_2@C at 0.5A/g, **i** rate performance comparison study and, **j** potassium ion battery cycling at 2 A/g. Reprinted with permission from [47] ©2023 Elsevier B.V.

3 MoS_2 Electrode for Multivalent-Ion Batteries

MoS_2 for Mg/Zn-ion batteries

Computational results on Mg/Zn-ion batteries

Recently, Mg/Zn-ion batteries are attracting greater attention than LIBs due to their low cost, abundance of Mg and Zn in the Earth's crust, and safety. In addition, being a divalent metallic cation, Mg/Zn can provide double charge transportation in comparison to monovalent cations such as Li, Na, and K. Mg and Zn show weak adsorption strength on the MoS_2 surface, making MoS_2 an unsuitable candidate for anode application [48]. The estimated energy barrier for Mg and Zn is 0.45 and 0.6 eV, respectively, showing poor electrochemical performance as an anode material for Mg/Zn-ion batteries. Moreover, 1H to 1T phase transition does not occur during the loading of Mg/Zn atoms [48]. Yang et al. [49] investigated zigzag nanoribbons of MoS_2 as an anode material for Mg-ion batteries. With double-sided adsorption, a maximum theoretical capacity of 223 mAh/g could be achieved. Mg prefers the Mo site for adsorption, and only partial charge transfer was observed during adsorption. The activation barrier estimated for MoS_2 nanoribbons was 0.48 eV, which is much lower than that for bulk inlayer migration (2.61 eV) [49]. Wu et al. [50] discussed the buckled square lattice of MoS_2, known as B-MoS_2, as a promising anode material for Mg-ion batteries, providing a high capacity of 921 mAh/g and low OCV of 0.15 eV. This high capacity could be explained in terms of the occurrence of sub-interfacial and interfacial charge transfer between Mg and B-MoS_2 layers [50]. B-MoS_2 offers fast charge/discharge rates due to a low migration barrier of 0.60 eV. Zn-ion batteries suffer from sluggish Zn-ion migration dynamics and severe structural degradation upon the insertion/extraction of Zn-ions. Recently, insertion of foreign species (e.g., crystal water, graphene, and NH_4^+) or heteroatoms in the MoS_2 frame successfully increased the interlayer spacing towards enhancing the specific capacity and reducing the migration barrier [51, 52]. Li et al. [53] designed D-MoS_2-O materials using a molecular engineering strategy, which increased the interlayer spacing from 6.2 Å to 9.6 Å and enhanced the electrical conductivity of the material. DFT calculations revealed that the Zn-ion migration barrier reduced drastically due to O-doping (0.59 eV) and S-vacancies (0.43 eV) compared with the pristine MoS_2 (0.99 eV).

Experimental results for Mg-ion batteries

Among the next-generation electrochemical storage systems, the rechargeable Mg-ion battery is considered to be promising due to its safety, abundance of Mg, and volumetric capacity [54]. The 2D-layered material is a promising candidate for Mg^{2+} intercalation due to its higher surface area and surplus active sites as well as the ion transport pathway. The ionic radius (0.072 nm) and high polarity of Mg^{2+}aid its interaction with host lattices. This leads to the limited storage ability of Mg in pristine MoS_2 (only 20 mAh/g) [55, 56]. Wu et al. [57] introduced lattice engineering to overcome the challenges of the van der Waals heterostructure; this helped lower the energy barrier and led to specific ion diffusion. Figure 6a–d shows the electrochemical performance of the MoS_2/Graphene heterostructure (MoS_2/GR), exhibiting a capacity of 90 mAh/g at a rate of 500 mA/g. Cycling at 500 mA/g maintains 83% retention for 500 cycles. Galvanostatic intermittent titration technique analysis at 10 mA/g resulted in a Mg^{2+} diffusion coefficient value of 3.24×10^{-9} cm^2 s^{-1}, whereas the value for bulk was 10^{-20} cm^2 s^{-1}, indicating the sluggish kinetics of Mg^{2+}. A hierarchical nanocomposite was fabricated and demonstrated for Mg-ion storage application. A hierarchical nanocapsule of hydrogen-substituted graphdiyne nanotube, encapsulated in Cu-doped MoS_2 nanopetals, was synthesized, which could successfully provide a buffer zone. This was investigated as a cathode material and showed a capacity of 148 mAh/g at a current density of 50 mA/g [58]. In another study, a combination of MoS_2 and MXene, which resembled a petal-like morphology, was synthesized using a hydrothermal method. An enhanced capacity of 165 mAh/g at a current density of 50 mA/g was obtained. The rate performance study of the material as a cathode for Mg-ion batteries showed a capacity of about 93 mAh/g at a rate of 200 mA/g [59].

Experimental Results for Zn-ion batteries

Rechargeable aqueous batteries have achieved great attention due to their safety, non-toxicity, and low cost. Among aqueous batteries, Zn-ion batteries are promising systems for large-scale energy storage applications. The unique properties of these batteries such as high theoretical capacity close to 820 mAh/g and −0.76 V redox potential are their main advantages. However, Zn-ion batteries face challenges in terms of a suitable cathode material. Various cathode materials including vanadium-based and manganese-based metal oxides and prussian blue analogs have been explored. The poor cyclability or low capacity of all these electrodes is a major issue [60–62]. In contrast, transition metal dichalcogenide-based electrodes have emerged as a potential electrode material for metal cations, specifically due to their high theoretical capacity and low cost. In an aqueous electrolyte, a high desolvation barrier must be overcome by the Zn ion [63]. The kinetics of Zn intercalation in these materials are mostly hindered by electrostatic interaction between ions of Zn^{2+} and the materials; this results in the sluggish kinetics of Zn intercalation. More specifically, at a high rate, the poor performance can be attributed to the difficulty of intercalating Zn^{2+} into the electrode [64]. Shao et al. [65] developed a mixed-phase MoS_2 with 66% of 1T phase and remaining with 2H phase using a hydrothermal method. The MoS_2 delivered a satisfactory performance with a capacity of 156 mAh/g and a capacity retention of 97.3% at 1 A/g for about 500 cycles. Another study reported the MoS_2@CNT hybrid for application in Zn-ion batteries; the comparison with MoS_2@CNT with different CNT concentrations at a rate of 1A/g and comparison with P-MoS_2 were also described. The rate performance of MoS_2@CNT showed a capacity of 180 mAh/g and 93.1 mAh/g at 0.1 A/g and 3.0 A/g, respectively [66]. Zhao et al. [67] explored MoS_{2-x}@NCNTs as a cathode for Zn-ion batteries and showed that the sulfur vacancies typically reduce the Zn^{2+} diffusion barrier. At 0.1 A/g, a specific capacity of 220 mAh/g was obtained, whereas at a high rate of 3 A/g, a specific capacity of 173 mAh/g was obtained. Figure 6e displays an MoS_{2-x}@NCNT

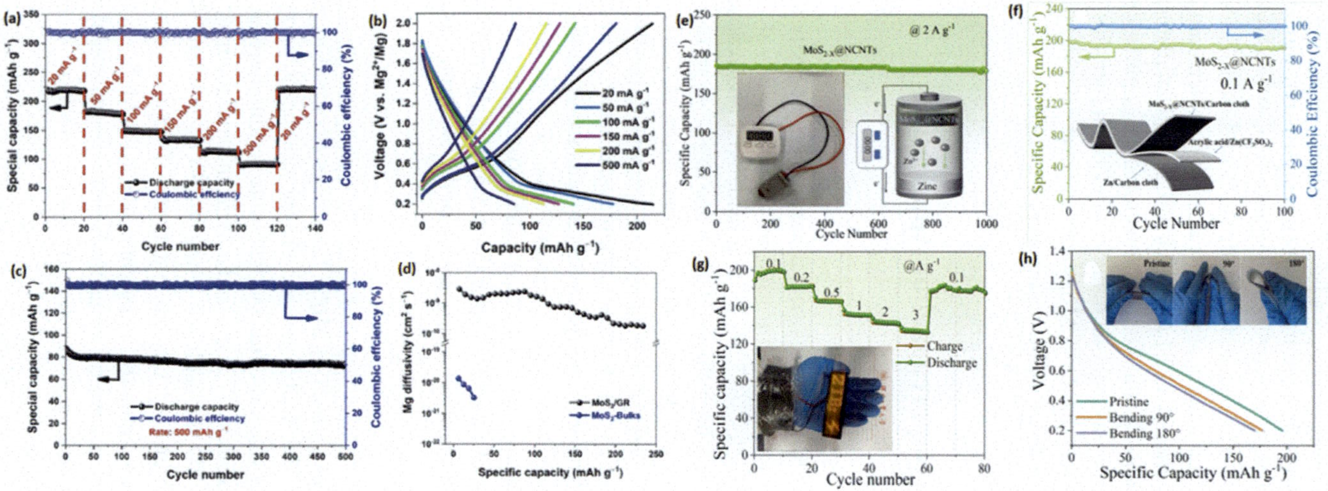

Fig. 6 **a–d** Electrochemical studies of MoS_2/GR, rate performance study, galvanostatic charge/discharge (GCD) profile, and cycling performance at 500 mA/g and comparison plot of the diffusion coefficient of MoS_2/GR and MoS_2/Bulk. Reprinted with permission from [57] ©2021 Elsevier B.V. **e–h** MoS_2-x@NCNT at 2A/g and optical and schematic of zinc battery powered timer, the cycling and CE of flexible Zn-ion battery, rate performance study of flexible zinc battery and voltage curve for Zn battery at 0°, 90°, and 180°. The electrochemical performance of MoS_2 for Zn-ion battery cycling studies, GCD profile, Reprinted with permission from [67] ©2023 Elsevier B.V.

at 2A/g and the optical and schematic representation of the Zn-ion battery-powered timer, which shows no capacity decay at 2A/g for about 1000 cycles and the series-connected Zn-ion battery shows that the timer functions properly. MoS_{2-x} flexible Zn-ion battery was also assembled, with a discharge capacity of 192 mAh/g at 0.1 A/g for about 100 cycles (Fig. 6f). The rate performance study of flexible Zn-ion battery and the voltage curve for Zn-ion battery at 0°, 90°, and 180°are depicted in Fig. 6g and h, showing a rate performance of 152.1 mAh/g at 3 A/g and a small energy decay under severe bending conditions. Li et al. [68] demonstrated the application of MoS_2 with expanded interlayer spacing as cathode materials in Zn-ion batteries. At 0.1 A/g, a capacity of 202.6 mAh/g was obtained. Moreover, an energy density of 148 Wh/kg and capacity retention of 98.6% for 600 cycles were reported.

4 MoS_2 Electrode for Supercapacitor Applications

Computational studies

Supercapacitors are energy storage devices with a lower energy density and higher power density than batteries and hence are used as complementary devices with batteries in hybrid electric vehicles [69]. MoS_2 is considered a good capacitor, as it has a layered structure that offers high charge storage due to a large surface area, and ions are stored between the layers through intercalation. Xu et al. [70] studied the effect of B, N, P, Al, Cu, Au, and Ag doping and S vacancies in pristine MoS_2 on properties such as stability and electrical and quantum capacitance. Al and B doping significantly increased the electronic and quantum capacitance of MoS_2 compared with pristine MoS_2. With an increase in the concentrations of Al from 1.3 to 8.3%, the quantum capacitance increased from 44.76 $\mu F/cm^2$ at 0.17 V to 227.85 $\mu F/cm^2$ at 0.53 V. This enhancement of quantum capacitance was mainly attributed to the local states (significant reduction in bandgap) arising in the electronic structure due to doping.

Experimental studies

The unique supercapacitive properties of MoS_2 have been explored in the field of supercapacitor applications. The double-layered charge storage property due to a large surface area of the sheet-like structure and fast ionic conductivity are the main advantages of MoS_2 [71]. Gao et al. [72] utilized a hydrothermal method to prepare MoS_2 nanosheet-assembled 3D nanospheres using an SiO_2 template. A specific capacitance of 683 F/g at a rate of 1A/g and a stability of 1000 cycles with 85% retention were reported. An asymmetric supercapacitor with MoS_2 as a positive electrode and activated carbon as a negative electrode were demonstrated to have an energy density of 20.42 Wh/kg and power density of 750.3 kW/kg. Recently, Lee et al. [73] investigated oxygen-incorporated 1T/2H hybrid phase MoS_2 nanoflowers on graphite foil. The hybrid phase provided excellent stability and conductivity due to the presence of 2H and 1T mixed phases. A high specific capacitance of about 280 F/

g at a current density of 1 A/g was obtained. A 10,000-cycle stable performance with 88% retention at a rate of 10 A/g was also demonstrated. An asymmetric supercapacitor was fabricated with MoS_2@GF, showing a power density of 450–1800 W/kg and energy density of 39.7–5.46 Wh/kg. Figure 7a–d illustrates the MoS_2 nanodots synthesized by dip coating onto a 3D nickel substrate. The specific capacitance at a rate of 1.5 A/g was recorded to be 395 F/g. For the symmetric supercapacitor, the rate performance retention was about 95% for a rate increase from 1.5 to 10 A/g [74]. Another study verified the importance of nanocomposites and prepared a biocarbon-based MoS_2 using a hydrothermal method. A specific capacitance of 945 F/g was obtained at a current density of 0.5 A/g. This material exhibited an excellent power density of about 3800–8000 W/kg and an energy density of 74.9–157 Wh/kg [75]. From Fig. 7e–h, a specific capacitance of 38.3 F/g at a rate of 1 A/g and 19 F/g at a rate of 10 A/g can be noted. The corresponding energy density was 7.6 Wh/kg, with a corresponding power density of 608 W/kg. Further, 92% retention was recorded for the symmetric supercapacitor at the end of 2300 cycles [76]. Moreover, doping has been found to be an effective strategy. Vikraman et al. [77] demonstrated a series of transition metal doping to activate more active sites like Ni, Cu, and Fe. The maximum capacitance was shown by the Cu-doped MoS_2 nanostructure, with a capacitance of 353 F/g at a current density of 1A/g and an exceptional retention of 94% after 5000 cycles.

5 MoS_2 Heterostructures for Battery and Supercapacitor Applications

Computational studies

The 2D van der Waals heterostructures have evolved as a new class of materials that can be used in battery applications as an effective electrode. This section discusses how the 2D heterostructures open the possibility of new battery electrode materials, taking advantage of favorable properties and overcoming the drawbacks of individual 2D layers. First-principle simulations play a crucial role in advancing this class of materials by offering insights into the process, which is difficult to investigate otherwise.

Pushparaj et al. [78] studied the coal-derived graphene/MoS_2 heterostructure as an anode for LIBs. The anode showed a capacity of 813 mAh/g with 99% Coulombic efficiency even after 150 cycles at a constant current density of 100 mA/g. The barrier was reduced significantly from that in bare MoS_2, as estimated using DFT. Barik et al. [79] explored the feasibility of graphene/blue P/MoS_2 vdW trilayer heterostructure for LIB applications. The specific capacity of this heterostructure was 818 mAh/g, which is much higher than that of MoS_2/VS_2 (584 mAh/g) and black P/graphene (626 mAh/g). The Li adsorption energy at a higher Li concentration was higher than that of the Li cohesive energy, which indicates that clustering does not take place at a higher Li concentration. At maximum Li coverage, the OCV was estimated to be

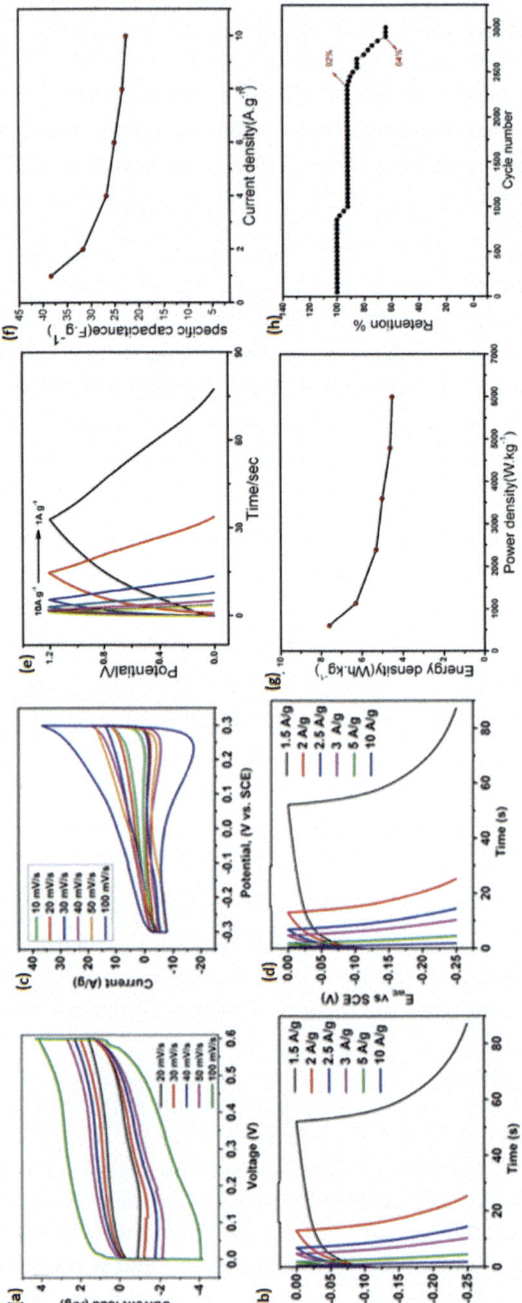

Fig. 7 **a–d** Electrochemical performance of MoS_2 nanodots, cyclic voltammetry (CV) plot, and galvanostatic charge/discharge (GCD) profile at different rates of symmetric supercapacitor. **c** and **d** Electrochemical performance of MoS_2 NDs in three-electrode configuration CV and GCD profiles. Reprinted with permission from [74] ©CC-BY-NC-NP. **e–h** Electrochemical performance of MoS_2/CB-C symmetric supercapacitor, GCD curves, current density variation as a function of specific capacitance, specific capacitance variation as a function of current density, Ragone plot and cycling at 2 A. Reprinted with permission from [76] ©2020 Elsevier B.V.

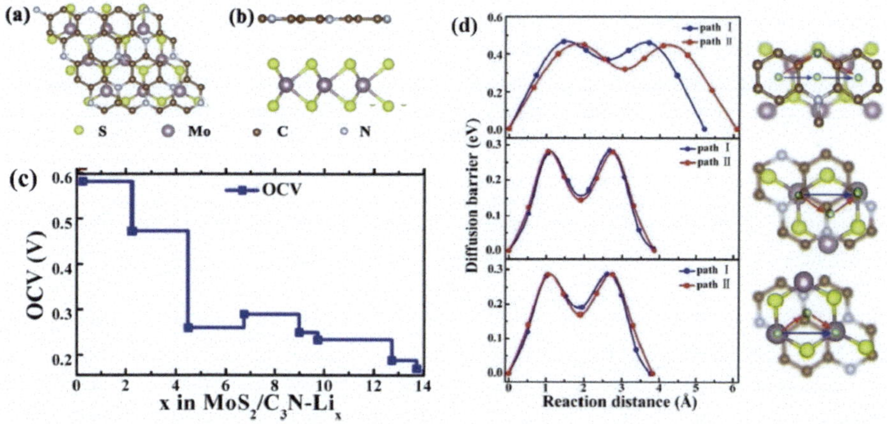

Fig. 8 Top **a** and side **b** views of MoS_2/C_3N heterostructure. Open circuit voltage of MoS_2/C_3N heterostructure versus Li content intercalated in the interlayer spacing of the heterostructure **c**. Diffusion barrier of Li migration on the outer C_3N and C_3N/MoS_2 heterostructure and outer MoS_2 surface is shown **d**. Pathways highlighted using red and blue arrows correspond to the red and blue color energy profiles. Reprinted with permission from [80] ©2022 with permissions from Elsevier

0.26 eV, suggesting that this material is an excellent anode for LIBs. The diffusion barrier varied between 0.06 and 0.22 eV depending on the layer or interlayer to which Li is transported. He et al. [80] studied the MoS_2/C_3N heterostructure for application as an anode in LIBs (Fig. 8a–b). The estimated OCV was in the range of 0.6–0.2 V, as shown in Fig. 8c. The energy barrier varied depending on the surface to which Li migration takes place (Fig. 8d); the lowest barrier was observed on MoS_2 and the interlayer spacing of the MoS_2-C_3N heterostructure. MoS_2 heterostructures for LIB/NIB applications are provided in Table 1.

These heterostructures can also be used effectively in supercapacitors for higher values of capacitance. Kapse et al. [81] studied the MoS_2/C heterostructure to increase the quantum capacitance for supercapacitor applications using first-principles analysis. The density of states near the Fermi level of 1T/1T' MoS_2 resulted in an increased value of quantum capacitance. The 1T' MoS_2/CNT heterostructure showed a quantum capacitance of 500 $\mu F/cm^2$ at an electrode potential of 0.6 V (Fig. 9a–c). The quantum capacitance for different combinations of the $MoS_2$and carbon heterostructures is shown in Fig. 9d.

Experimental studies

The 2D heterostructures have garnered interest in various applications in the field of material science and engineering. The distinctive chemical and physical properties with tunable compositions, morphologies, and structural properties have gained significant attention. The functional applications of 2D materials in energy storage can be attributed to their charge transport properties and favorable interlayer spacing. The reduced number of electrochemical active sites, associated with the numerous stacking and agglomeration of layers, resulting in a low Coulombic efficiency and

Table 1 Battery characteristic of 2D MoS_2 heterostructures as an electrode in Li/Na-ion batteries estimated using first-principle calculations

Two-dimensional heterostructure	Bandgap (eV)	Binding atom	Binding energy or adsorption energy (eV)	Open circuit voltage (V)	Theoretical specific capacity (mAh/g)	Diffusion barrier (eV)	References
$MoS_2/TiCO_2$	0.00	Li	−2.00	0.2.0–0.75	470	0.15	[82]
MoS_2/N-doped carbon	–	Li	−2.48	–	815	1.49	[83]
MoS_2/C_3N	0.00	Li	−2.22	0.17	743	0.3	[80]
$C_3N/MoS_2/C_3N$	0.00	Li	−2.82	–	814	–	[80]
MoS_2/borophene	0.00	Li	−1.68	0.64	539	0.3	[84]
1T MoS_2/Ti_2C	0.00	Li	−0.5	0.62–0.25	362	0.03	[82]
MoS_2/VS_2	0.00	Li	2.52	1.8–0.5	584	0.24	[85]
1T MoS_2/graphene	0.00	Na	−1.22	–	–	0.7	[86]
3R MoS_2/graphene	0.00	Na	−0.065	–	–	0.25	[86]
MoS_2/WS_2	1.45	Na	−2.41	–	–	0.36	[87]
MoS_2/VS_2	0.00	Na	1.90	1.4–0.4	584	0.45	[85]

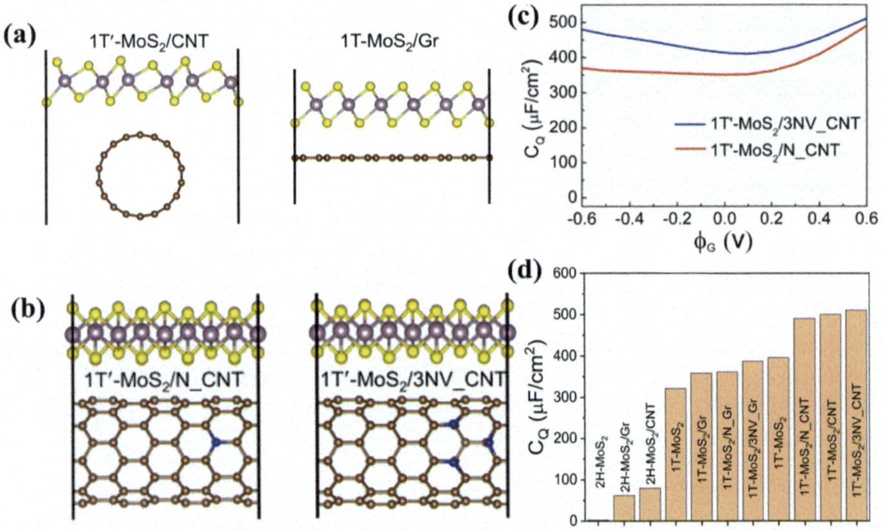

Fig. 9 Side view of 1T'-MoS_2/CNT and 1T-MoS_2/graphene heterostructures **a**. Optimized structure of 1T'-MoS_2/N_CNT and 1T'-MoS_2/3NV_CNT heterostructure **b** and their quantum capacitance as a function of voltage in the range of −0.6–0.6 V for both heterostructures **c**. Quantum capacitance of 2H MoS_2 and various heterostructures of MoS_2 **d**. Reprinted with permission from [81] ©2022 with permissions from Elsevier

capacity fading, which are considered major drawbacks of 2D MoS_2 [88]. In contrast, the synergistic effect of heterostructures leads to cyclability and higher rate capability. Another important property of heterostructures is the presence of a supporting framework for freestanding electrodes. Moreover, the 2D framework is capable of accommodating insulated active materials like sulfur, red phosphorus, and silicon to stabilize and suppress the volume expansion [89]. Pham-Cong et al. [90] developed a hierarchical heterostructure with $TiNb_2O_7$ along with ultrathin nanosheets of MoS_2 using a simple electrospinning and hydrothermal method. This entangled structure revealed a synergy of both 1D and 2D combinations. This heterostructure was studied for Li storage application and delivered a capacity of 740 mAh/g at 1 A/g after 200 cycles. Hussain et al. [91] synthesized $MoS_2@X_2C$ (X = Mo or W) hybrid supercapacitors with excellent cyclability. The hybrid $MoS_2@Mo_2C$ supercapacitor showed a specific capacitance of 1040 F/g at a current rate of 0.5 A/g, whereas $MoS_2@W_2C$ delivered a capacitance of 681 F/g at a current density of 0.5 A/g. A symmetric supercapacitor with $MoS_2@Mo_2C$ was fabricated and showed a capacitance of 349 F/g at a current density of 0.5 A/g, with an energy density maximum of 48 Wh/kg and a power density of 0.25 kW/kg. Tiwari et al. [92] produced chemical vapor deposition-synthesized atomically thin MoS_2 nanoflakes on carbon nanotubes by magnetron sputtering. A symmetric MoS_2@CNT supercapacitor was fabricated and delivered an aerial capacitance of 131 mF/cm^2 and a volumetric capacitance of 2.9 F/cm^3. This hybrid electrode showed a cycling stability of 97.6% even after 2500 cycles. Pan et al. [93] developed a hierarchical tubular heterostructure of MoS_2/C using a templated method. This hetero-interface had a vital role in achieving a stable performance as well as tackling the volume expansion and improvisation of Na-ion diffusivity. This heterostructure delivered a capacity of 484.9 mAh/g at a current rate of 2 A/g even after 1500 cycles. Lutz et al. [94] explored the MoS_{2+x} nanosheet/nanocarbon heterostructure. The nanocarbon was classified into 1D-carbon nanotube (CNT) and 2D-graphene nanoplatelet (GNP). Using a hydrothermal reaction, a 3D binder-free porous electrode with a high loading of 16.2–21.0 mg/cm^2 was obtained. MoS_{2+x}-GNP showed a capacity of 78 mAh/g for 100 cycles at a rate of 3.2 mA/cm^2. However, the MoS_{2+x}-CNT structure could not deliver a high capacity or stability.

6 Conclusion

In this chapter, we highlighted that MoS_2 is a prospective material for energy storage applications. The energy storage applications of MoS_2 range from monovalent-to multivalent-ion batteries and even supercapacitors. MoS_2 can be easily synthesized, is cheap, and can be effectively used in energy storage devices. Weak van der Waals forces between the layers make MoS_2 a suitable candidate for metal-ion batteries, as it allows easy intercalation of metal atoms between the layers. In addition, owing to its large surface area, superconductivity, and easy restacking, it is a suitable material for supercapacitor applications. This chapter also provides a brief note on each application with examples from both experimental and computational point-of-views.

Of note, MoS_2 faces certain challenges including large-scale production, controlling defects, high conductivity, and stability. We also discussed how nanocomposites and heterostructures can help improve some of its properties and make it suitable for energy storage applications. Although further research is needed to replace carbon-based materials in real-time energy storage applications, it is worth investigating this material. MoS_2 has the potential to replace other 2D materials in energy storage applications because it is economic, efficient, and an easily available solution.

Acknowledgements AS, DS, and SN are thankful to the Department of Science and Technology (Ref: DST/TMD/MES/2k18/225) for the financial support.

References

1. Wang T, Chen S, Pang H, Xue H, Yu Y (2017) MoS_2-based nanocomposites for electrochemical energy storage. Adv Sci 4:1600289
2. Yang T, Liang J, Sultana I, Rahman MM, Monteiro MJ, Chen YI et al (2018) Formation of hollow MoS_2/carbon microspheres for high capacity and high-rate reversible alkali-ion storage. J Mater Chem A 6:8280–8288
3. Singh A, Shirodkar SN, Waghmare UV (2015) 1H and 1T polymorphs, structural transitions and anomalous properties of $(Mo,W)(S,Se)_2$ monolayers: first-principles analysis. 2D. Mater 2:035013. https://doi.org/10.1088/2053-1583/2/3/035013
4. Xu X, Liu W, Kim Y, Cho J (2014) Nanostructured transition metal sulfides for lithium ion batteries: progress and challenges. Nano Today 9:604–630
5. Jayabal S, Saranya G, Wu J, Liu Y, Geng D, Meng X (2017) Understanding the high-electrocatalytic performance of two-dimensional MoS_2 nanosheets and their composite materials. J Mater Chem A 5:24540–24563. https://doi.org/10.1039/C7TA08327K
6. Chhowalla M, Shin HS, Eda G, Li L-J, Loh KP, Zhang H (2013) The chemistry of two-dimensional layered transition metal dichalcogenide nanosheets. Nat Chem 5:263–275. https://doi.org/10.1038/nchem.1589
7. Kalin M, Kogovšek J, Remškar M (2012) Mechanisms and improvements in the friction and wear behavior using MoS_2 nanotubes as potential oil additives. Wear 280–281:36–45. https://doi.org/10.1016/j.wear.2012.01.011
8. Wang D, Zhu W, Yuan Y, Du G, Zhu J, Zhu X et al (2018) Kelp-like structured $NiCo_2S_4$-C-MoS_2 composite electrodes for high performance supercapacitor. J Alloys Compd 735:1505–1513. https://doi.org/10.1016/j.jallcom.2017.11.249
9. Laffleur F, Keckeis V (2020) Advances in drug delivery systems: work in progress still needed? Int J Pharm 590:119912. https://doi.org/10.1016/j.ijpharm.2020.119912
10. Zhang H, Wang Y, Guo S, Hui Y, Wei H, Song J et al (2023) Advances in MoS_2 based hollow structural materials for high-performance metal-ion batteries. Batteries Supercaps
11. Meng YS, Arroyo-de Dompablo ME (2009) First principles computational materials design for energy storage materials in lithium ion batteries. Energy Environ Sci 2:589. https://doi.org/10.1039/b901825e
12. Whittingham MS (2004) Lithium batteries and cathode materials. Chem Rev 104:4271–4302. https://doi.org/10.1021/cr020731c
13. Samaras I, Saikh SI, Julien C, Balkanski M (1989) Lithium insertion in layered materials as battery cathodes. Mater Sci Eng B 3:209–214. https://doi.org/10.1016/0921-5107(89)90203-1

14. Julien C, Saikh SI, Nazri GA (1992) Electrochemical studies of disordered MoS_2 as cathode material in lithium batteries. Mater Sci Eng B 15:73–77. https://doi.org/10.1016/0921-5107(92)90034-7
15. Manthiram A (2020) A reflection on lithium-ion battery cathode chemistry. Nat Commun 11:1550. https://doi.org/10.1038/s41467-020-15355-0
16. Santiago Y, Cabrera CR (1994) Surface analysis and electrochemistry of MoS_2 thin films prepared by intercalation-exfoliation techniques. J Electrochem Soc 141:629–635. https://doi.org/10.1149/1.2054782
17. Li Y, Wu D, Zhou Z, Cabrera CR, Chen Z (2012) Enhanced Li adsorption and diffusion on MoS_2zigzag nanoribbons by edge effects: a computational study. J Phys Chem Lett 3:2221–2227. https://doi.org/10.1021/jz300792n
18. Seifert G, Terrones H, Terrones M, Jungnickel G, Frauenheim T (2000) Structure and electronic properties of MoS_2 nanotubes. Phys Rev Lett 85:146–149. https://doi.org/10.1103/PhysRevLett.85.146
19. Enyashin AN, Seifert G (2012) Density-functional study of Li_xMoS_2 intercalates ($0<x<1$). Comput Theor Chem 999:13–20. https://doi.org/10.1016/j.comptc.2012.08.005
20. Liu T, Jin Z, Liu D-X, Du C, Wang L, Lin H et al (2020) A density functional theory study of high-performance pre-lithiated MS_2 (M = Mo, W, V) Monolayers as the anode material of lithiumion batteries. Sci Rep 10:6897. https://doi.org/10.1038/s41598-020-63743-9
21. Kabiraj A, Mahapatra S (2018) High-throughput first-principles-calculations based estimation of lithium ion storage in monolayer rhenium disulfide. Commun Chem 1:81. https://doi.org/10.1038/s42004-018-0082-3
22. Chang K, Chen W (2011) L-Cysteine-assisted synthesis of layered MoS_2/graphene composites with excellent electrochemical performances for lithium-ion batteries. ACS Nano 5:4720–4728. https://doi.org/10.1021/nn200659w
23. Pereira AO, Miranda CR (2015) First-principles investigation of transition metal dichalcogenide nanotubes for Li and Mg ion battery applications. J Phys Chem C 119:4302–4311. https://doi.org/10.1021/jp510182u
24. Sukhanova EV, Bereznikova LA, Manakhov AM, Al Qahtani HS, Popov ZI (2022) A novel membrane-like 2D A'-MoS_2 as anode for lithium- and sodium-ion batteries. Membranes 12:1156. https://doi.org/10.3390/membranes12111156
25. Yang T, Liang J, Sultana I, Rahman MM, Monteiro MJ, Chen YI et al (2018) Formation of hollow MoS_2/carbon microspheres for high capacity and high rate reversible alkali-ion storage. J Mater Chem A 6:8280–8288
26. He H, Zhen L, Peng C, Jun L, Wang X, Jing Z (2022) 3D MoS_2/graphene nanoflowers as anode for advanced lithium-ion batteries. Trans Nonferrous Met Soc China 32:4041–4049
27. He J, Li P, Lv W, Wen K, Chen Y, Zhang W et al (2016) Three-dimensional hierarchically structured aerogels constructed with layered MoS_2/graphene nanosheets as free-standing anodes for high-performance lithium-ion batteries. Electrochim Acta 215:12–18
28. Wei X, Lin C-C, Wu C, Qaiser N, Cai Y, Lu A-Y et al (2022) Three-dimensional hierarchically porous MoS_2 foam as high-rate and stable lithium-ion battery anode. Nat Commun 13:6006
29. Huang G, Cai B, Zhan C, Sun P (2020) Two-dimensional material as anode for lithium ion batteries: recent progress. Int J Electrochem Sci 15:5416–5429
30. Sun C, Zhao K, He Y, Zheng J, Xu W, Zhang C et al (2019) Interconnected vertically stacked 2D-MoS_2 for ultrastable cycling of rechargeable Li-ion battery. ACS Appl Mater Interfaces 11:20762–20769
31. Hao X, Zhang J, Wang J, Zhao B, Qian M, Wang R et al (2022) Metallothermic-synchronous construction of compact dual-two-dimensional MoS_2-graphene composites for high-capacity lithium storage. Nano Energy 103:107850
32. Mateti S, Rahman MM, Cizek P, Chen Y (2020) In situ production of a two-dimensional molybdenum disulfide/graphene hybrid nanosheet anode for lithium-ion batteries. RSC Adv 10:12754–12758
33. Jing L, Lian G, Niu F, Yang J, Wang Q, Cui D et al (2018) Few-atomic-layered hollow nanospheres constructed from alternate intercalation of carbon and MoS_2 monolayers for

sodium and lithium storage. Nano Energy 51:546–555. https://doi.org/10.1016/j.nanoen.2018.06.084
34. Wang Y, Yu L, Lou XW (2016) Synthesis of highly uniform molybdenum–glycerate spheres and their conversion into hierarchical MoS_2 hollow nanospheres for lithium-ion batteries. Angew Chem 128:7549–7552
35. Yazami R, Touzain Ph (1983) A reversible graphite-lithium negative electrode for electrochemical generators. J Power Sources 9:365–371. https://doi.org/10.1016/0378-7753(83)87040-2
36. Park J, Kim J-S, Park J-W, Nam T-H, Kim K-W, Ahn J-H et al (2013) Discharge mechanism of MoS_2 for sodium ion battery: electrochemical measurements and characterization. Electrochim Acta 92:427–432. https://doi.org/10.1016/j.electacta.2013.01.057
37. Hao J, Zheng J, Ling F, Chen Y, Jing H, Zhou T et al (2018) Strain-engineered two-dimensional MoS_2 as anode material for performance enhancement of Li/Na-ion batteries. Sci Rep 8:2079. https://doi.org/10.1038/s41598-018-20334-z
38. Jian Z, Luo W, Ji X (2015) Carbon electrodes for K-ion batteries. J Am Chem Soc 137:11566–11569. https://doi.org/10.1021/jacs.5b06809
39. Rehman J, Fan X, Laref A, Dinh VA, Zheng WT (2021) Potential anodic applications of 2D MoS_2 for K-ion batteries. J Alloys Compd 865:158782. https://doi.org/10.1016/j.jallcom.2021.158782
40. Kulka A, Hanc A, Walczak K, P\lotek J, Sun J, Lu L et al (2022) Direct evidence of an unanticipated crystalline phase responsible for the high performance of few-layered-MoS_2anodes for Na-ion batteries. Energy Storage Mater 48:314–324
41. Fagiolari L, Versaci D, Di Berardino F, Amici J, Francia C, Bodoardo S et al (2022) An exploratory study of MoS_2 as anode material for potassium batteries. Batteries 8:242
42. Chen J, Chua DH, Lee PS (2020) The advances of metal sulfides and in situ characterization methods beyond Li ion batteries: sodium, potassium, and aluminum ion batteries. Small Methods 4:1900648
43. Wu J, Liu J, Cui J, Yao S, Ihsan-Ul-Haq M, Mubarak N et al (2020) Dual-phase MoS_2 as a high-performance sodium-ion battery anode. J Mater Chem A 8:2114–2122
44. Yao K, Xu Z, Huang J, Ma M, Fu L, Shen X et al (2019) Bundled defect-rich MoS_2 for a high-rate and long-life sodium-ion battery: achieving 3D diffusion of sodium ion by vacancies to improve kinetics. Small 15:1805405
45. Dong X, Xing Z, Zheng G, Gao X, Hong H, Ju Z et al (2020) MoS_2/N-doped graphene aerogles composite anode for high performance sodium/potassium ion batteries. Electrochim Acta 339:135932
46. Li J, Rui B, Wei W, Nie P, Chang L, Le Z et al (2020) Nanosheets assembled layered MoS_2/MXene as high performance anode materials for potassium ion batteries. J Power Sources 449:227481
47. Dong G, Yu H, Li L, Zhang R, Yang X, Zhu K et al (2023) Surface engineering of core-shell MoS_2@N-doped carbon spheres as stable and ultra-long lifetime anode for sodium-ion batteries. J Colloid Interface Sci 647:395–405
48. Zhang J, Lu X, Zhang J, Li H, Huang B, Chen B et al (2021) Metal-ions intercalation mechanism in layered anode from first-principles calculation. Front Chem 9:677620. https://doi.org/10.3389/fchem.2021.677620
49. Yang S, Li D, Zhang T, Tao Z, Chen J (2012) First-principles study of zigzag MoS_2 nanoribbon as a promising cathode material for rechargeable Mg batteries. J Phys Chem C 116:1307–1312. https://doi.org/10.1021/jp2097026
50. Wu D, Yang B, Zhang S, Ruckenstein E, Chen H (2021) Reshaping two-dimensional MoS_2 for superior magnesium-ion battery anodes. J Colloid Interface Sci 597:401–408. https://doi.org/10.1016/j.jcis.2021.04.002
51. Liang H, Cao Z, Ming F, Zhang W, Anjum DH, Cui Y et al (2019) Aqueous zinc-ion storage in MoS_2 by tuning the intercalation energy. Nano Lett 19:3199–3206. https://doi.org/10.1021/acs.nanolett.9b00697

52. Luo X, Peng W, Li Y, Zhang F, Fan X (2022) Understanding of the electrochemical behaviors of aqueous zinc–manganese batteries: reaction processes and failure mechanisms. Green Energy Environ 7:858–899. https://doi.org/10.1016/j.gee.2021.08.006
53. Li S, Liu Y, Zhao X, Cui K, Shen Q, Li P et al (2021) Molecular engineering on MoS_2enables large interlayers and unlocked basal planes for high-performance aqueous Zn-ion storage. Angew Chem Int Ed 60:20286–20293. https://doi.org/10.1002/anie.202108317
54. Ji X, Chen J, Wang F, Sun W, Ruan Y, Miao L et al (2018) Water-activated $VOPO_4$ for magnesium ion batteries. Nano Lett 18:6441–6448
55. Wu Y, Yu Y (2019) 2D material as anode for sodium ion batteries: recent progress and perspectives. Energy Storage Mater 16:323–343
56. Tan S, Xiong F, Wang J, An Q, Mai L (2020) Crystal regulation towards rechargeable magnesium battery cathode materials. Mater Horiz 7:1971–1995
57. Wu C, Zhao G, Yu X, Liu C, Lyu P, Maurin G et al (2021) MoS_2/graphene heterostructure with facilitated Mg-diffusion kinetics for high-performance rechargeable magnesium batteries. Chem Eng J 412:128736
58. Zhuo S, Huang G, Sougrat R, Guo J, Wei N, Shi L et al (2022) Hierarchical nanocapsules of Cu-doped MoS_2@H-substituted graphdiyne for magnesium storage. ACS Nano 16:3955–3964
59. Xu M, Bai N, Li H-X, Hu C, Qi J, Yan X-B (2018) Synthesis of MXene-supported layered MoS_2 with enhanced electrochemical performance for Mg batteries. Chin Chem Lett 29:1313–1316
60. Blanc LE, Kundu D, Nazar LF (2020) Scientific challenges for the implementation of Zn-ion batteries. Joule 4:771–799
61. Liu J, Peng W, Li Y, Zhang F, Fan X (2021) A VS 2@ N-doped carbon hybrid with strong interfacial interaction for high-performance rechargeable aqueous Zn-ion batteries. J. Mater. Chem. C. 9:6308–6315
62. Xiong T, Zhang Y, Lee WSV, Xue J (2020) Defect engineering in manganese-based oxides for aqueous rechargeable zinc-ion batteries: a review. Adv Energy Mater 10:2001769
63. Deng Y-P, Liang R, Jiang G, Jiang Y, Yu A, Chen Z (2020) The current state of aqueous Zn-based rechargeable batteries. ACS Energy Lett 5:1665–1675
64. Zhang Y, Wang Y, Lu L, Sun C, Denis Y (2021) Vanadium hexacyanoferrate with two redox active sites as cathode material for aqueous Zn-ion batteries. J Power Sources 484:229263
65. Shao F, Huang Y, Wang X, Li Z, Huang X, Huang W et al (2022) MoS_2 with high 1T phase content enables fast reversible zinc-ion storage via pseudocapacitance. Chem Eng J 448:137688
66. Huang M, Mai Y, Zhao L, Liang X, Fang Z, Jie X (2020) Hierarchical MoS_2@CNTs hybrid as a long-life and high-rate cathode for aqueous rechargeable Zn-ion batteries. ChemElectroChem 7:4218–4223
67. Zhao P, Shen B, Shi G, Yang J, Lu F, Shi Q et al (2023) Vacancy engineering of MoS_{2-x}@NCNTs for efficient storage of zinc ions. J Alloys Compd 171114
68. Li H, Yang Q, Mo F, Liang G, Liu Z, Tang Z et al (2019) MoS_2 nanosheets with expanded interlayer spacing for rechargeable aqueous Zn-ion batteries. Energy Storage Mater 19:94–101
69. Saha D, Kruse P (2020) Editors' choice—review—conductive forms of MoS_2 and their applications in energy storage and conversion. J Electrochem Soc 167:126517. https://doi.org/10.1149/1945-7111/abb34b
70. Xu Q, Yang GM, Zheng WT (2020) DFT calculation for stability and quantum capacitance of MoS_2 monolayer-based electrode materials. Mater Today Commun 22:100772. https://doi.org/10.1016/j.mtcomm.2019.100772
71. Bello IT, Oladipo AO, Adedokun O, Dhlamini SM (2020) Recent advances on the preparation and electrochemical analysis of MoS_2-based materials for supercapacitor applications: a mini-review. Mater Today Commun 25:101664
72. Gao Y-P, Huang K-J, Wu X, Hou Z-Q, Liu Y-Y (2018) MoS_2 nanosheets assembling three-dimensional nanospheres for enhanced-performance supercapacitor. J Alloys Compd 741:174–181
73. Lee S, Hwang J, Kim D, Ahn H (2021) Oxygen incorporated in 1T/2H hybrid MoS_2 nanoflowers prepared from molybdenum blue solution for asymmetric supercapacitor applications. Chem Eng J 419:129701

74. Abraham AM, Lonkar SP, Pillai VV, Alhassan SM (2020) Three-dimensional MoS_2 nanodot-impregnated nickel foam electrodes for high-performance supercapacitor applications. ACS Omega 5:11721–11729
75. Mahajan H, Mohanan KU, Cho S (2022) Facile synthesis of biocarbon-based MoS_2 composite for high-performance supercapacitor application. Nano Lett 22:8161–8167
76. Wang F, Ma J, Zhou K, Li X (2020) MoS_2/corncob-derived activated carbon for supercapacitor application. Mater Chem Phys 244:122215
77. Vikraman D, Hussain S, Karuppasamy K, Kathalingam A, Jo E-B, Sanmugam A et al (2022) Engineering the active sites tuned MoS_2 nanoarray structures by transition metal doping for hydrogen evolution and supercapacitor applications. J Alloys Compd 893:162271
78. Pushparaj RI, Cakir D, Zhang X, Xu S, Mann M, Hou X (2021) Coal-derived Graphene/ MoS_2heterostructure electrodes for Li-ion batteries: experiment and simulation study. ACS Appl Mater Interfaces 13:59950–59961. https://doi.org/10.1021/acsami.1c18993
79. Barik G, Pal S (2021) Two-dimensional graphene/BlueP/MoS_2 van der Waals multilayer heterostructure as a high-performance anode material for LIBs. J Phys Chem C 125:8980–8992. https://doi.org/10.1021/acs.jpcc.1c02238
80. He J, Jiao Z (2022) MoS_2/C_3N heterostructure: a promising anode material for lithium-ion batteries. Appl Surf Sci 580:152371. https://doi.org/10.1016/j.apsusc.2021.152371
81. Kapse S, Benny B, Mandal P, Thapa R (2021) Design principle of MoS_2/C heterostructure to enhance the quantum capacitance for supercapacitor application. J. Energy Storage 44:103476. https://doi.org/10.1016/j.est.2021.103476
82. Shao Y, Gong P, Pan H, Shi X (2019) H-/dT-MoS_2-on-MXene heterostructures as promising 2D anode materials for lithium-ion batteries: insights from first principles. Adv Theory Simul 2:1900045. https://doi.org/10.1002/adts.201900045
83. Zhen M, Wang J, Guo S, Shen B (2019) Vertically aligned nanosheets with MoS_2/N-doped-carbon interfaces enhance lithium-ion storage. Appl Surf Sci 487:285–294. https://doi.org/10.1016/j.apsusc.2019.05.110
84. Xiang P, Chen X, Liu J, Xiao B, Yang L (2018) Borophene as conductive additive to boost the performance of MoS_2-based anode materials. J Phys Chem C 122:9302–9311. https://doi.org/10.1021/acs.jpcc.8b00768
85. Samad A, Shin Y-H (2017) MoS_2@VS_2 nanocomposite as a superior hybrid anode material. ACS Appl Mater Interfaces 9:29942–29949. https://doi.org/10.1021/acsami.7b07161
86. Massaro A, Pecoraro A, Muñoz-García AB, Pavone M (2021) First-principles study of Na intercalation and diffusion mechanisms at 2D MoS_2/graphene interfaces. J Phys Chem C 125:2276–2286. https://doi.org/10.1021/acs.jpcc.0c10107
87. Xiao J, Zhou J, Chen L-N, Chen J (2022) Na transport in bilayer MoS_2 and MoS_2-WS_2 heterojunction with S vacancy defect: first-principles study. AIP Adv 12:065218. https://doi.org/10.1063/5.0096098
88. Zhao Y, Xu L, Yan J, Yan W, Wu C, Lian J et al (2017) Facile preparation of $NiFe_2O_4$/ MoS_2 composite material with synergistic effect for high performance supercapacitor. J Alloys Compd 726:608–617
89. Hu Z, Liu Q, Chou S-L, Dou S-X (2021) Two-dimensional material-based heterostructures for rechargeable batteries. Cell Rep Phys Sci 2
90. Pham-Cong D, Choi JH, Yun J, Bandarenka AS, Kim J, Braun PV et al (2017) Synergistically enhanced electrochemical performance of hierarchical $MoS_2/TiNb_2O_7$ hetero-nanostructures as anode materials for Li-ion batteries. ACS Nano 11:1026–1033
91. Hussain S, Rabani I, Vikraman D, Feroze A, Ali M, Seo Y-S et al (2021) MoS_2@X_2C (X= Mo or W) hybrids for enhanced supercapacitor and hydrogen evolution performances. Chem Eng J 421:127843.
92. Tiwari P, Jaiswal J, Chandra R (2019) Hierarchal growth of MoS_2 @ CNT heterostructure for all solid state symmetric supercapacitor: insights into the surface science and storage mechanism. Electrochim Acta 324:134767
93. Pan Q, Zhang Q, Zheng F, Liu Y, Li Y, Ou X et al (2018) Construction of MoS_2/C hierarchical tubular heterostructures for high-performance sodium ion batteries. ACS Nano 12:12578–12586. https://doi.org/10.1021/acsnano.8b07172

94. Lutz DM, Dunkin MR, King ST, Stackhouse CA, Kuang J, Du Y et al (2022) Hybrid MoS_{2+x} nanosheet/nanocarbon heterostructures for lithium-ion batteries. ACS Appl Nano Mater 5:5103–5118.

Chapter 13
MoS_2 for Hydrogen Evolution Reaction

Siby Thomas, Minu Mathew, K. P. Priyanka, and Dickson D. Babu

1 Introduction

While fossil fuels maintain their prominent position in the global energy sector, mounting apprehensions regarding their detrimental effects on our planet has made it imperative to develop green, renewable, and efficient fuels to fulfill the world's rising energy demands [1, 2]. However, in the case of renewable sources like solar, tidal and wind energy, their intermittent nature coupled with low energy density severely limit widespread practical applications. To address these shortcomings, electro/photochemical water electrolysis driven by renewable energy sources to obtain hydrogen is a viable alternative as hydrogen is an energy vector with tremendous energy density (gravimetric energy density of 120–142 MJ kg^{-1}) and almost no harmful emissions [3]. The technique of water electrolysis involves the utilization of a suitable catalyst to expedite the rate of hydrogen production. Among the catalysts employed for hydrogen evolution reaction (HER), platinum (Pt)-based catalysts still continue to be the most efficient. However, the exorbitant price and scarcity of these precious metals have circumscribed their large-scale application. Keeping this in

S. Thomas (✉)
Department of Chemical Engineering, University of Florida, Gainesville, FL 32611, USA
e-mail: siby.thomas@ufl.edu

M. Mathew
Department of Chemistry, University of Florida, Gainesville, FL 32611, USA

K. P. Priyanka
Department of Physics, M S Ramaiah University of Applied Sciences, Bengaluru 560058, Karnataka, India

D. D. Babu
Department of Chemistry, St. Thomas College, Kozhencherry 689641, Kerala, India

D. J. Late and C. S. Rout (eds.), *Science and Technology of 2D MoS_2*,
Materials Horizons: From Nature to Nanomaterials,
https://doi.org/10.1007/978-981-97-7367-1_13

view, scientists have designed and developed numerous catalysts with varying backbones and functionalities that are cheaper and sustainable while also being reasonably efficient [4, 5].

Among these precious metal-free catalysts, layered transition metal dichalcogenides (TMDs), have engrossed a lot of traction owing to some key advantages over noble metal-based counterparts. Some of the prominent members of the aforementioned group include molybdenum disulfide (MoS_2), molybdenum diselenide ($MoSe_2$), tungsten disulfide (WS_2), tungsten diselenide (WSe_2), etc. [6–9]. Among the aforementioned TMDs, MoS_2 has emerged as a material of choice, owing to its high catalytic performance, good stability, low-cost, higher abundance, excellent in-plane carrier mobility as well as favorable mechanical properties. Notably, the bulk form of MoS_2 depicts negligible catalytic performance for HER, whereas nanoparticulate MoS_2 has been established as a more proficient system. Furthermore, theoretical studies have identified the edges of MoS_2 to be the active site for HER catalysis with the free energy of H adsorption (ΔG_H*) close to 0 eV [10]. However, appropriate address of bottlenecks such as the limited number of edge sites, poor interlayer conductivity and catalytically inert basal planes in layered MoS_2 must be undertaken to further ameliorate its catalytic activities. In order to mitigate these drawbacks, various strategies, such as phase engineering, interfacial electronic coupling, structural modulation, unsaturated strain, and defects generation, have been employed with varying degrees of success [11].

Taking all this into consideration, this chapter focuses on multifarious strategies for synthesizing MoS_2 to obtain desired structures, which render enhanced HER performance. Additionally, explicit discussions regarding the various performance metrics for HER have also been incorporated. Furthermore, discussions regarding the mechanisms of HER based on theoretical investigations with special focus on hydrogen adsorption-free energy (ΔG_H*), energy level matching and d-band theory have been presented. Finally, challenges and potential strategies to overcome the bottlenecks in developing efficient yet sustainable catalysts for HER have been discussed.

2 Structure, Properties, and Synthesis of MoS_2

2.1 *Structure and Properties*

Depending on the synthetic approach undertaken, MoS_2 materials with diverse structures, properties as well as applications can be obtained. This is particularly important as the catalytic activity of MoS_2, and associated materials relies mainly on its layered structure and number of active edge sites. Bulk MoS_2 entails vertically stacked sheets with hexagonal structure, whereas, the monolayer MoS_2 structure contains an S-Mo-S unit with a layer of Mo atoms sandwiched by two layers of Sulphur atoms [12]. Notably, the MoS_2 sheets are held together via weak van der Waals interactions,

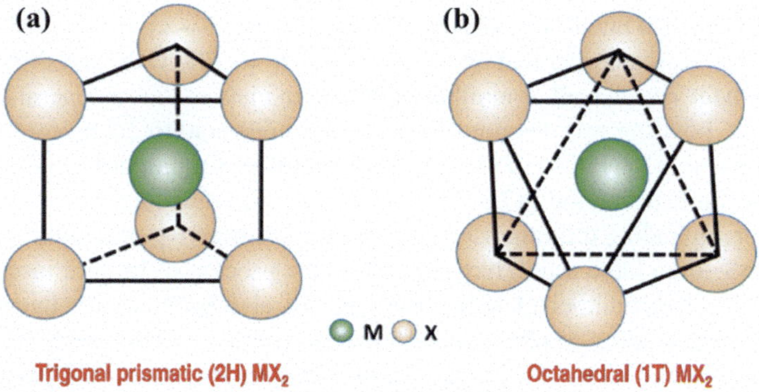

Fig. 1 Unit cell structures of 2H-MoS_2 and 1T-MoS_2 (Reproduced with permission from Ref. [14]; *Acc. Chem. Res.* 2015, 48, 1, 56–64; https://pubs.acs.org/doi/10.1021/ar5002846)

permitting the exfoliation of bulk MoS_2 into monolayers. Furthermore, bulk 2H MoS_2 has a bandgap of ~1.29 eV and is an indirect semiconductor, whereas, in monolayer MoS_2, energy gap can go up to ~1.90 eV, making it a direct bandgap material which is suitable for application in electronic devices[13]. Primarily, layered MoS_2 exists in two different polymorphs: stable 2H (hexagonal) phase semiconducting state and metastable 1T(trigonal) metallic state. In 2H-MoS_2, every Mo atom is prismatically surrounded by six Sulphur atoms, creating a thermodynamically stable structure (Fig. 1a), whereas in metallic 1T-MoS_2 phase, six Sulphur atoms coordinate to form a twisted octahedron around one Mo atom (Fig. 1b) [14]. Interestingly, phase engineering can convert the stable semiconducting 2H-MoS_2 phase into the metallic 1T-structure [15]. For instance, 2H- MoS_2 can be transformed into 1T- MoS_2 via the intercalation of Li or K and since 1T -MoS_2 is thermodynamically unstable it slowly reverts back to the trigonal prismatic configuration (2H- MoS_2) at ambient temperature. It is worth noting that 1T-MoS_2 seems to be a better HER catalyst than 2H-MoS_2. This can be attributed to the significantly ameliorated conductivity of 1T-MoS_2 (~105 times) than that of 2H -MoS_2, which promotes the transfer of charge carriers in the electrode material.

2.2 *Synthesis*

Due to the structural similarity of MoS_2 with graphene, most approaches employed to synthesize graphene are also applicable for obtaining MoS_2 nanosheets. The synthesis and exfoliation of MoS_2 layers can be classified into two broad categories (1) top-down methods, which include mechanical and chemical exfoliation [16] and (2) bottom-up methods, such as chemical syntheses and vapor deposition techniques. Exfoliation methods came to the fore, when bulk graphite was successfully exfoliated to yield graphene by the simple Scotch tape method, which paved the way for the synthesis of other 2D materials like MoS_2 nanosheets. However, this method is not suitable for large-scale practical applications due to its low yield and difficulties in regulating sheet size as well as number of layers. In view of this, a general technique called anodic bonding was developed for making high-quality 2D layered materials and it was possible to control the size and layer number for the MoS_2 obtained using the aforementioned method [17]. Another method to realize exfoliation is via ion intercalation, including lithium insertion, which is an effective strategy to synthesize layered MoS_2 on a large scale. Notably, while the process of lithium insertion will minimize the layer number of bulk MoS_2, it will also bring some changes in the crystal structure leading to phase transformation (typically from inert 2H phase to metallic 1T phase), thereby further enhancing the corresponding HER performance. Another novel method to exfoliate bulk MoS_2 was reported by Zhang et al. by means of an electrochemical lithiation discharge, which resulted in lithium intercalation. The subsequent ultrasonication yielded monolayer MoS_2 in copious amounts and a large percentage of the MoS_2 obtained by the aforementioned method was metallic 1T-MoS_2 [18]. Liquid exfoliation is another method for synthesizing large amounts of single as well as few-layered MoS_2 nanoflakes [19].

On the other hand, bottom-up wet chemical approaches have emerged as an alternative to exfoliation methods for synthesizing MoS_2 nanosheets. Numerous efforts have been undertaken to grow monolayer MoS_2 using chemical vapor deposition (CVD) methods. By the strategic manipulation of various parameters such as reaction conditions and surface stabilizing ligands, one can modulate the shape, size, and uniformity of MoS_2 nanosheets. For instance, a novel synthetic method based on a vapor–solid growth mechanism was developed in 2013, wherein, optical quality monolayer MoS_2 films were prepared on various insulating substrates (SiO_2, glass, and sapphire) (Fig. 2a) [20]. Wang et al. reported an approach for obtaining rhomboid-shaped MoS_2 flakes having controlled the number of layers via the layer-by-layer sulfurization using MoO_{21} microcrystallites as templates [21]. The process involves annealing the MoO_2 microsheets in sulfur vapor and depending on the annealing duration MoS_2 layers with varied numbers can be synthesized (Fig. 2b). Another effective method to synthesize large-area MoS_2 films with exceptional electrical performance was realized by the thermal decomposition of ammonium thiomolybdate, which yielded bi- or tri-layer continuous films on insulating substrates (Fig. 2c). Furthermore, the importance of the second-step high temperature sulfurization process in improving the crystallization was also established [22]. Additionally, wet chemical synthetic methods such as hydrothermal method and thermal injection method have also been developed in order to synthesize MoS_2 nanosheets. However, in spite of these developments, the wet chemical synthesis of MoS_2 nanosheets needs to overcome few synthetic challenges including the amelioration of yield as well as purity while also coming up with solutions to truncate vertical growth and enable lateral growth.

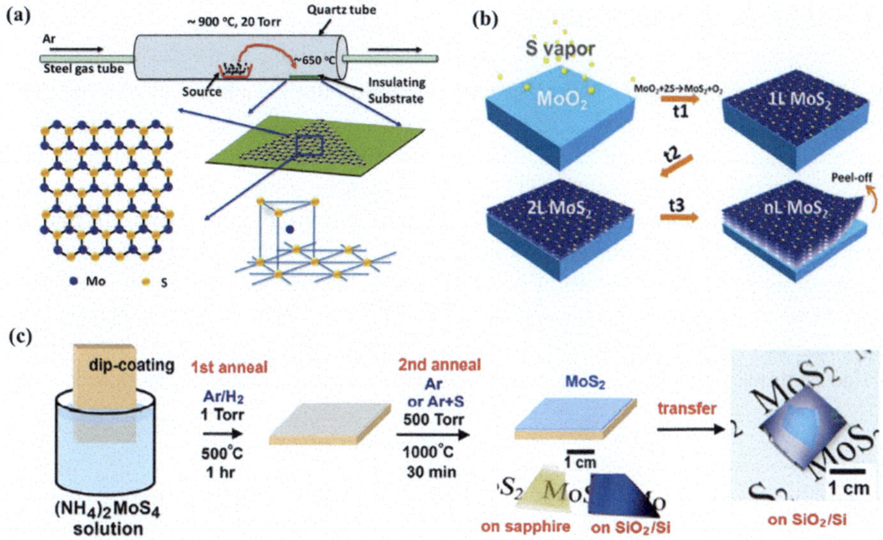

Fig. 2 **a** Vapor–solid growth from MoS$_2$ powder (Reproduced with permission from Ref. [20]; *ACS Nano* 2013, 7, 3, 2768–2772; https://pubs.acs.org/doi/10.1021/nn4002038). **b** Schematics for creating and breaking down MoS$_2$ involve the synthesis of MoO$_2$ microplates through the reduction of MoO$_3$, serving as a framework for the incremental growth of MoS$_2$ via surface sulfurization. The resulting MoS$_2$ coating is isolated from MoO$_2$ and relocated to a different substrate, orienting the top layer upwards for subsequent analyses. The notation of 1L, 2L, and nL denotes single-layer, bilayer, and multi-layer configurations, respectively (Reproduced with permission from Ref. [21]; *J. Am. Chem. Soc.* 2013, 135, 14, 5304–5307; https://pubs.acs.org/doi/10.1021/ja4013485). **c** Diagram depicting the two-stage thermal decomposition method used to produce thin layers of MoS$_2$ on insulating substrates (Reproduced with permission from Ref. [22]; *Nano Lett.* 2012, 12, 3, 1538–1544; https://pubs.acs.org/doi/10.1021/nl2043612)

3 Understanding of HER Mechanism

It is well understood that theoretical understanding of HER mechanism is very crucial when detailing the electrocatalytic HER process. In general, HER mechanisms basically take place at the interface of electrode and electrolyte. Throughout the electrochemical process, H_2 is produced through the reduction of either water (H_2O) or proton (H^+), or both which encompass a sequence of basic steps and also depend on the pH value of the electrolyte. The most celebrated procedure for the evolution of hydrogen involves different reaction steps including Volmer reaction, Heyrovsky, and the Tafel reaction. However, in a broad sense, the HER process can be classified based on their occurrence in acidic and alkaline media, respectively [23].

3.1 HER in Acidic Media

The HER in acidic media includes two sequential steps such as: (1) the catalyst adsorbs a proton or H^+ to its surface and forms an adsorbed hydrogen atom (H*), where * denotes the most probable active site of H on the surface of the catalyst, and (2) the adsorbed hydrogen atom (H*) combines with a proton (H^+) and an electron (e^-) to form an H_2 molecule. This former step is known as the Volmer reaction, and the latter one is widely called the electrochemical desorption step or Heyrovsky reaction. On the other hand, the formation of an H_2 molecule can also be possible through a chemical desorption step, known as the Tafel reaction. In the Tafel reaction, the formation of H_2 molecules on the surface of the catalyst is governed by the grouping of two protons. The Volmer, Heyrovsky, and Tafel reactions are shown below.

$$H^+ + e^- \rightarrow H* \qquad \text{(Volmer reaction)} \tag{1}$$

$$H* + H^+ + e^- \rightarrow H_2 \qquad \text{(Heyrovsky reaction)} \tag{2}$$

$$2H* \rightarrow H_2 \qquad \text{(Tafel reaction)} \tag{3}$$

Then, the overall HER mechanism in the acidic media is: $2H^+ + 2e^- \rightarrow H_2$. Nonetheless, the Sabatier principle reveals that HER kinetics strongly axes on the interaction between H* and the catalyst [24, 25]. Thus, keeping the hydrogen atom (H*) on various locations of the catalyst surface can provide the hydride formation energy and that can be used to describe the adsorption behavior and adsorption energy of hydrogen atom (ΔG_H^0).

3.2 HER in Alkaline Media

For describing the HER in alkaline media, similar to the case of acidic media, ΔG_H^0 is considered as a base parameter in which the adsorption behavior of hydrogen on the surface of the catalyst material can be computed. However, it is observed that the HER exchange current densities of the majority of metal catalysts in alkaline solutions are smaller compared to acidic media [26, 27]. This low current density and HER activity in alkaline media is associated with the distinct pathway of acidic media. However, due to the absence of H^+, the HER in alkaline solutions begins from the dissociation of H_2O molecules and will generate protons. Similar to the case of HER in acidic media, the HER in alkaline media is also associated with the Volmer, Heyrovsky, and Tafel reactions, as detailed below.

$$H_2O + e^- \rightarrow H* + OH^- \qquad \text{(Volmer reaction)} \tag{4}$$

$$H* + H_2O + e^- \rightarrow OH^- + H_2 \quad \text{(Heyrovsky reaction)} \tag{5}$$

It is interesting to notice that the Tafel process in alkaline media is similar to that of acidic media ($2H^* \rightarrow H_2$) and the overall HER process in alkaline media is described as:

$$2H_2O + 2e^- \rightarrow 2OH^- + H_2 \tag{6}$$

The HER mechanism in solutions generally takes place the conversion of water molecules to Hydrogen molecules with the help of a suitable catalyst. The reaction pathway of HER in acidic and alkaline medium is shown in Fig. 3. The reaction energy is dependent on the pathway of reaction. The efficiency of HER in acidic medium is greater than in an alkaline medium as it may be associated with the dissociation of water molecules in the Volmer step in an alkaline medium. The rate-determining step of overall HER includes the transfer of adsorbed hydrogen to the consecutive steps via a strong H-O-H bond [29]. There are several reports that the catalyst activity towards HER only relies on the pH of the solution. In fact, the binding energy of hydrogen is lower in a solution with lower pH, which leads to faster reaction kinetics of HER in an acidic medium. Nonetheless, some researchers consider that this model is inappropriate to detail the HER mechanism as it depends on the adsorption of intermediates like OH^- and this should be taken into consideration [29].

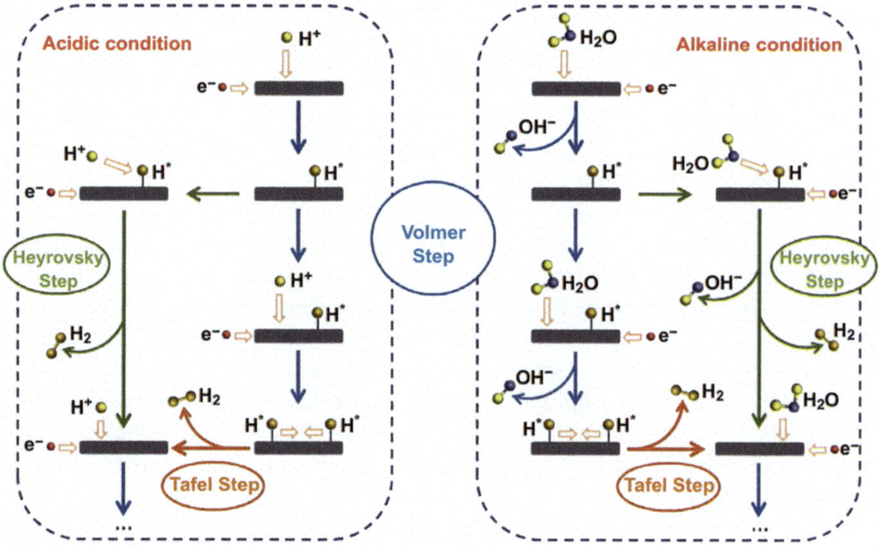

Fig. 3 Schematic pathways for hydrogen evolution reaction under acidic and alkaline conditions (Reproduced with permission from Ref. [28]; Nano-Micro Lett. (2018) 10:75; 10.1007/s40820-018–0229-x)

Based on the insights from earlier studies, both experimental and theoretical studies confirmed high HER activity of the edge sites of MoS_2 [10, 30–33]. It should be noticed that the advancement of computational techniques, especially density functional theory (DFT) plays an indispensable role in computing the close-to-neutral value of ΔG_H^0, much needed for the modern catalysis technologies. Even so, since additional energy is needed for the generation of protons during the HER process in alkaline media, the kinetics on the surface of most catalysts are slow in the alkaline solutions. Previous studies also noticed that the HER activity in alkaline media is governed by a gentle equilibrium between the energy required to dissociate H_2O and ΔG_H^0 [34]. Although different HER catalysts including Co_2P and MoS_2 are not ideal for the water dissociation process, endorsing this process while remaining a moderate ΔG_H^0 is a feasible approach to propose efficient catalysts for alkaline HER.

4 General Approaches for Catalyst Evaluation

4.1 Overpotential

Oxygen evolution reaction (OER) and HER are the two major steps of electrochemical water splitting that occurs at the anode and cathode of the electrolyzer, respectively. These reactions are thermodynamically not viable under standard conditions of temperature and pressure and require additional energy. Overpotential is known as the difference between experimentally obtained potential and the thermodynamically determined potential of an electrochemical reaction. It is one of the major parameters for evaluating the performance of an electrocatalyst because a large overpotential leads to the low energy conversion efficiency of an electrochemical water splitting system. Linear sweep voltammetry (LSV) with small sweep rates is usually performed to minimize the no faradaic current. The presence of overpotential points out the activation of reaction, diffusion of charge carriers, and series resistance. Exactly, the activation overpotential directly connects to the catalytic activity, whereas the overpotential originated by the charge carrier diffusion and series resistance mainly come from the electrolysis system. So, the catalytic activity should be accessed through the evaluation of activation overpotential. Besides, the diffusion overpotential can be minimized effectively using a rotating disk electrode (RDE) system because it employs a continuously rotating electrode for recording LSV curves. Meantime, IR compensation should be done to correct the resistance overpotential as provided in the equation (R_s = series resistance and I = current flowing in the circuit).

$$E_{\text{corrected}} = E_{\text{uncorrected}} - IR_S \tag{7}$$

The catalytic performance is usually evaluated based on the overpotential obtained at a current density of 10 $mAcm^{-2}$, with regard to the working current density of photoelectrochemical water splitting system. The current density is commonly measured based on catalytic mass, the specific geometric area of the catalyst, or the electrochemical active surface area of the catalyst.

4.2 Exchange Current Density and Tafel Slope

The exchange current density is another important factor to assess the performance of a catalyst in which it is measured at the equilibrium where the anodic and cathodic currents are the same. The exchange current density plays a major role in catalytic activity as it essentially reflects the intrinsic charge transfer between the electrolyte and electrode. Also, previous studies noticed that the exchange current density is larger on the surface of most of the catalysts which also ensures a better catalytic activity[35]. The exchange current can be extracted from the Tafel slope, which is the extrapolated liner part of the Tafel plot (overpotential vs. log |current density|). Remarkably, the Tafel plot can be obtained from the equivalent LSV curve and the Tafel slope can provide a detailed understanding about the HER reaction mechanisms on the surface of the catalyst.

4.3 Turnover Frequency

The turnover frequency (TOF) is defined as how many product molecules are generated per active site in unit time and is expressed as:

$$TOF = \frac{\text{numberofproductmolecules}}{\text{numberofactivesites}} \times \frac{1}{\text{unittime}} \quad (8)$$

The above expression clearly indicates that for the evaluation of the number of H_2 molecules, H_2 should be collected and theoretically, H_2 is the charge flowing through the circuit of the electrochemical system which is governed by Faraday's law of electrolysis and is given below.

$$n = \frac{It}{zF} \quad (\text{Faraday'slaw}) \quad (9)$$

where n is the amount of constituent matter, I is constant current applied during a time t, z is the electron number transferred per molecule, and F is the Faraday constant. A realistic TOF versus overpotential curve is necessary for evaluating the catalytic performance and can be obtained by evaluating the number of active sites on the surface of the catalyst using numerous ways including copper underpotential deposition method [36] and cyclic voltammetry (CV) [37, 38]. It should be noticed that the TOF versus overpotential graph can be achieved by utilizing Eqs. (8) and (9) and becomes:

$$TOF = \frac{I \times N_A}{zF \times \text{numberofactivesites}} \quad (10)$$

4.4 *Durability*

Durability or stability is one of the important parameters to assess the performance of a catalyst for practically viable applications. This mainly involves the ability to preserve the original activity of the catalyst over a long period of time without ruining their stability. In general, the stability is evaluated by monitoring the rate of change in overpotential with respect to current density. Besides, a CV approach can also be used to evaluate stability by analyzing the cyclic performance. However, until now, most of the methods for evaluating stability are limited to working situations and more effective approaches are much needed to evaluate stability under idle conditions as well.

5 Structure-Performance and Active HER Sites of MoS_2

The impediment in improving the catalytic performance of MoS_2 is due to the lack of explicit knowledge about the active sites. Ergo, in order to design and synthesize high-performance HER catalysts, it is imperative to have an unambiguous understanding about the catalytically active sites. Based on the available studies, it is widely accepted that only the edge sites of MoS_2 are catalytically active, while the basal plane remains inert for HER. The underwhelming activity on MoS_2 basal plane is ascribed to the exceptionally high hydrogen adsorption-free energy ($\Delta G_H = 1.92$ eV), whereas, on the edge sites, the value of ΔG_H is close to zero (~0.08 eV). Keeping this in view, myriads of strategies have been explored to ameliorate the catalytic performance by augmenting the number of edge sites. Additionally, it must be noted that monolayer MoS_2 entails two types of edges; sulfur rich edge or molybdenum rich edge and based on studies, it has been established that, between the two types of edges, Mo-terminated edges are the predominant contributor towards the HER activity of MoS_2 [39]. DFT studies have revealed that HER performance of Mo-terminated edges is attributed to their considerably lower hydrogen binding energy. In addition to the edge sites, DFT simulation as well as experimental results have established the sulfur vacancies of MoS_2 as another key active site for HER. However, unlike the direct dependence of the catalytic activity on the length of the edge sites, maximum catalytic performance is revealed when the density of sulfur vacancies is around 7–10%. Interestingly, the grain boundary has also displayed some catalytic performance. Li and co-workers measured the turnover frequency (TOF) of the edge sites, sulfur vacancies as well as grain boundaries to be7.5 ± 1.5, 3.2 ± 0.4, and around 0.1 s^{-1}, respectively [40]. Besides, the study also indicated that the catalytic performance of sulfur vacancy is also reliant on the local crystalline structure around the vacancy and a higher crystalline quality delivers higher catalytic activity. Hong et al. utilized scanning electrochemical microscopy (SECM) for gaining kinetic information about both strained and unstrained S vacancies on the basal plane of MoS_2 monolayers to comprehend how strain can modulate the activity of sulfur vacancy (Fig. 4a). The experiment revealed that the strained S vacancy has 4 times higher electron-transfer rate than that the corresponding unstrained S vacancy [41].

Some of the strategies adopted to increase the edge sites include incorporation of transition metal and varying conductive supports [41–43]. Notably, theoretical studies have demonstrated that the choice of support plays a vital role in modulating the HER activity of supported MoS_2 and with judicious selection of support material, the catalytic activity of Mo-edge can be enhanced by achieving $\Delta G_H \sim 0$. The comparison of the theoretical hydrogen binding energy of MoS_2 and the experimental HER exchange current density with that of several pure metals, it is evident that supported MoS_2 exhibits an HER performance, which is on par with non-precious metals (Fig. 4b). Keeping this view, the ΔG_H* on the Mo-edge could be improved either by selecting ideal catalyst support or by synthesizing catalysts where the active edge sites are far-off from the support to mitigate the effect [43].

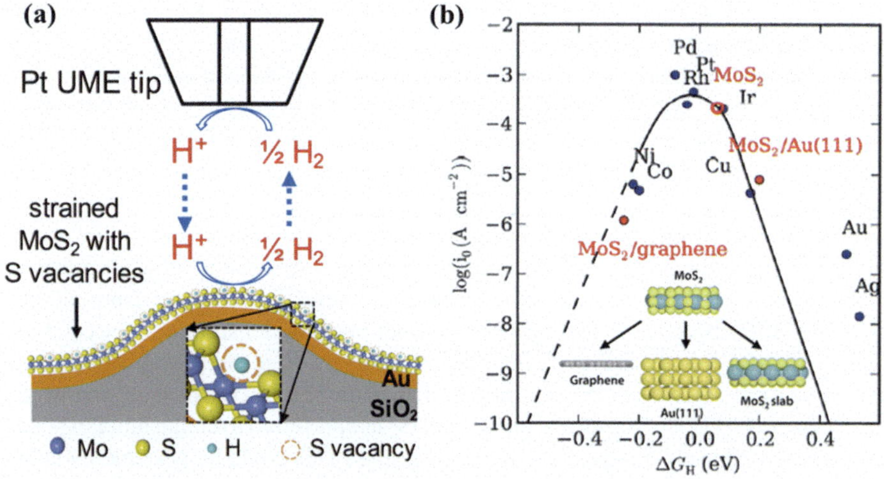

Fig. 4 **a** Methodology for quantifying the HER kinetics of strained/unstrained S vacancies on the basal plane of MoS_2 monolayers. Schematic of the SECM measurement setup (Reproduced with permission from Ref. [41]; *Am. Chem. Soc.* 2016, 138, 15, 5123–5129; https://pubs.acs.org/doi/10.1021/jacs.6b01377).). **b** Activity map for the hydrogen evolution reaction (HER) showing the exchange current density as a function of the calculated free energy of hydrogen adsorption, ΔG_H (Reproduced with permission from Ref. [43]; *Nano Lett.* 2014, 14, 3, 1381–1387; https://pubs.acs.org/doi/10.1021/nl404444k)

6 MoS_2-Based Nanocomposites for Efficient HER Activity

Among the TMDs, MoS_2 shows significant catalytic activity and is known as a promising catalyst due to its larger surface area, channels for transportation of ions, and low adsorption energy between hydrogen atoms and unsaturated sulfur atoms [10]. The coating of S atoms on the Mo edges of Mo-S-Mo nanosheets makes them more active because of their low coordination number and high electron density [44]. The electrocatalytic activity of MoS_2 is mainly determined by the unsaturated coordinate bonds of MoS_2, which helps reactants to coordinate with the Mo atom [45]. So, the effective usage of the electrocatalytic activity of MoS_2 is possible through the modification of MoS_2 with other materials but this remains a big challenge. Nonetheless, there are several factors that inhibit the HER activity of MoS_2 such as intrinsic low conductivity, few active sites, and insufficient stability [46]. These issues have been addressed using various methods such as element doping, structural design, and defect and interface engineering for improving the electrocatalytic activity of MoS_2. Taking into consideration that the active site was originated from the Mo-S edge of MoS_2 catalyst, surface defect or interfaces are crucial to enhance the electrocatalytic activity and various defect-rich MoS_2 electrocatalysts have been investigated by doping heteroatom to form various M-Mo-S structures [47], the addition of excessive sulfur sources [48], interlayer expanded MoS_2 nanosheet [49]and N-doped MoS_2 nanosheet [50]. These modified MoS_2 structures resulted in poor electrochemical stability due to their low crystallinity, which can be easily dissolved in acidic solutions [45]. Moreover, owing to less stability and slow HER kinetics in the alkaline medium, the industrial requirements of MoS_2-based catalysts still remain a big challenge. The kinetics of HER in an acidic medium is much faster than in an alkaline solution because of the large adsorption energy of OH^- on the catalyst surface in an alkaline medium and the high dissociation energy barrier between water and MoS_2 [42].

Platinum-based electrocatalysts are popular for HER owing to their favorable binding energies toward protons and the highly expensive platinum limits its practical applications. This leads to the search for new cost-effective catalytic materials like MoS_2 for HER instead of Pt for hydrogen production via water splitting. The catalytic activity of MoS_2 is related to its number of exposed edge sites whose electronic structure, local stoichiometry, and physical structure are different from the MoS_2 basal planes (001) [31]. Nanostructures of MoS_2 are preferred to create more exposed sites on their edges in comparison to the bulk sites. This could be possible by using suitable supports or substrates (MoS_2 nanocomposites) [51]. The presence of more exposed sulfur edge sites results in more active MoS_2 nanoparticles than the inert bulk form and the MoS_2 reduction of protons occurs at low overpotentials with a high current density via its coordinated sulfur edge sites [51]. Table 1 shows the comparison of the electrocatalytic HER activity of MoS_2-containing compounds in acidic and alkaline electrolytes.

Table 1 Comparison table showing HER activity of MoS_2-containing compounds

Catalysts	Medium	H (mV)	Tafel slope (mV dec^{-1})	Stability	References
Ni_3S_2@MoS_2	1.0 M KOH	60	48	5000	[52]
MoS_2-g-$CuNi_2S_4$	0.5 M H_2SO_4	150	29.3	1000	[53]
MoS_2/rGO/NiS	0.5 M H_2SO_4	152	54.3	1000	[45]
MoS_2@Ag/AuNP	0.5 M H_2SO_4	225	57	5000	[54]
ZnS/C-MoS_2	1.0 M KOH	118	55.4	500	[55]
MoS_2@prGO	0.5 M H_2SO_4	263	60	500	[56]
Bi_2S_3/MoS_2	1.0 M KOH	124	123	–	[57]

Electrocatalytic efficiency of 3D hierarchical Ni_3S_2/MoS_2 heterostructures was investigated by Cui et al. In the alkaline solution, MoS_2 typically deals with unfavorable desorption of hydrogen, and its positive HER kinetics mainly depends on the large overpotentials. Taking everything into consideration, they showed a synergistically electronic and structure modulation strategy to improve the performance of MoS_2. A heterogeneous hollow nanoframe of Ni_3S_2@MoS_2 core–shell structure supported on Ni foam was synthesized by three steps such as hydrothermal, sulfurization, and calcination. The optimized composite displays an overpotential of only 60 mV to attain a current density of -10 mA/cm^2 and Tafel slope of 48 mV/dec in 1.0 M KOH. The catalyst proved excellent stability towards HER after 5000 cycles of continuous operation with a slight decrease in electrocatalytic efficiency. The enhanced performance can be associated with kinetic and thermodynamic acceleration originated from the synergistic effect of smooth heterojunction and rich active edges (morphological peculiarity) (Fig. 5a). The electrocatalytic HER activities of the as-prepared Ni_3S_2@MoS_2NF electrocatalyst measured in 1 M KOH are shown in Fig. 5b and c. The design of the prepared catalyst offers new directions for re-engineering electrocatalysts by including promoters to optimize the electrocatalytic efficiency of these particular materials and beyond [52].

An advanced electrocatalyst for HER was designed by synthesizing a novel MoS_2-g-$CuNi_2S_4$ nanocomposite in which the intrinsic HER activity of MoS_2 was augmented by raising the affinity of the active edge sites for H^+ adsorption via dopants such as Cu and Ni. Simultaneously, the edge sites were increased by anchoring them to the graphene framework. The optimized nanocomposite was immobilized on SPEs and showed HER activity a small Tafel slope and onset potential value of 29.3 mV dec^{-1} and -0.05 V, respectively, a near Pt-like electrocatalytic behavior toward HER [53]. Liu et al. synthesized a defect-rich heterogeneous ternary nanocomposite based on MoS_2, r-GO, and NiS using ultrathin α-$Ni(OH)_2$ nanowires (Fig. 6a). The composite showed exceptional catalytic activity and stability over a wide range of pH due to the presence of heterogeneous interfaces and defects. The outstanding performance can be related to the following points: (1) heterogeneous interface can

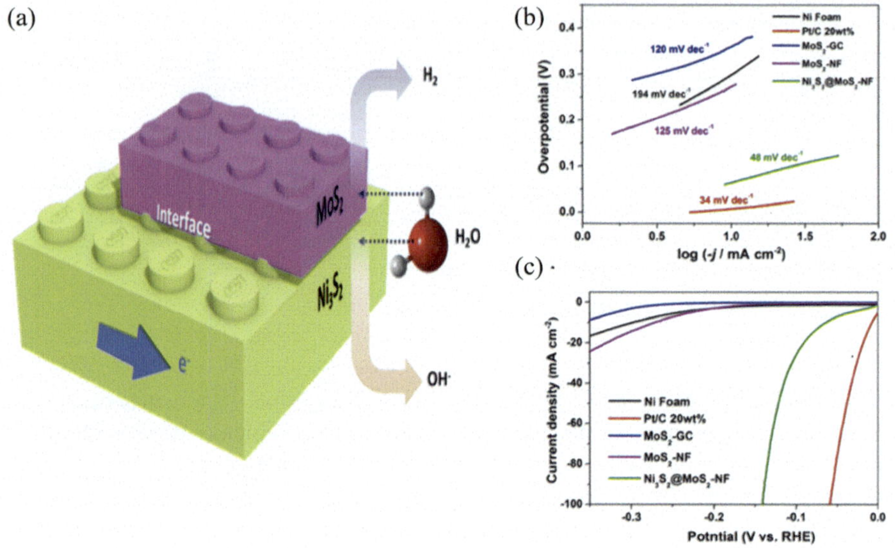

Fig. 5 **a** Schematic diagram of hybridization between MoS_2 and Ni_3S_2 for HER. Electrocatalytic HER activities of the as-prepared Ni_3S_2@MoS_2NF electrocatalyst measured in 1 M KOH. **b** LSV curves and **c** Tafel plots (Reproduced with permission from Ref. [52]; *Materials Today Energy,* 2019, 13:85–92. 10.1016/j.mtener.2019.05.001)

produce dense catalytic sites and accelerate electron transfer; (2) the enlarged interlayer spacing of MoS_2 creates many active edge sites; (3) the high intrinsic conductivity of rGO promotes the charge transfer between active catalytic site and electrode. A small Tafel slope of only 54.3 mV/dec was obtained with an overpotential of 152 mV at 10 mA/cm^2 (Fig. 6b and c), describing a high electrocatalytic efficiency. The high current efficiency could be maintained after 1000 cycles of reaction (Fig. 6d), indicating excellent stability [45].

Wang et al. synthesized three distinct MoS_2 formulations with varying proportions of the 1T phase using a cost-effective methodology: 2H-MoS_2 (13.6% 1T phase), 1T-MoS_2 (80.1% 1T phase), and 1T/2H-MoS_2 (57.6% 1T phase) nanosheets. Following this, they fabricated three types of nanocomposites by modifying Ag and Au nanoparticles onto the MoS_2 nanosheets employing diverse strategies. The hydrogen evolution reaction (HER) activity of these composites was subsequently investigated. The outcomes revealed that the crystalline phase exerted a pivotal influence on dictating the catalytic activity. To elaborate, the 1T-MoS_2 showed better electrocatalytic hydrogen activity than the 2H-MoS_2 because of the decreased charge transfer resistance in the metallic phase and the higher content of the 1T phase in the samples indicated improved performance. Furthermore, with the similar 1T phase concentration, the improvement of the HER of the MoS_2@Ag/AuNP nanocomposite compared to that of pristine MoS_2 is due to the greater enhancement of charge transfer, as pointed out by the lower charge-transfer impedance. The composite gives an excellent overpotential of 225 mV at 10 mA/cm^2 and a smaller Tafel slope of 57 mV/dec

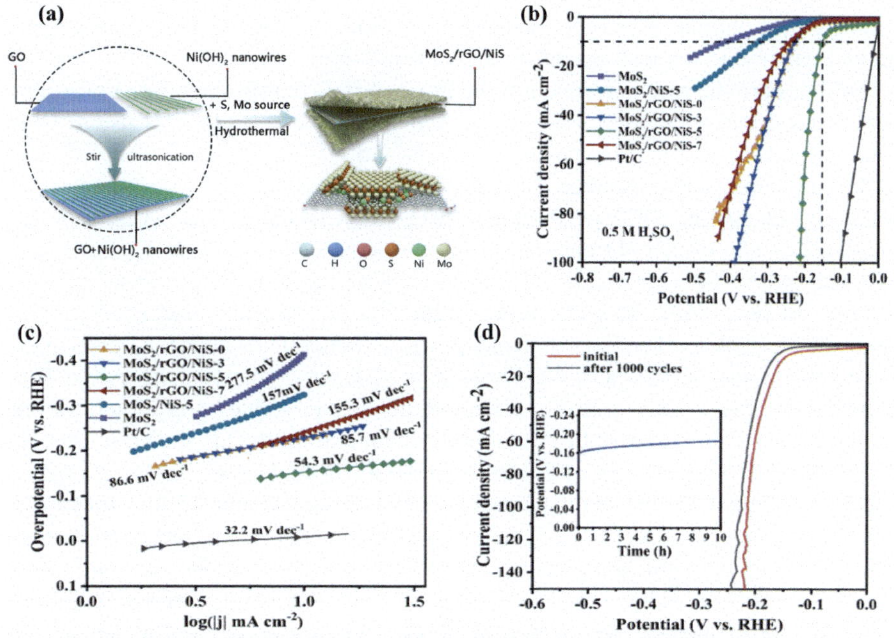

Fig. 6 **a** Schematic of synthesized process and structure of MoS_2/rGO/NiS-X (where X is the volume of $Ni(OH)_2$ added, X = 0,3,5,7).HER catalytic activities of the as-prepared samples. **b** Polarization curves in 0.5 M H_2SO_4, **c** Tafel plots under 0.5 M H_2SO_4, and **d** the polarization curves of MoS_2/rGO/NiS-5 before and after 1000 cycles of voltammetry in 0.5 M H_2SO_4 (Reproduced with permission from Ref. [45]; *Nanomaterials,* 2021, 11:662; 10.3390/nano11030662)

[54]. To date, various strategies have been explored for hydrogen evolution reaction, but only a few studies were concentrated on the effect of structural morphology of MoS_2-based composites on electrocatalysis [55–57]. So, the effect of edge sites, adsorption, and structural defects need to be investigated.

7 Comparison of MoS_2 with Other HER Catalysts

Based on the aforementioned details, it is clear that MoS_2 and MoS_2-based nanocomposites are potential candidates for HER activities in both alkaline and acidic media. However, in recent years, numerous catalytic materials have been proposed and studied for utilizing them in HER activities and are including 2D metal oxides, MXenes, alloys, metal phosphides, heteroatom-doped carbon materials, etc. Similar to TMDs, layered structure of metal oxides and (oxy) hyroxides such as δ-MnO_2 and V_2O_5 are widely used for catalytic activities. One of the advantages of these structures is their large surface area that helps high surface exposure for electrocatalysis. However, it should be noticed that although these materials are promising for oxygen evolution and oxygen reduction reactions, their HER performance is not good compared to the well-established TMDs [58–60].

MXenes are the recent addition of the 2D materials family and consist of a class of transition metal carbides and nitrides, which have been studied for use in HER catalysts. Except a few MXenes (W_2N, V_2N and Hf_2N), the basal plane of most MXens are HER active and their ΔG_H^* close to 0 eV also become one of the reasons for their outstanding HER performance [61]. Most recently, Djire et al. reported the development of mixed transition metal nitride MXenes, M-$Ti_4N_3T_X$ (M = Mo, Mn, Cr, and V; T_X = O and/or OH) by modifying the solution-processed pristine $Ti_4N_3T_X$MXenes [62]. They also reported the first experimental evidence by modifying the basal plane chemistry of the MXenes. To better understand the trends in the HER performance of MXenes, Wei She et al. used density functional theory (DFT) to compute the ΔG_H^*. They, at first, considered the most probable adsorption locations on the MXene surface including fcc, hcp, bridge, and on top, as shown in Fig. 7a. Then, the adsorption of 1 ML of O, H, OH, and H_2O on the MXene surface of pristine and surface functional group (T_x) terminated MXenes are computed for a total of 16 distinct adlayers per MXene (four categories of sites with four probable adsorbates). The DFT-computed volcano plots for carbide and nitride-based MXenes indicate a wide range of HER activity, as shown in Fig. 7b. Although ΔG_H^* varies largely from −1.5 to 1.5 eV, most of the reported MXenes fall on the weak-binding side of the volcano plot (the right leg), as shown in Fig. 7b and this indicated that the interaction between H and MXenes is weak. In addition to this, numerous metal-free HER catalysts have been developed and in 2006, Greeley et al. employed high-throughput computational screening and computed the ΔG_H^* of many pure metals and their bimetallic alloys and noticed that most of the alloys exhibit high HER activity [63]. Nonetheless, they noticed that only a small fraction of the studied alloys including IrRe, BiPt, PtRh, and PtRu are suitable for acidic HER activities [63], but they are expensive and unsuitable for industry-scale implementations. In addition to these types of metals and alloys, metal-free alloys such as NiMo have also been reported as a suitable HER catalyst in alkaline media, but their existence in acidic media is still a challenging issue due to the corrosive nature [64, 65].

Furthermore, although the tri-metallic type catalysts have also been tried for efficient HER activities, their challenging fabrication routes and poor catalytic performance restrict use in efficient HER activities. In this respect, TMDs play a significant role as an efficient catalyst for use in both acidic and alkaline media due to their promising physical and chemical properties including the high surface area as well as the exposure of the surface. In order to overcome the many of the disadvantages associated with the alloy-based catalysts, quite a lot of heteroatoms (O, N, P, and S atoms) doped carbon-based materials have been tested for HER activities. However, they also possess poor catalytic performance compared to TMDs. In addition to this, it should be noticed that the electronic conductivity of most TMDs including 1T MoS_2 (10–100 cm^{-1} at room temperature) are superior compared to the most celebrated HER catalyst, reduced graphene oxide [66]. On the other hand, the poor charge transfer and conductivity issues of 2H MoS_2can be solved by making a stable contact with the conductive surface of graphene [67]. Nevertheless, it is much exciting to

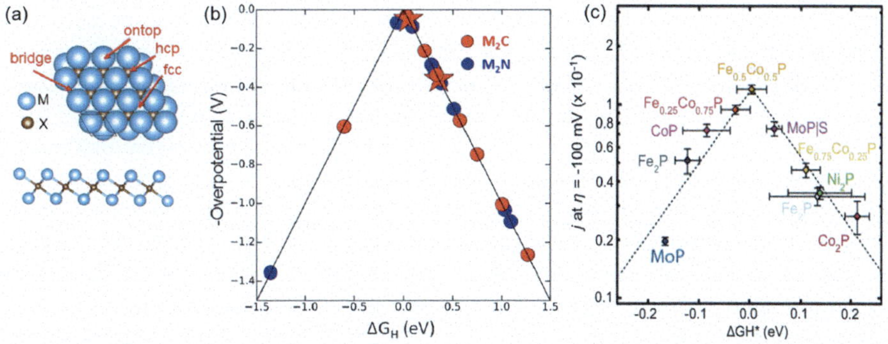

Fig. 7 **a** Top and side views of the MXene structure and the unit cell is marked using black rhombus structure. Most probable adsorption sites are also provided. **b** The HER volcano plots and theoretical overpotential of titanium carbide (red circles) and nitride (blue circles) MXenes $\Delta G^*_H 0$ (Reproduced with permission from Ref. [71]; *ACS Energy Lett.* 2016, 1, 589–594; 10.1021/acsenergylett.6b00247). **c** Activity volcano for the HER revealing the electrochemical active surface area and normalized current density at $\eta = 100$ mV as a function of(Reproduced with permission from Ref. [71]; *Energy Environ. Sci.*, 2015, 8, 3022–3029; 10.1039/C5EE02179K)

notice that metal phosphides such as FeCoP, Co_2P, and MoP show superior HER performance compared to many TMDs and is further supported by their observed 0 eV value of $|\Delta G^*_H|$, as shown in Fig. 7c, i.e., for the case of FeP nanoparticles in 0.5 M H_2SO_4 electrolytic solution, the Tafel slope is 32 mV dec^{-1}[68], and for the case of Ni_5P_4 nanocrystals, the Tafel slope is 33 mV $dec^{-1}$with an overpotential of 23 mV at j = 10 mA cm^{-2} [69]. Previous studies also reported that some of the studied metal phosphides exhibited decent HER performance in the alkaline media and also showed superior oxidation resistance in both alkaline and acidic media due to the fairly robust chemical bonds between the phosphide and metal atoms [70]. Even so, the non-layered nature of these materials further hinders the extensive use for HER applications. All these indicate that the polymorphs of MoS_2 can be a most suitable candidate for efficient HER activities.

8 Strategies to Improve the Electrocatalytic Performance of MoS_2

Although most MoS_2 polymorphs and their nanocomposites proved their potential for efficient HER activity and performance, the semiconducting electronic property as well as the low concentration of exposed catalytic sites of 2H-MoS_2 restricts their extensive HER activity. In order to increase the catalytic performance, there are methods for enhancing the catalytic activity and conductivity as well as constructing more catalytic sites. Crystallization modulation, structure and phase engineering, and nanostructure fabrication, modulating current density distribution using a conducting substrate, etc. are some of the approaches to enhance the reactivity and electronic properties including conductivity of the catalysts. Besides, introducing defect engineering by locating possible defect locations as well as creating different types of defects, and doping are also considered as some of the efficient strategies for creating large numbers of catalytic locations in MoS_2 for enhanced HER performance [72].

Structure and phase engineering have been observed as an effective method to improve the catalytic performance of MoS_2 polymorphs. This, in general, can be done by increasing the number of active sites for catalytic activity and/or elevating the mass transfer and electron transportation. Earlier studies reported that among the 2H, 1T and 3R polymorphs of MoS_2, 1T phase shows superior HER performance due to the charge transfer kinetics and improved H adsorption properties. Although phase engineering changes the polymorphic phase from semiconducting 2H to metallic 1T phase by improving the electronic conductivity, the thermodynamically unstable nature of metallic 1T phase is challenging for superior HER activity and can be overcome by slowly changing to semiconducting 2H phase impulsively with the help of n-butyllithium [36, 73, 74]. In addition to this, numerous strategies have been applied to stabilize the metallic 1T phase, including synthesis under high magnetic field, making heterojunctions on the basal plane, donating electrons to the d orbitals of Mo atom, and intercalation, etc. [72]. When compared to crystalline MoS_2, amorphous MoS_2 exhibits improved HER performance and is attributed to the shorter Mo-Mo bonds as well as the presence of a large number intrinsic catalytic sites in the disordered amorphous structure [75]. These sites resemble defects and the main drawback of these types of amorphous structures is their poor stability. Due to this stability issue, the amorphous structure spontaneously crystallizes to form 2H $MoS_{2.}$

Nanostructure fabrication is considered as a simple and powerful technique to enhance the electrochemical performance of MoS_2 polymorphs. Different research groups have investigated this possibility using different materials and methods including the development of three-dimensional superstructure of MoS_2, synthesis of MoS_2quantum dots using one-step hydrothermal reaction [76], analyze the HER performance using vertically grown carbon-cloth-MoS_2 heterostructures [77], and coupling the few-layered MoS_2 to Ti metal substrate, etc. All these approaches claimed an elevated HER performance of MoS_2 with the presence of these nanostructures. Defects, especially atomic vacancies are considered as universally accepted catalytic sites for HER. Typical atomic vacancies in MoS_2 are Mo and S vacancies and the concentration and structure of vacancies could significantly increase the HER performance [78]. Among these, sulfur vacancies are found to be more appropriate for HER due to the presence of active sites. Using density functional theory (DFT), Charlie Tsai et al. investigated the HER activity of 2H MoS_2 basal plane with different adsorbate species at a fixed sulfur vacancy of 3.1% and reported the most stable hydrogen adsorption site within the MoS_2 lattice, as depicted in Fig. 8a. From their analysis, they also noticed that edge vacancies do not increase the HER activity. The computed relative surface energies with the presence of sulfur vacancy as a function of applied potential revealed that higher defect concentrations tend to form in clusters to stabilize the system as shown in Fig. 8b and c. The electrochemical analysis also showed that the ideal $\Delta G_H^* = 0$ eV for the highest HER performance happens for a sulfur vacancy concentration between 12.5 and 15.6%, as shown in Fig. 8d [79].

9 Conclusions and Perspectives

Hydrogen fuel is considered to be the primary alternative to fossil fuels for future energy supply and it is one of the cleanest renewable energy sources available now. Extensive studies have been reported over the past decades on the mechanism of production of hydrogen through electro-catalytic hydrogen evolution reaction. In this chapter, we summarized the fundamentals of HER and reviewed the recent advances in the HER production using transition metal dichalcogenides especially using MoS_2. It is considered that the mechanism behind HER in acidic and alkaline media incorporates two major steps, namely, Volmer reaction and the Heyrovsky reaction. The formation of an H_2 molecule can also be possible through a chemical desorption step, called the Tafel reaction in an acidic medium. It has to be noted that, the equilibrium existing between the energy required to dissociate H_2O and ΔG_H^0 governs the HER activity in alkaline media. Even though the chemical reactions involved in both acidic and alkaline media are slightly different, these three reactions effectively describe the activity behind HER under normal conditions.

The important factors that depend on evaluating the performance of an electrocatalyst include overpotential, exchange current density, turnover frequency, and durability. These factors affect differently from one another for producing an effective HER method. For example, a large overpotential leads to the low energy conversion efficiency of an electrochemical water splitting system. So, the catalytic activity should be accessed through the evaluation of activation overpotential. Usually, the catalytic performance is evaluated based on the overpotential obtained at a current density of 10 mA cm^{-2}, with regard to the working current density of photoelectrochemical water splitting system. Similarly, the exchange current density and turnover frequency also play major roles in catalytic activity of a reaction. The former gives an idea about the intrinsic charge transfer mechanism between the electrolyte and electrode and the latter presents how many product molecules are generated per active site in unit time. As per the reported literature, to obtain a better catalytic efficiency, the exchange current density should be large on the surface of the catalysts. Apart from these, for practical applications, durability or stability of the catalysts also should be examined in a crucial way. Unfortunately, most of the methods for evaluating stability are limited to working situations and more effective approaches are much needed to evaluate stability under idle conditions as well.

There are many catalytic materials such as 2D metal oxides, MXenes, alloys, metal phosphides, heteroatom-doped carbon materials, etc. similar to TMDs have been reported for their HER activities in both alkaline and acidic media. As per the reports, their large surface area helps in exposing the surface for more effective electrocatalytic reactions. Although these materials are promising for oxygen evolution and oxygen reduction reactions, they are expensive, their production requires challenging

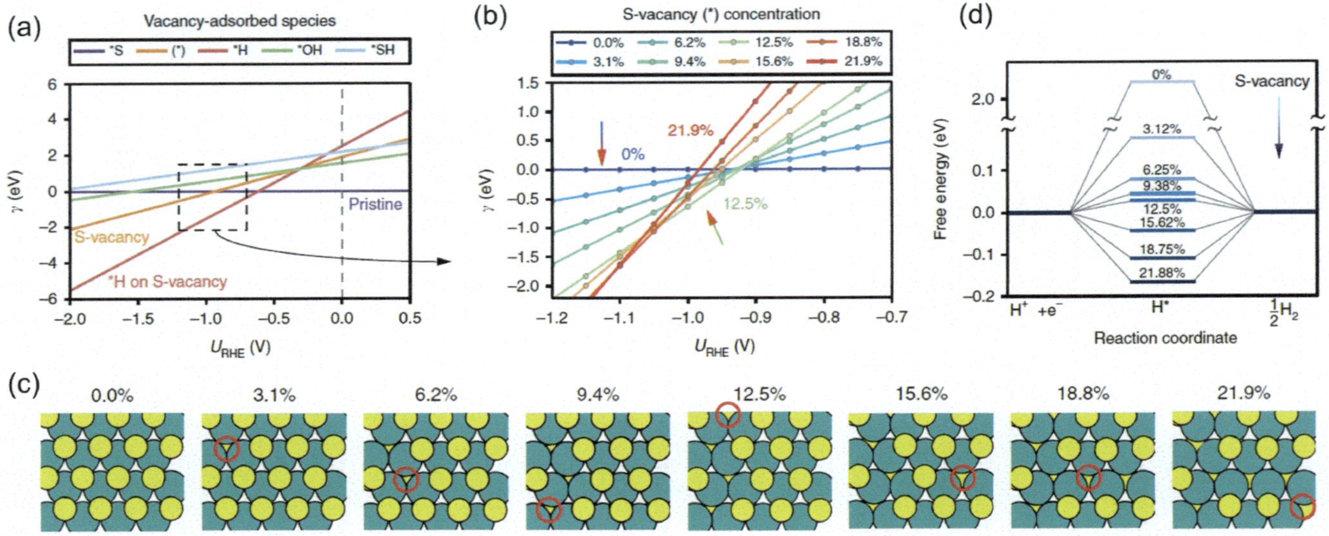

Fig. 8 **a** DFT computed surface energy of various species in 2H MoS_2 basal plane configuration as a function of applied potential at a fixed sulfur vacancy of 3.1%. The (*) and * prefix denote the surface vacancy and the adsorbed species in the sulfur vacancy, respectively. *S denotes sulfur adsorbed in the vacancy, that is, a pristine basal plane without sulfur vacancies. **b** Without the presence of any adsorbates, surface energy per unit cell for a range of sulfur vacancy concentrations (0–21.9%) within a reversible hydrogen electrode of −1.0 to 1.1 V with respect to the basal plane ($\gamma = 0$ eV) of pristine MoS_2. **c** Sulfur vacancy locations in MoS_2 with varying defect concentration, and (d) the free energy diagram for the HER activity of 2H-MoS_2 with the presence of varying sulfur vacancy concentration (Reproduced with permission from Ref. [79]; *Nature Communications*, 2017, 8, 15,113; doi: 10.1038/ncomms15113)

fabrication routes and hence, unsuitable for industry-scale implementations. In this scenario, MoS_2 and MoS_2-based nanocomposites can be potential candidates for HER production. There are different strategies to enhance the HER mechanism and further reactivity for MoS_2 and MoS_2-based polymorphs. This includes crystallization modulation, structure and phase engineering, nanostructure fabrication, modulating current density distribution using a conducting substrate, etc. Besides, introducing defect engineering by locating possible defect locations as well as creating different types of defects, and doping are also considered as some of the efficient strategies for creating large numbers of catalytic locations in MoS_2 for efficient HER performance. In addition to that, structure and phase engineering which employs increasing the number of active sites for catalytic activity and/or elevating the mass transfer and electron transportation can also improve the catalytic performance of MoS_2 polymorphs on a large scale. Despite the numerous challenges existing in the production of a novel renewable energy resource, worldwide ongoing research is still looking into this domain in a subtle way. Hence, the scope for developing an efficient, economical, and sustainable production of clean energy mechanisms is still open for the scientific community to explore using new materials and methodologies.

References

1. Babu DD, Huang Y, Anandhababu G et al (2017) Mixed-metal–organic framework self-template synthesis of porous hybrid oxyphosphides for efficient oxygen evolution reaction. ACS Appl Mater Interfaces 9:38621–38628. https://doi.org/10.1021/acsami.7b13359; Huang Y, Babu DD, Wu M, Wang Y (2018) Synergistic supports beyond carbon black for polymer electrolyte fuel cell anodes. ChemCatChem 23;10(20):4497–508; Huang Y, Liu Q, Lv J, Babu DD, Wang W, Wu M, Yuan D, Wang Y (2017) Co-intercalation of multiple active units into graphene by pyrolysis of hydrogen-bonded precursors for zinc–air batteries and water splitting. J Mater Chemistry A 5(39):20882–20891
2. Anandhababu G, Babu DD, Manimuthu V, Baskaran S (2021) Profuse surface activation of Ir-dispersed titanium nitride bifunctional electrocatalysts. Adv Energy Sustain Res 2:2000054. https://doi.org/10.1002/aesr.202000054; Abbas SC, Wu J, Huang Y, Babu DD, Anandhababu G, Ghausi MA, Wu M, Wang Y (2018). Novel strongly coupled tungsten-carbon-nitrogen complex for efficient hydrogen evolution reaction. Int J Hydrog Energy 43(1):16–23
3. Nikolaidis P, Poullikkas A (2017) A comparative overview of hydrogen production processes. Renew Sustain Energy Rev 67:597–611. https://doi.org/10.1016/j.rser.2016.09.044
4. Yin Y, Zhang Y, Gao T et al (2017) Synergistic phase and disorder engineering in 1T-MoSe2 nanosheets for enhanced hydrogen-evolution reaction. Adv Mater 29:1700311. https://doi.org/10.1002/adma.201700311
5. Zhou G, Li M, Li Y et al (2020) Regulating the electronic structure of CoP nanosheets by O incorporation for high-efficiency electrochemical overall water splitting. Adv Func Mater 30:1905252. https://doi.org/10.1002/adfm.201905252
6. Najafi L, Bellani S, Oropesa-Nuñez R et al (2018) Engineered MoSe2-based heterostructures for efficient electrochemical hydrogen evolution reaction. Adv Energy Mater 8:1703212. https://doi.org/10.1002/aenm.201703212
7. Li L, Qin Z, Ries L et al (2019) Role of sulfur vacancies and undercoordinated Mo regions in MoS2 nanosheets toward the evolution of hydrogen. ACS Nano 13:6824–6834. https://doi.org/10.1021/acsnano.9b01583

8. Henckel DA, Lenz OM, Krishnan KM, Cossairt BM (2018) Improved HER catalysis through facile, aqueous electrochemical activation of nanoscale WSe2. Nano Lett 18:2329–2335. https://doi.org/10.1021/acs.nanolett.7b05213
9. Cheng L, Huang W, Gong Q et al (2014) Ultrathin WS2 nanoflakes as a high-performance electrocatalyst for the hydrogen evolution reaction. Angew Chem Int Ed 53:7860–7863. https://doi.org/10.1002/anie.201402315
10. Hinnemann B, Moses PG, Bonde J et al (2005) Biomimetic hydrogen evolution: MoS2 nanoparticles as catalyst for hydrogen evolution. J Am Chem Soc 127:5308–5309. https://doi.org/10.1021/ja0504690
11. Yu Y, Nam G-H, He Q et al (2018) High phase-purity 1T′-MoS2- and 1T′-MoSe2-layered crystals. Nature Chem 10:638–643. https://doi.org/10.1038/s41557-018-0035-6
12. Zhang X, Jia F, Yang B, Song S (2017) Oxidation of molybdenum disulfide sheet in water under in situ atomic force microscopy observation. J Phys Chem C 121:9938–9943. https://doi.org/10.1021/acs.jpcc.7b01863
13. Splendiani A, Sun L, Zhang Y et al (2010) Emerging photoluminescence in monolayer MoS2. Nano Lett 10:1271–1275. https://doi.org/10.1021/nl903868w
14. Lv R, Robinson JA, Schaak RE et al (2015) Transition metal dichalcogenides and beyond: synthesis, properties, and applications of single- and few-layer nanosheets. Acc Chem Res 48:56–64. https://doi.org/10.1021/ar5002846
15. Voiry D, Mohite A, Chhowalla M (2015) Phase engineering of transition metal dichalcogenides. Chem Soc Rev 44:2702–2712. https://doi.org/10.1039/C5CS00151J
16. Li H, Yin Z, He Q et al (2012) Fabrication of single- and multilayer MoS2 film-based field-effect transistors for sensing no at room temperature. Small 8:63–67. https://doi.org/10.1002/smll.201101016
17. Gacem K, Boukhicha M, Chen Z, Shukla A (2012) High quality 2D crystals made by anodic bonding: a general technique for layered materials. Nanotechnology 23:505709. https://doi.org/10.1088/0957-4484/23/50/505709
18. Zeng Z, Yin Z, Huang X et al (2011) Single-layer semiconducting nanosheets: high-yield preparation and device fabrication. Angew Chem Int Ed 50:11093–11097. https://doi.org/10.1002/anie.201106004
19. Ghasemi F, Mohajerzadeh S (2016) Sequential solvent exchange method for controlled exfoliation of MoS2 suitable for phototransistor fabrication. ACS Appl Mater Interfaces 8:31179–31191. https://doi.org/10.1021/acsami.6b07211
20. Wu S, Huang C, Aivazian G et al (2013) Vapor-solid growth of high optical quality MoS2 monolayers with near-unity valley polarization. ACS Nano 7:2768–2772. https://doi.org/10.1021/nn4002038
21. Wang X, Feng H, Wu Y, Jiao L (2013) Controlled synthesis of highly crystalline MoS2 flakes by chemical vapor deposition. J Am Chem Soc 135:5304–5307. https://doi.org/10.1021/ja4013485
22. Liu K-K, Zhang W, Lee Y-H et al (2012) Growth of large-area and highly crystalline MoS2 thin layers on insulating substrates. Nano Lett 12:1538–1544. https://doi.org/10.1021/nl2043612
23. Zhao G, Rui K, Dou SX, Sun W (2018) Heterostructures for electrochemical hydrogen evolution reaction: a review. Adv Func Mater 28:1803291. https://doi.org/10.1002/adfm.201803291
24. Quaino P, Juarez F, Santos E, Schmickler W (2014) Volcano plots in hydrogen electrocatalysis – uses and abuses. Beilstein J Nanotechnol 5:846–854. https://doi.org/10.3762/bjnano.5.96
25. Parsons R (1958) The rate of electrolytic hydrogen evolution and the heat of adsorption of hydrogen. Trans Faraday Soc 54:1053–1063. https://doi.org/10.1039/TF9585401053
26. Danilovic N, Subbaraman R, Strmcnik D et al (2012) Enhancing the alkaline hydrogen evolution reaction activity through the bifunctionality of Ni(OH)2/metal catalysts. Angew Chem Int Ed 51:12495–12498. https://doi.org/10.1002/anie.201204842
27. Schmidt TJ, Ross PN, Markovic NM (2002) Temperature dependent surface electrochemistry on Pt single crystals in alkaline electrolytes: Part 2. The hydrogen evolution/oxidation reaction. J Electroanal Chem 524–525:252–260. https://doi.org/10.1016/S0022-0728(02)00683-6

28. Wei J, Zhou M, Long A et al (2018) Heterostructured electrocatalysts for hydrogen evolution reaction under alkaline conditions. Nano-Micro Lett 10:75. https://doi.org/10.1007/s40820-018-0229-x
29. Ahmaruzzaman Md, Gadore V (2021) MoS2 based nanocomposites: an excellent material for energy and environmental applications. J Environ Chem Eng 9:105836. https://doi.org/10.1016/j.jece.2021.105836
30. Nørskov JK, Bligaard T, Logadottir A et al (2005) Trends in the exchange current for hydrogen evolution. J Electrochem Soc 152:J23. https://doi.org/10.1149/1.1856988
31. Jaramillo TF, Jørgensen KP, Bonde J et al (2007) Identification of active edge sites for electrochemical H2 evolution from MoS2 nanocatalysts. Science. https://doi.org/10.1126/science.1141483
32. Li Y, Wang H, Xie L et al (2011) MoS2 nanoparticles grown on graphene: an advanced catalyst for the hydrogen evolution reaction. J Am Chem Soc 133:7296–7299. https://doi.org/10.1021/ja201269b
33. Liao L, Zhu J, Bian X et al (2013) MoS_2 formed on mesoporous graphene as a highly active catalyst for hydrogen evolution. Adv Func Mater 23:5326–5333. https://doi.org/10.1002/adfm.201300318
34. Chen L, Zhang J, Ren X et al (2017) A Ni(OH)2–CoS2 hybrid nanowire array: a superior non-noble-metal catalyst toward the hydrogen evolution reaction in alkaline media. Nanoscale 9:16632–16637. https://doi.org/10.1039/C7NR06001G
35. Cao C (2008) Principles of electrochemistry of corrosion. Chemical Industry Press, Peking, China
36. Voiry D, Yamaguchi H, Li J et al (2013) Enhanced catalytic activity in strained chemically exfoliated WS2 nanosheets for hydrogen evolution. Nature Mater 12:850–855. https://doi.org/10.1038/nmat3700
37. Yan Y, Ge X, Liu Z et al (2013) Facile synthesis of low crystalline MoS2 nanosheet-coated CNTs for enhanced hydrogen evolution reaction. Nanoscale 5:7768–7771. https://doi.org/10.1039/C3NR02994H
38. Mukherjee D, Austeria PM, Sampath S (2016) Two-dimensional, few-layer phosphochalcogenide, FePS3: a new catalyst for electrochemical hydrogen evolution over wide pH range. ACS Energy Lett 1:367–372. https://doi.org/10.1021/acsenergylett.6b00184
39. Jaramillo TF, Jørgensen KP, Bonde J et al (2007) Identification of active edge sites for electrochemical H2 evolution from MoS2 nanocatalysts. Science 317:100–102. https://doi.org/10.1126/science.1141483
40. Li G, Zhang D, Qiao Q et al (2016) All the catalytic active sites of MoS2 for hydrogen evolution. J Am Chem Soc 138:16632–16638. https://doi.org/10.1021/jacs.6b05940
41. Li H, Du M, Mleczko MJ et al (2016) Kinetic study of hydrogen evolution reaction over strained MoS2 with sulfur vacancies using scanning electrochemical microscopy. J Am Chem Soc 138:5123–5129. https://doi.org/10.1021/jacs.6b01377
42. Zhang J, Wang T, Liu P et al (2016) Engineering water dissociation sites in MoS2 nanosheets for accelerated electrocatalytic hydrogen production. Energy Environ Sci 9:2789–2793. https://doi.org/10.1039/C6EE01786J
43. Tsai C, Abild-Pedersen F, Nørskov JK (2014) Tuning the MoS2 edge-site activity for hydrogen evolution via support interactions. Nano Lett 14:1381–1387. https://doi.org/10.1021/nl404444k
44. Winchester A, Ghosh S, Feng S et al (2014) Electrochemical characterization of liquid phase exfoliated two-dimensional layers of molybdenum disulfide. ACS Appl Mater Interfaces 6:2125–2130. https://doi.org/10.1021/am4051316
45. Liu G, Thummavichai K, Lv X et al (2021) Defect-rich heterogeneous MoS2/rGO/NiS nanocomposite for efficient ph-universal hydrogen evolution. Nanomaterials 11:662. https://doi.org/10.3390/nano11030662
46. Yu XY, (David) Lou XW (2018) Mixed metal sulfides for electrochemical energy storage and conversion. Adv Energy Mater 8:1701592. https://doi.org/10.1002/aenm.201701592

47. Yu X-Y, Feng Y, Jeon Y et al (2016) Formation of Ni–Co–MoS2 nanoboxes with enhanced electrocatalytic activity for hydrogen evolution. Adv Mater 28:9006–9011. https://doi.org/10.1002/adma.201601188
48. Xie J, Zhang H, Li S et al (2013) Defect-rich MoS2 ultrathin nanosheets with additional active edge sites for enhanced electrocatalytic hydrogen evolution. Adv Mater 25:5807–5813. https://doi.org/10.1002/adma.201302685
49. Gao M-R, Chan MKY, Sun Y (2015) Edge-terminated molybdenum disulfide with a 9.4-Å interlayer spacing for electrochemical hydrogen production. Nat Commun 6:7493. https://doi.org/10.1038/ncomms8493
50. Xiao W, Liu P, Zhang J et al (2017) Dual-functional N dopants in edges and basal plane of MoS2 nanosheets toward efficient and durable hydrogen evolution. Adv Energy Mater 7:1602086. https://doi.org/10.1002/aenm.201602086
51. Laursen AB, Kegnæs S, Dahl S, Chorkendorff I (2012) Molybdenum sulfides—efficient and viable materials for electro - and photoelectrocatalytic hydrogen evolution. Energy Environ Sci 5:5577–5591. https://doi.org/10.1039/C2EE02618J
52. Cui B, Zhang M, Zhao Y, Hu S (2019) Heterogenization of few-layer MoS2 with highly crystalline 3D Ni3S2 nanoframes effectively synergizes the electrocatalytic hydrogen generation in alkaline medium. Mater Today Energy 13:85–92. https://doi.org/10.1016/j.mtener.2019.05.001
53. Adarakatti PS, Mahanthappa M, Hughes JP et al (2019) MoS2-graphene-CuNi2S4 nanocomposite an efficient electrocatalyst for the hydrogen evolution reaction. Int J Hydrog Energy 44:16069–16078. https://doi.org/10.1016/j.ijhydene.2019.05.004
54. Wang J, Fang W, Hu Y et al (2020) Different phases of few-layer MoS2 and their silver/gold nanocomposites for efficient hydrogen evolution reaction. Catal Sci Technol 10:154–163. https://doi.org/10.1039/C9CY02158B
55. Liu S, Li S, Sekar K et al (2019) Hierarchical ZnS@C@MoS2 core-shell nanostructures as efficient hydrogen evolution electrocatalyst for alkaline water electrolysis. Int J Hydrog Energy 44:25310–25318. https://doi.org/10.1016/j.ijhydene.2019.08.048
56. Yu C, Cao Z, Yang F et al (2019) MoS2 confined on graphene by triethanolamine for enhancing electrocatalytic hydrogen evolution performance. Int J Hydrog Energy 44:28151–28162. https://doi.org/10.1016/j.ijhydene.2019.09.052
57. Zhou L, Yang T, Chen S et al (2020) Tunably fabricated nanotremella-like Bi2S3/MoS2: an excellent and highly stable electrocatalyst for alkaline hydrogen evolution reaction. Int J Hydrog Energy 45:9535–9545. https://doi.org/10.1016/j.ijhydene.2020.01.168
58. Burke MS, Enman LJ, Batchellor AS et al (2015) Oxygen evolution reaction electrocatalysis on transition metal oxides and (oxy)hydroxides: activity trends and design principles. Chem Mater 27:7549–7558. https://doi.org/10.1021/acs.chemmater.5b03148
59. Browne MP, Sofer Z, Pumera M (2019) Layered and two dimensional metal oxides for electrochemical energy conversion. Energy Environ Sci 12:41–58. https://doi.org/10.1039/C8EE02495B
60. Lu Z, Wang H, Kong D et al (2014) Electrochemical tuning of layered lithium transition metal oxides for improvement of oxygen evolution reaction. Nat Commun 5:4345. https://doi.org/10.1038/ncomms5345
61. Seh ZW, Fredrickson KD, Anasori B et al (2016) Two-dimensional molybdenum carbide (MXene) as an efficient electrocatalyst for hydrogen evolution. ACS Energy Lett 1:589–594. https://doi.org/10.1021/acsenergylett.6b00247
62. Djire A, Wang X, Xiao C et al (2020) Basal plane hydrogen evolution activity from mixed metal nitride MXenes measured by scanning electrochemical microscopy. Adv Func Mater 30:2001136. https://doi.org/10.1002/adfm.202001136
63. Greeley J, Jaramillo TF, Bonde J et al (2006) Computational high-throughput screening of electrocatalytic materials for hydrogen evolution. Nat Mater 5:909–913. https://doi.org/10.1038/nmat1752
64. Fang M, Gao W, Dong G et al (2016) Hierarchical NiMo-based 3D electrocatalysts for highly-efficient hydrogen evolution in alkaline conditions. Nano Energy 27:247–254. https://doi.org/10.1016/j.nanoen.2016.07.005

65. Callejas JF, Read CG, Roske CW et al (2016) Synthesis, characterization, and properties of metal phosphide catalysts for the hydrogen-evolution reaction. Chem Mater 28:6017–6044. https://doi.org/10.1021/acs.chemmater.6b02148
66. Acerce M, Voiry D, Chhowalla M (2015) Metallic 1T phase MoS2 nanosheets as supercapacitor electrode materials. Nat Nanotech 10:313–318. https://doi.org/10.1038/nnano.2015.40
67. Lin L, Lei W, Zhang S et al (2019) Two-dimensional transition metal dichalcogenides in supercapacitors and secondary batteries. Energy Stor Mater 19:408–423. https://doi.org/10.1016/j.ensm.2019.02.023
68. Tian J, Liu Q, Liang Y et al (2014) FeP nanoparticles film grown on carbon cloth: an ultrahighly active 3D hydrogen evolution cathode in both acidic and neutral solutions. ACS Appl Mater Interfaces 6:20579–20584. https://doi.org/10.1021/am5064684
69. Laursen AB, Patraju KR, Whitaker MJ et al (2015) Nanocrystalline Ni5P4: a hydrogen evolution electrocatalyst of exceptional efficiency in both alkaline and acidic media. Energy Environ Sci 8:1027–1034. https://doi.org/10.1039/C4EE02940B
70. Dey KK, Jha S, Kumar A et al (2019) Layered vanadium oxide nanofibers as impressive electrocatalyst for hydrogen evolution reaction in acidic medium. Electrochim Acta 312:89–99. https://doi.org/10.1016/j.electacta.2019.04.185
71. Kibsgaard J, Tsai C, Chan K et al (2015) Designing an improved transition metal phosphide catalyst for hydrogen evolution using experimental and theoretical trends. Energy Environ Sci 8:3022–3029. https://doi.org/10.1039/C5EE02179K
72. Cao Y (2021) Roadmap and direction toward high-performance MoS2 hydrogen evolution catalysts. ACS Nano 15:11014–11039. https://doi.org/10.1021/acsnano.1c01879
73. Lukowski MA, Daniel AS, Meng F et al (2013) Enhanced hydrogen evolution catalysis from chemically exfoliated metallic MoS2 nanosheets. J Am Chem Soc 135:10274–10277. https://doi.org/10.1021/ja404523s
74. Voiry D, Salehi M, Silva R et al (2013) Conducting MoS2 nanosheets as catalysts for hydrogen evolution reaction. Nano Lett 13:6222–6227. https://doi.org/10.1021/nl403661s
75. Merki D, Fierro S, Vrubel H, Hu X (2011) Amorphous molybdenum sulfide films as catalysts for electrochemical hydrogen production in water. Chem Sci 2:1262–1267. https://doi.org/10.1039/C1SC00117E
76. Ren X, Pang L, Zhang Y et al (2015) One-step hydrothermal synthesis of monolayer MoS2 quantum dots for highly efficient electrocatalytic hydrogen evolution. J Mater Chem A 3:10693–10697. https://doi.org/10.1039/C5TA02198G
77. Zhang N, Gan S, Wu T et al (2015) Growth control of MoS2 nanosheets on carbon cloth for maximum active edges exposed: an excellent hydrogen evolution 3D cathode. ACS Appl Mater Interfaces 7:12193–12202. https://doi.org/10.1021/acsami.5b02586
78. Bai F, Xu L, Zhai X et al (2020) Vacancy in ultrathin 2D nanomaterials toward sustainable energy application. Adv Energy Mater 10:1902107. https://doi.org/10.1002/aenm.201902107
79. Tsai C, Li H, Park S et al (2017) Electrochemical generation of sulfur vacancies in the basal plane of MoS_2 for hydrogen evolution. Nat Commun 8:15113. https://doi.org/10.1038/ncomms15113

Chapter 14
MoS_2: The New Avenue in Biomedical Applications

Snehal Gagare, Ratna Chauhan, Dattatray J. Late, and Suresh W. Gosavi

1 Introduction

The common infectious infections are now less widespread, and antibiotic medications have become crucial for many medical procedures [1]. Nevertheless since bacteria are constantly changing and because antibiotics are misused, drug resistance develops quickly and widely making it harder to treat bacterial diseases [2, 3]. Pathogens that are resistant to antibiotics have proliferated across human and animal populations all over the world. The pace of bacterial evolution is outpacing the effectiveness of antibiotics in treating bacterial infections [4]. The prevalence of pathogens including carbapenem-resistant Enterobacteriaceae (CRE) [5] and meticillin-resistant Staphylococcus aureus (MRSA) [6] has increased significantly. The rise in antibiotic-resistant bacteria (ARB) in recent years has had a noteworthy effect on public health [7]. Animal excrement, water, and air are all potential carriers for ARB. In particular, samples taken from livestock and poultry breeding facilities contain more than a hundred different types of resistance genes [8]. By gene

S. Gagare
Department of Biotechnology, Modern College of Arts Science and Commerce, Ganeshkhind, Pune 411016, India
e-mail: snehal@moderncollegegk.org

R. Chauhan
Department of Environmental Science, Savitribai Phule Pune University, Pune 411007, India

D. J. Late
Department of Physics, Federal University of Lavras, Campus Universitário, PO Box 3037, Lavras, Minas Gerais, Brazil

S. W. Gosavi (✉)
Department of Physics, Savitribai Phule Pune University, Pune 411007, India
e-mail: swg@physics.unipune.ac.in

D. J. Late and C. S. Rout (eds.), *Science and Technology of 2D MoS_2*,
Materials Horizons: From Nature to Nanomaterials,
https://doi.org/10.1007/978-981-97-7367-1_14

horizontal transfer, the antibiotic-resistant bacteria might obtain their antibiotic-resistance genes either from pathogens or from environmental pathogens to the bacteria that serve as their primary host. Bacterial infections persist to be one of the biggest threats to human health since there are an estimated 600–700 types of bacteria on the globe and they continue to evolve and mutate [9]. The loss of efficacy against common bacteria has led to higher rates of morbidity and mortality in low- and middle-income countries due to the inability to afford second-line medications. Conversely, in high-income nations, this has also resulted in a shift toward further costly antibiotic medications [10].

It is a concerning situation and an urgent need to investigate and develop non-antibiotic antibacterial agents to address the aforementioned issues.

Developing an innovative and effective antimicrobial approach that meets the necessary requirements of being biocompatible and free of antibiotics is crucial.

The photodynamic antibacterial approach has received a lot of attention in microbiological applications due to the low risk of developing drug resistance and the fact that it depends on photo-responsive materials to carry out the operation [11, 12].

In order to meet the needs of a wide range of biomedical applications, researchers have been working on the development and analysis of nanocomposites with customizable characteristics, excellent targeting ability, low toxicity, and multiple functions [13–15]. Inspired by graphene's success in biomedical applications, transition-metal dichalcogenides (TMDs) [16], a new class of nanomaterials and prospective replacement for graphene, share similarities with graphene in their shape and physical characteristics [17]. TMDs have piqued the interest of scientists due to their high conductivity, exceptional catalytic activity, and favourable optical characteristics. The most well-known type of TMD is molybdenum disulphide (MoS_2). Several dimensions MoS_2-based nanostructures, including quantum dots (QDs), two-dimensional (2D) nanosheets (NSs), flower-like particles, nanospheres, and nanotubes [18], have sparked the curiosity amongst the scientific community. Specifically, 2D MoS_2 NSs have garnered interest as they are the most typical of 2D TMDs. Due to their outstanding electronic and optical characteristics, which include strong near-infrared absorbance, thickness-dependent bandgaps, large surface areas, and easily functionalized surface sites [19], it has tremendous nanoplatforms for a diversity of biomedical mechanisms like targeted drug delivery, phototherapy for tumors [20], sensing biological materials [21], bioimaging, theranostics [22], development of antibacterial agents [23].

2 The Emergence of 2D MoS_2 Nanomaterials as a Novel Nanomedicine

Numerous graphene-based, polymer-, and metal-based nanocarriers have been explored for both diagnostic and therapeutic purposes [24]. The emergence of 2D MoS2 nanomaterials as a novel nanomedicine alternative to graphene and its derivatives in the field of biomedical research can be attributed to the unique structural, optical, electrical, thermal, and biocompatible properties of these nanomaterials. MoS_2 has become a promising contender for use in cancer theranostics within the class of 2D nanomaterials being investigated as nanomedicine platforms [25]. Because of their superior physicochemical and biocompatible qualities, MoS_2-based nanomaterials like nanosheets, quantum dots, and nanocomposites are being extensively researched for use in the field of nanomedicine [26]. Potential applications include using them as nanocarriers for the delivery of therapeutic cargo and as nanoprobes for bioimaging of cells and tissue MoS_2 and its nanocomposites can be potential candidates as future antibiotics. The MoS_2 has been utilized in the departments of catalysis [27], drug delivery [28], and biomedicine [19, 29] for the sake of its superior biocompatibility, high specific surface area, ultrathin atomic layer structure in two dimensions (2D) [30]. The MoS_2 nanoparticles offer a very wide range of potential biological applications owing to its outstanding photothermal conversion efficiency, strong near-infrared (NIR) absorbance, and peroxidase-like activity [31, 32]. (Fig. 14.1). The layers of MoS_2 are stacked together by a weak van der Waals attraction, and molybdenum and sulphur atoms are joined by covalent bonds to create an S-Mo-S sandwich structure. The S-Mo configuration has high adsorption [33]. For the first time, Chou et al. reported that MoS_2 has a superb near-infrared photothermal effect because of its remarkable surface-area to mass ratio and ability to load with large cargo concentrations. When exposed to specific light wavelengths, they can form electron hole pairs. Reactive oxygen species (ROS), which can lead to the death of pathogenic or unhealthy organisms, can be created when the expelled electrons are subsequently absorbed by the oxygen in the external environment. Photodynamic treatment (PDT), in particular, is able to prevent the development of drug resistance by primarily acting as a biocide through the oxidative degradation of biological macromolecules such lecithin, enzymes, nutrients, and DNA in cell membranes.

3 Construction of Functionalized Molybdenum Disulphide Nanoflowers

In order to scavenge antibiotic-resistant bacteria (ARB), Liu et al. 2021 created PEG-MoS_2 NFs, which are polyethylene glycol functionalized molybdenum disulphide nanoflowers [35]. PEG-MoS_2 NFs exhibit good photothermal performance around

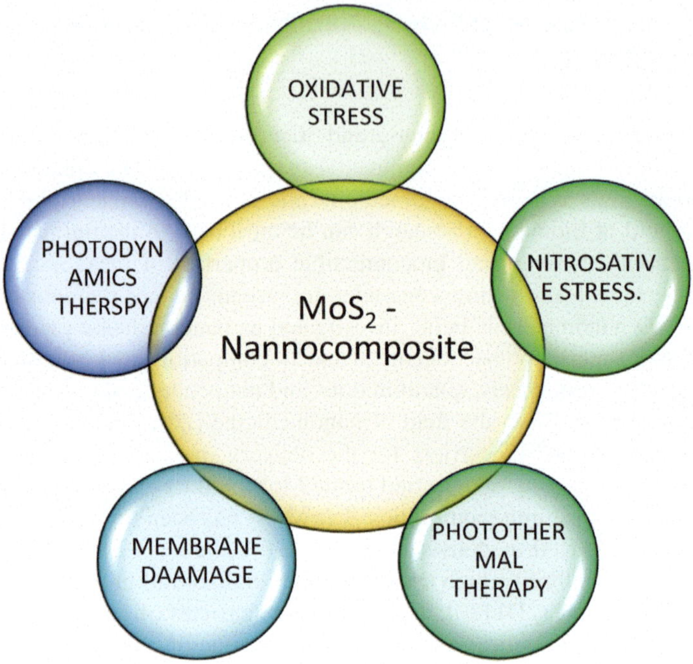

Fig. 14.1 Biological properties of MoS_2 nanocomposites

808 nm NIR illumination, which was utilized to study the inhibition of antibiotic-resistant bacteria. The nanoflowers of PEG-MoS_2 are biocompatible with both Gram positive and negative bacterial cell membrane can convert the light energy of NIR into heat energy, which increase the temperature in that localized area to effectively work as a photothermal lysing agent. This nanocomposite attacks Glutathione, the intercellular protection machinery that prevents the oxidative damage caused by reactive oxygen species and oxidative stress. Glutathione is an antioxidant that guards the cellular biomolecules from the oxidative damage triggered by external oxidative stress on the cell's constituent parts [36]. When exposed to reactive oxygen or other oxidants, the sulfhydryl group on cysteine, which is its active group, can quickly oxidise into disulphide bonds, destroying the proteins and enzymes and causing internal cell dysfunction that results in cell death. PEG-MoS_2 nanoflowers were constructed by one-step hydrothermal approach to synthesize a potent bacterial inhibitor with outstanding photothermal conversion capability and photothermal stability. Unlike the narrow spectrum antibiotics the nanocomposite of PEG-MoS_2 can effectively eradicate both Gram-positive and Gram-negative antibiotic-resistant bacteria. The schematic representation in Fig. 14.2 depicts the mechanism of PEG-MoS_2 NFs, the nanocomposites when prompted by high temperature NIR laser at 808 nm resulted in oxidation of Gluthathione that collapsed the cellular antioxidant and protection system and abrupt cell lysis of antibiotic resistant bacteria. The

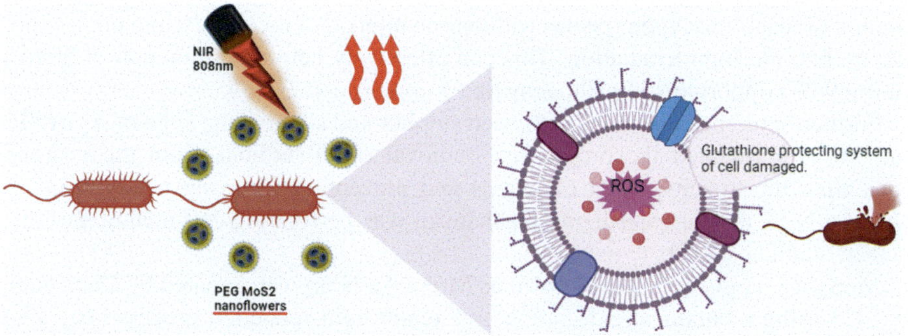

Fig. 14.2 Removal of antibiotic-resistant bacteria from the contaminated environment using photothermal potential of PEG-MoS_2 nanoflowers

study illustrates the tremendous potential of PEG-MoS_2 nanoflowers to commendably function in the environment contaminated with antibiotic-resistant bacteria and act instantly to eradicate them.

MoS_2 is a suitable candidate in the fields of energy storage [37] and hydrogen evolution owing to its properties like two dimensioned layer structure, huge specific surface area and most significantly the surface modifications can be easily done [38], and robust compatibility to biological systems. It can be used in photothermal therapy (PTT) due to its excellent near-infrared absorption in the range of [39] and the human body-friendly nature of Mo and S. Nevertheless, MoS_2's narrow band gap makes it easier for electron–hole pairs to join, and less suitable for photodynamic activity [40]. Thus, finding other suitable alternative materials to amalgamate with MoS_2 is essential this can be achieved by increasing the flow of electrons, exquisite metals and semiconductor materials can significantly augment the photodynamic characteristics [41] of semiconductor materials. The Ag is an excellent precious metal material due to its superior conductivity and innate antibacterial qualities. More pointedly, the Ag also exhibits a phenomenon, which facilitates electrons to leak from the outermost surface and be absorbed by the surrounding oxygen to form reactive oxygen species [42]. The ability of MoS_2 to produce ROS when reacting with hydrogen peroxide (H_2O_2), which functions as an enzyme similar to peroxidase, is another benefit. This ability speeds up the death of bacteria [43].

4 Construction of Functionalized MoS_2/Ag Nanosheets

A remarkable success was achieved by Zhu et al. in 2019 by functionalizing surface of two-dimensional MoS_2 nanosheets with AgBrNPs significantly suppressing the recombination of photoelectrons and holes [44]. The composite AgBr@MoS_2 coating displayed good photocatalytic and had the ability to generate substantial amount of reactive oxygen species within time interval of meagre 20 min upon exposure to 660 nm light irradiation. This can effectively help in elimination of almost all (99.94%) opportunistic pathogens like *S. aureus* and *E. coli.*

Furthermore, the close bonds between sulphur and silver at the edge of AgBrNPs and MoS_2 stimulated the progressive removal of Silver ions from the coating, preventing the toxicity effect of metals and preventing recurring bacterial infection. Ag NPs' primary sterilizing mechanism was perforating cell membranes due to their small size [45].

MoS_2/Ag nanosheets (also known as MoS_2/Ag NSs) were created by Chen et al. in 2023 using a chemical exfoliation and xenon light reduction process [46]. Zhu and his coworkers demonstrate that on irradiation by a 660 nm laser, the valence electrons of MoS_2 were stimulated to change into the conduction band and produce electron–hole pairs. The resultant nanocomposite of MoS_2 with Ag could expedite electron transport and produce a substantial amount of photoelectrons and holes as it would inhibit the loss caused due to compounding of electron–hole pairs. Collectively Ag NPs and MoS_2 improved the energy evolution pathways and electron transport at the interface, significantly boosting MoS_2's photodynamic activity [47] and, consequently, the quantity of ROS generated when exposed to 660 nm near-infrared light. The intrinsic peroxidase-like ability of MoS_2/Ag NSs could convert a modest quantity of H_2O_2 into harmful ROS, which was crucial in damaging the bacterial membranes. This work thus showed exceptional potential in the clinical management of inflammatory illnesses and provided a viable approach for rapid and effective sterilization.

The nanozymes displayed good antibacterial activity when used in disinfection treatments that used OH to oxidise bacteria's membranes. Nonetheless, there is still much need for improvement in many areas of the efficient bacteriotoxic application of nanozymes [48]. For instance, the inadequate capacity of most synthetic enzymes to bind bacteria as well as the short activity span and meagre diffusivity of ROS significantly affected the interaction between bacteria and the nanozyme tremendously affecting the efficiency of disinfection. So, coming up with original solutions to the aforementioned problems and constraints continues to be a very difficult task. The interaction of bacteria and antibacterial compounds during the disinfection process is widely acknowledged as one of the essential processes for determining the disinfection performance [49]. The antibacterial agents' capacity to stick to surfaces in this regard plays a crucial role in determining the fate of bacterial survival. The painstaking efforts to engineer nanoenzymes continued, which led to create Specific adhesive materials. These designs are effectively binding to bacteria, owing to the unexpected adherence from distinctive rough surfaces, which were inspired by the fascinating biological occurrences [50].

Remarkable works have been done by F. Cao et al. 2019 in designing a group of defect-rich sticky nanozymes that naturally capture bacteria and have powerful antibacterial properties. Using MoS_2 as a model nanozymes were synthesized with an ability to vertically align on various substrates, forming abundant coarse exteriors to trap bacteria and simultaneously exposing more sites to improve peroxidase-like activity [51]. The removal of the natural antibacterial barriers of nanozymes, the scientific community has been successful in improving in situ sterilization to counter attack antibiotic-resistant *Staphylococcus aureus* and *E. coli*. To overcome the problems that nanozymes faced, these defect-rich sticky nanozymes were developed. They could quickly trap bacteria within the range of elevated ROS destruction, demonstrating noticeably higher bacterial binding capacity and enhanced in situ antibacterial efficacy in comparison to pristine nanozyme. This was made possible by the formation of numerous rough surfaces and the exposure of defect-rich edge sites. Also, the addition of the innate photothermal effect produced better therapeutic results both in vitro and in vivo. This work integrates nanotopology, defecting chemistry, and catalytic performance to create effective nanozymes with the capacity to capture alternative antibiotics. This work also offers fresh insight into the design of biomaterials that surmounted the problem in mode of action of artificial enzymes like less bacteria-binding capacity short lifespan and poor diffusivity of ROS. This had adversely affected the interaction with bacteria and restricted the disinfection performance.

5 Construction of Recoverable Peroxidase like Fe_3O_4@MoS_2-Ag Nanocomposites

A two-step hydrothermal process and in-situ Ag nanoparticle development, a simple artificial nanozyme of Fe3O4@MoS2-1%Ag was created. By wrapping a sharp MoS_2 sheet around the Fe_3O_4 surface, which demonstrated strong adhesion properties for bacteria, the composite displayed an appealing rough surface [52]. The local hyperthermia and improved peroxidase mimicking properties can significantly increase the disinfectant activity when subjected to 808 nm NIR radiation. According to antibacterial tests conducted on E. coli, Fe_3O_4@ MoS_2-1%Ag can effectively inactivate bacteria by encasing the pathogen, releasing poisonous OH and Ag+, and then using localized hyperthermia to assault the membranes. The composite is reusable and promising nanoenzyme owing to the magnetism conferred by Fe_3O_4 [53].

The schematic representation in Fig. 14.3 gives a detailed insight into the underlying mechanism of bactericidal properties of composite. Furthermore, it possesses antibacterial activity against antibiotic-resistant Gram-positive and Gram-negative bacteria. The creation of this promising adhesive nanozyme could significantly reduce the diffusive distance of OH, opening up the possibility of a quick and efficient antibacterial therapy.

Silver nanoparticles were effectively added to the surface of MoS_2 nanosheets by means of UV light-induced reduction, this generated a photo-responsive antibacterial coating for implants. The photo-generated electrons utilized the quick electron transport capabilities of silver and are moved from MoS_2, this lowers the rate at which electron–hole couples recombine. The photocatalytic activity of this system is dramatically increased when exposed to 660 nm light. Moreover, CS in the system

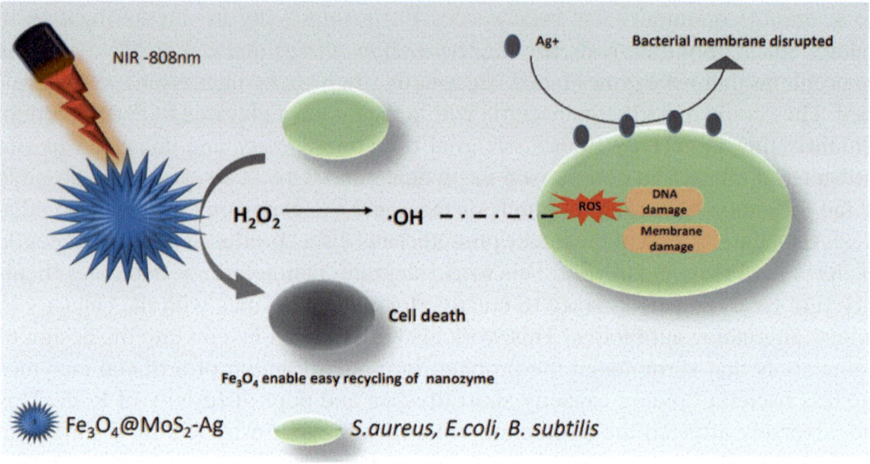

Fig. 14.3 Nanozymes and essential bactericidal mechanism

keeps cells from coming into interface with Ag NPs, lowering Ag NPs' cytotoxicity [40].

6 MoS_2 Nanocomposites in Drug Delivery and Cancer Therapy

The potential of various nanoagents is being explored as effective drug delivery vehicles, to function as a potential agent in drug delivery a composite should have a strong NIR-responsive photothermal transducing ability, and planar structure and high surface-area-to-volume ratio. The various possible areas of cancer therapy where MoS_2 can effectively work are depicted in the schematic representation in Fig. 14.4. Leong and colleagues created a novel drug delivery system made of multi-layered MoS_2 nanosheets connected by DNA [54].

Enormous surface-to-volume ratios of 2D materials allow for a variety of anchoring sites for the active drug molecules to be loaded into drug delivery carriers, resulting in high drug loading capacities. To release the medication in a sustainable manner, the carrier can use 2D materials because they have a comparatively higher chemical or physical adsorption capacity with drug molecules than organic liposomes [55].

Additionally, the diverse chemical compositions of 2D materials and the specific interactions between drug molecules enable external triggers for on-demand intelligent drug release. Due to its straight band gap, high carrier mobility, and quick optical response, MoS_2, one type of sophisticated 2D material, has a wide range of potential applications in electronics and optoelectronics. Since they are transparent

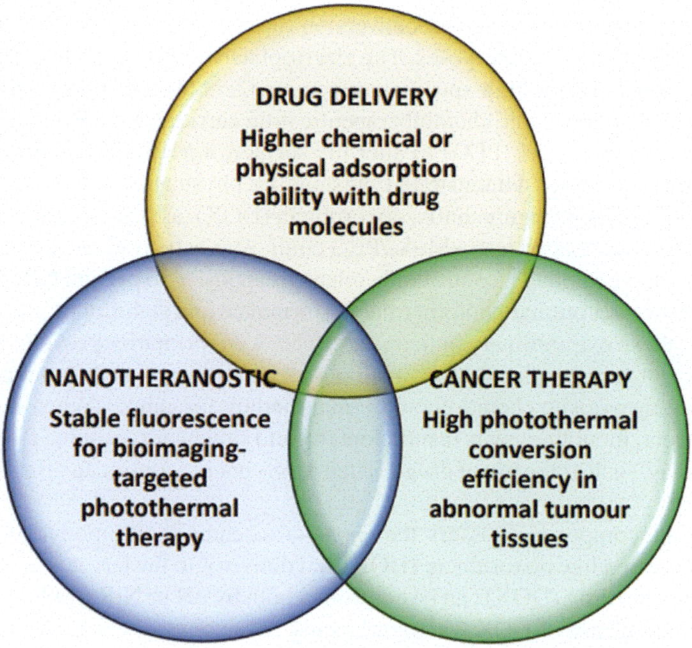

Fig. 14.4 Schematic representation of 2D MoS_2—potential candidate for drug delivery and anticancer treatments

to normal biological tissues and have a high photothermal conversion efficiency in abnormal tumour tissues, their superiority in biological applications also extends to biomedical ones. These qualities are attributed to their high loading capacity and strong light absorption in the near-infrared (NIR) window of 700–1300 nm [56].

These nanoscale 2D materials' biological behaviour, including their cytotoxicity, biodegradability, and dangerous by-products, is also being carefully examined in order to test their biocompatibility and biosafety [57]. Research on the biological interaction of two-dimensional materials is a promising area that could lead to innovative developments in the field of biomedicine. The front-line anti-tuberculosis (TB) drugs isoniazid (INH) and pyrazinamide (PZA) adsorbed on functionalized 2D materials (MoS_2 and phosphorene) were studied for their electrical and optical properties. It is anticipated that these functional 2D complexes will remain mechanically stable while retaining their band gaps, band topologies, and density of states for phosphorene and MoS_2. On their basal plane, INH and PZA both facilitate the physical adsorption of 2D compounds. In the INH/PZA/MoS_2 complexes, shallow gap states were observed close to the valence band maximum [58]. It is found that the formation energy at high temperatures and an acidic environment is smaller than that at low temperatures and a neutral environment. This finding offers a novel way to deliver and release drugs on the platform of biocompatible 2D materials at specific positions [59].

Due to its superior photothermal conversion capabilities, the MoS_2 has been used extensively for biomedical purposes. The electroplated one-dimensional (1D) MoS_2 porous nanofibers had a high specific surface area, controlled shape, and uniform size, making them a suitable chemotherapeutic drug carrier after post-annealing. By using polyethylene glycol (PEG) to alter the surface, a MoS_2/PEG composite was created. This composite demonstrated outstanding physical/chemical stability and biocompatibility. More significantly, doxorubicin (DOX), a chemotherapy drug used to treat cancer, was loaded into a MoS_2/PEG composite, which showed precise DOX dose release and exhibited a controllable release responsive to pH and NIR radiation. Because of the exceptional photothermal performance and stability of porous MoS_2 nanofibers, these extraordinary anticancer effects were primarily possible [60].

An efficient functionalized MoS_2 nanoplatform was developed for targeted drug delivery and synergistic chemo-photothermal therapy for cancer. When it concerns treating cancer, monotherapy is insufficient to yield satisfactory results; thus, combination therapy, such as the use of drug-loaded MoS_2 nanocarriers, can offer intriguing alternatives.

These nanocomposites possess the capacity to carry high amount of the anti-cancer medication like doxorubicin (DOX) and deliver it to the target site. Moreover, the release of the drug (DOX) can be controlled both by pH or NIR irradiation. It was found that the release of DOX was significantly accelerated at pH 5.0 as opposed to pH 7.4. In addition, the application of near-infrared laser (NIR) radiation enhances release. MDA-MB-231 is a type of highly invasive breast cancer cell line in humans was used to assess the efficacy of nanocomposites of MoS_2 functionalized with folic acid and the drug. The various techniques like confocal microscopy and cell vitality assay demonstrated that the noncancerous cells were unaffected whereas the cancerous cell will be specifically binding to nanocomposites by endocytosis using folic acid targeting. The nanocomposites are biocompatible delivery drug to the specific target cell where certain immunotherapies fail due to the absence of specific receptors and upon binding induce localized heating owing to the near infrared light irradiation. Thus, the designed nanocomposite holds future for a holistic combination therapy in cancer treatments [61].

By integrating ZnO into 2D-MoS_2, multifunctional nanocomposites based on MoS_2 demonstrate its application as an effective anti-angiogenic and cancer therapy agent. ZnO is an ideal nanocomposite structure with MoS_2 for cancer therapy because of its natural cytotoxicity against cancer cells that targets them specifically. This combination exhibits impressive down- and up-conversion photoluminescence that is excitation wavelength dependent. The cytotoxic effects of the MoS_2-ZnO nanocomposite were evident, and its safety index was as high as B2. By specifically activating caspase-3 and hence triggering cellular death, this nanocomposite slows the growth of tumours, according to an in ovo xenograft assay done with fertilized chick eggs. Additionally, the MoS_2-ZnO nano-composite therapy for xenografts helped restraining vascularization into interiors of the tumour by limiting the expression of crucial pro-angiogenic genes that produce vascular endothelial growth factor. MoS_2-ZnO nanocomposite treatment of tumour xenografts resulted in declined expression of genes specific for mesenchymal cell development and stimulated expression of genes specific for epithelial cell development, proposing a role for the material in delaying the enhanced migratory capacity of cell which may play critical role in delaying the onset of tumour growth. This study shows that incorporating ZnO into MoS_2 nanostructures provides a novel approach for designing effective MoS_2-based functionalized nanocarriers that open up possibilities in pioneering combinatorial applications of light detecting systems and biomedical sciences.

The MoS_2 nanosheets quench or reduce the emission of fluorescence dyes when they are combined with these dyes, which is another reason why fluorescence-imaging-guided nanotheranostic systems are designed. In order to build these systems, mesoporous silica nanoparticles (MSNs) were modified chemically to produce the Aggregation Induced Emission (AIE) fluorogen $PhENH_2$ on their surfaces. When compared to other MSNs with physically absorbed luminescent molecules, this construct establishes a more stable fluorescence emission. For targeted bioimaging and photothermal treatment, folic acid (FA) was chemically loaded on the nanoparticles as well. Due to the overexpression of folate receptors on human breast cancer cell line MDA-MB-231 cells, the produced composite containing folate, silica were more effectively taken up by human breast cancer epithelial cells (MDA-MB 231) than by non-tumorigenic epithelium-like hepatic cells with high proliferation rates (HepG2 cells). Meanwhile, these MDA-MB-231 cells were found to be efficiently killed by an 808 nm laser irradiation. The multi-functional MSNs synthesized, illustrated a stable fluorescence owing to the explicit chemical moiety added essential for bioimaging-based photothermal therapy that can kill only tumorogenic cell expressing precise receptors [62–64].

From nanosheets (MoS_2-NS) that were controllably functionalized with sequence-based DNA oligonucleotides and possessed thiol groups at the end terminals with strong binding to sulphur atom, a MoS_2 superstructure loaded with cancer drugs was created. DNA's thiol groups initially bound to sulphur atom vacancies on layered MoS_2 nanostructures, forming an anchor for DNA. A linker aptamer that caused inter-layer assembly is what gives the DNA/MoS_2-NS superstructure its driving force to assemble. These nanostructures were created as a reaction to cancer cells' sensitivity to ATP metabolism. These multilayer superstructures decomposed due to greater ATP molecule interaction with the linking aptamers when target ATP molecules were present. These superstructures have a protective MoS_2-NS shell that resembles an armour, yet it still responds to small, penetrating ATP molecules, which improves the stimuli-responsive drug release system for targeted chemotherapy [65].

MoS_2 nanocarriers widen the horizon of cancer therapy by opening up a new avenue of drug delivery in response to change in stimuli. This therapy allows timely release of anticancerous drug-like doxorubicin to specifically targeted tumorogenic cell. In preparation for use as a medication carrier in the treatment of cancer, bio-responsive polymeric MoS_2 nanocomposites were created to function as nanocomposite which is a self-assembly of MoS_2 and cystamine-glutathione complex. The nanocarrier released the drug in the glutathione-sensitive environment and assisted in the targeted killing of HeLa cells [66]. Overall, the polymer-coated MoS_2 nanocomposite has a huge potential to be successfully used as a GSH-sensitive drug carrier in the near future.

7 Conclusions

A new generation of antibacterial nanoagents and therapies is urgently needed due to the increasing risks of bacterial infections. We can functionalize MoS_2 with various moieties to generate a wide variety of MoS_2-based nanocomposites for extensive biological applications by taking inspiration from the functionalization strategies and procedures. This specific type of photo-responsive material showed strong antibacterial efficacy with low cytotoxicity, making it a suitable candidate for the treatment of bacterial infections without the use of antibiotics. The safe synergetic therapy strategy can also effectively repair wounds, elimination of inflammation and help to curb the increasing menace of antimicrobial resistance. The MoS_2-based nanocomposites will open up a new era of next-generation nanoantibiotics to curtail the threat of bacterial disease and existing antimicrobials to mankind, the environment and prospective of photodynamic agents as potent surface sterilizing sprays effective against antibiotic-resistant bacteria. These findings pave up a new avenue for designing novel drug-delivery systems that assist in molecular level understanding of nanomaterial with the onset of tumour, carrier responsive therapy and targeted multimodal therapy.

References

1. Van Boeckel TP, Gandra S, Ashok A, Caudron Q, Grenfell BT, Levin SA, Laxminarayan R (2014) Global antibiotic consumption 2000 to 2010: an analysis of national pharmaceutical sales data. Lancet Infect Dis 14:742–750. https://doi.org/10.1016/S1473-3099(14)70780-7
2. Fisher RA, Gollan B, Helaine S (2017) Persistent bacterial infections and persister cells. Nat Rev Microbiol 15:453–464. https://doi.org/10.1038/nrmicro.2017.42
3. Davies J, Davies D (2010) Origins and evolution of antibiotic resistance. Microbiol Mol Biol Rev 74:417–433
4. Qi B, Zhang D, Liu F-H, Qiao ZY, Wang H (2017) An "on-site transformation" strategy for treatment of bacterial infection. Adv Mater 29:1703461. https://doi.org/10.1002/adma.201703461
5. Nordmann P, Naas T, Poirel L (2011) Global spread of carbapenemase producing Enterobacteriaceae. Emerg Infect Dis 17:1791–1798
6. Grundmann H, Aires-de-Sousa M, Boyce J, Tiemersma E (2006) Emergence and resurgence of methicillin-resistant Staphylococcus aureus as a public-health threat. Lancet 368:874–885
7. Chen J, Sun R, Pan C, Sun Y, Li Q (2020) Antibiotics and food safety in aquaculture. J Agric Food Chem 68:11908–11919
8. Finley RL, Collignon P, Larsson DGJ, Mcewen SA, Li XZ, Gaze WH, Reid-Smith R, Timinouni M, Graham DW, Topp E (2013) The scourge of antibiotic resistance: the important role of the environment. Clin Infect Dis 57:704–710
9. Bakkeren E, Diard M, Hardt WD (2020) Evolutionary causes and consequences of bacterial antibiotic persistence. Nat Rev Microbiol 18:479–490. https://doi.org/10.1038/s41579-020-0378-z
10. Laxminarayan R, Duse A, Wattal C et al (2013) Antibiotic resistance—the need for global solutions. Lancet Infect Dis 13:1057–1098
11. He X, Hou J, Sun X, Jangili P, An J, Qian Y et al (2022) NIR-II photoamplified sonodynamic therapy using sodium molybdenum bronze nanoplatform against subcutaneous Staphylococcus aureus infection. Adv Funct Mater 32:2203964. https://doi.org/10.1002/adfm.202203964

12. Chauhan M, Sharma B, Kumar R, Chaudhary GR, Hassan AA, Kumar S (2019) Green synthesis of CuO nanomaterials and their proficient use for organic waste removal and antimicrobial application. Environ Res 168:85–95. https://doi.org/10.1016/j.envres.2018.09.024
13. Tan C, Cao X, Wu XJ, He Q, Yang J, Zhang X, Chen J, Zhao W, Han S, Nam GH, Sindoro M, Zhang H (2017) Recent advances in ultrathin two-dimensional nanomaterials. Chem Rev 117:6225–6331
14. Kim H, Beack S, Han S, Shin M, Lee T, Park Y, Kim KS, Yetisen AK, Yun SH, Kwon W, Hahn SK (2018) Multifunctional photonic nanomaterials for diagnostic, therapeutic, and theranostic applications. Adv Mater 30:1701460
15. Jiang S, Win KY, Liu S, Teng CP, Zheng Y, Han MY (2013) Surface-functionalized nanoparticles for biosensing and imaging guided therapeutics. Nanoscale 5:3127–3148
16. Gong L, Gu Z (2019) Transition metal dichalcogenides for biomedical applications. In: Two dimensional transition metal dichalcogenides, pp 241−292
17. Arul NS, Nithya VD (2016) Molybdenum disulfide quantum dots: synthesis and applications. RSC Adv 6:65670–65682
18. Cao X, Ding C, Zhang C, Gu W, Yan Y, Shi X, Xian Y (2018) Transition metal dichalcogenide quantum dots: synthesis, photoluminescence and biological applications. J Mater Chem B 6:8011–8036
19. Yadav V, Roy S, Singh P, Khan Z, Jaiswal A (2019) 2D MoS_2-based nanomaterials for therapeutic, bioimaging, and biosensing applications. Small 15:1803706
20. Shao T, Wen J, Zhang Q, Zhou Y, Liu L, Yuwen L, Tian Y, Zhang Y, Tian W, Su Y, Teng Z, Lu G, Xu J (2016) NIR photoresponsive drug delivery and synergistic chemo-photothermal therapy by monodispersed-MoS_2-nanosheets wrapped periodic mesoporous organosilicas. J Mater Chem B 4:7708–7717
21. Liu J, Zheng J, Nie H, Chen H, Li B, Jia L (2020) Co-delivery of erlotinib and doxorubicin by MoS_2 nanosheets for synergetic photothermal chemotherapy of cancer. Chem Eng J 381:122541
22. Wu J, Bremner DH, Niu S, Wu H, Wu J, Wang H, Li H, Zhu L-M (2018) Functionalized MoS_2 nanosheet-capped periodic mesoporous organosilicas as a multifunctional platform for synergistic targeted chemo-photothermal therapy. Chem Eng J 342:90–102
23. Feng ZZ, Liu XM, Tan L, Cui ZD, Yang XJ, Li ZY, Zheng YF, Yeung KWK, Wu SL (2018) Electrophoretic deposited stable chitosan@MoS_2 coating with rapid in situ bacteria-killing ability under dual-light irradiation. Small 14:1704347
24. Chen Y, Tan C, Zhang H, Wang L (2015) Two-dimensional graphene analogues for biomedical applications. Chem Soc Rev 44:2681–2701
25. Xu Y, Wang X, Zhang WL, Lv F, Guo S (2018) Recent progress in two-dimensional inorganic quantum dots. Chem Soc Rev 47:586–625
26. Sethulekshmi AS, Saritha A, Joseph K, Aprem AS, Sisupal SB (2022) MoS_2 based nanomaterials: advanced antibacterial agents for future. J Control Release 348:158–185. https://doi.org/10.1016/j.jconrel.2022.05.047
27. Zhao M, Yang R, Wei Y, Su J, Wang X, Zhang N et al (2022) Dual isolated bimetal single-atom catalysts for tumor ROS cycle and parallel catalytic therapy. Nano Today 44:101493. https://doi.org/10.1016/j.nantod.2022.101493]
28. Ganatra R, Zhang Q (2014) Few-layer MoS_2: a promising layered semiconductor. ACS Nano 8:4074–4099
29. Zhou X, Sun H, Bai X (2020) Two-dimensional transition metal dichalcogenides: synthesis, biomedical applications and biosafety evaluation. Front Bioeng Biotechnol 8:236
30. Wang X, Chang J, Wu C (2019) MoS_2-based biomaterials for cancer therapy. Biomater Transl Med 141–161
31. Gan X, Zhao H, Quan X (2017) Two-dimensional MoS_2: a promising building block for biosensors. Biosens Bioelectron 89:56–71
32. Yin W, Yu J, Lv F, Yan L, Zheng LR, Gu Z, Zhao Y (2016) Functionalized nano-MoS_2 with peroxidase catalytic and near-infrared photothermal activities for safe and synergetic wound antibacterial applications. ACS Nano 10:11000–11011

33. Dey S, Matte HSSR, Shirodkar SN, Waghmare UV, Rao CNR (2013) Charge-transfer interaction between few-layer MoS_2 and tetrathiafulvalene. Chem Asian J 8:1780–1784
34. Chou SS, Kaehr B, Kim J, Foley BM, De M, Hopkins PE, Huang J, Brinker CJ, Dravid VP (2013) Chemically exfoliated MoS_2 as near-infrared photothermal agents. Angew Chem 52:4160–4164
35. Liu L, Wu W, Fang Y, Liu H, Chen F, Zhang M, Qin Y (2021) Functionalized MoS_2 nanoflowers with excellent near-infrared photothermal activities for scavenging of antibiotic resistant bacteria. Nanomaterials 11:2829. https://doi.org/10.3390/nano11112829
36. Dicastillo C, Patio C, Galotto MJ, Vásquez-Martínez Y, Escrig J (2019) Novel hollow titanium dioxide nanospheres with antimicrobial activity against resistant bacteria. Beilstein J Nanotechnol 10:1716–1725
37. Zhang X, Zhang W, Liu L, Yang M, Huang L, Chen K, Wang R, Yang B, Zhang D, Wang J (2017) Antibiotic-loaded MoS_2 nanosheets to combat bacterial resistance via biofilm inhibition. Nanotechnology 28:225101
38. Hu Z, Wang L, Zhang K, Wang J, Cheng F, Tao Z, Chen J (2014) MoS_2 nanoflowers with expanded interlayers as high-performance anodes for sodium-ion batteries. Angew Chem Int Edit 53:12794–12798
39. Wang C, Su Y, Zhao X, Tong S, Han X (2018) MoS2@HKUST-1 flower-like nanohybrids for efficient hydrogen evolution reactions. Chem Eur J 24:1080
40. Zhu M, Liu X, Tan L, Cui Z, Liang Y, Li Z et al (2020) Photo-responsive chitosan/Ag/MoS2 for rapid bacteria-killing. J Hazard Mater 383:121122. https://doi.org/10.1016/j.jhazmat.2019.121122
41. Xia D, Shen Z, Huang G, Wang W, Yu JC, Wong PK (2015) Red phosphorus: an earth-abundant elemental photocatalyst for "green" bacterial inactivation under visible light. Environ Sci Technol 49:6264–6273. https://doi.org/10.1021/acs.est.5b00531]
42. Li X, Zhang X, Xue N, Li J, Wu T, Xu Z et al (2022b). Hierarchical self assembly of Ag-coordinated motifs on Ag (111). Acta Phys-Chim Sin 38:2011060. https://doi.org/10.3866/PKU.WHXB202011060
43. Zhao K, Gu W, Zheng S, Zhang C, Xian Y (2015) SDS–MoS_2 nanoparticles as highly-efficient peroxidase mimetics for colorimetric detection of H_2O_2 and glucose. Talanta 141:47–52. https://doi.org/10.1016/j.talanta.2015.03.055
44. Zhu W, Liu X, Tan L, Cui Z, Yang X, Liang Y, Li Z, Zhu S, Yeung KWK, Wu S (2019) AgBr nanoparticles in situ growth on 2D MoS2 nanosheets for rapid bacteriakilling and photodisinfection. ACS Appl Mater Inter 11:34364–34375
45. Akter M, Sikder MT, Rahman MM, Ullah AA, Hossain KFB, Banik S et al (2018) A systematic review on silver nanoparticles-induced cytotoxicity: physicochemical properties and perspectives. J Adv Res 9:1–16. https://doi.org/10.1016/j.jare.2017.10.008
46. Chen H, Zhao X, Cui B, Cui H, Zhao M, Shi J, Li J, Zhou Z (2023) Peroxidase like MoS_2/Ag nanosheets with synergistically enhanced NIR-responsive antibacterial activities. Front Chem 11:1148354. https://doi.org/10.3389/fchem.2023.1148354
47. Ma S, Zhan S, Jia Y, Shi Q, Zhou Q (2016) Enhanced disinfection application of Ag-modified g-C_3N_4 composite under visible light. Appl Catal B Environ 186:77–87. https://doi.org/10.1016/j.apcatb.2015.12.051
48. Liang M, Yan X (2019) Nanozymes: from new concepts, mechanisms, and standards to applications. Accounts Chem Res 52:2190–2200
49. Shan J, Li X, Yang K, Xiu W, Wen Q, Zhang Y, Yuwen L, Weng L, Teng Z, Wang L (2019) Efficient bacteria killing by Cu_2WS_4 nanocrystals with enzyme-like properties and bacteria-binding ability. ACS Nano 13:13797–13808
50. Deng Y, Yang F, Cocco E, Song E, Zhang J, Cui J, Mohideen M, Bellone S, Santin AD, Saltzman WM (2016) Improved i.p. drug delivery with bioadhesive nanoparticles. Proc Natl Acad Sci USA 113:11453–11458
51. Cao F, Zhang L, Wang H, You Y, Wang Y, Gao N, Ren J, Qu X (2019) Defect-rich adhesive nanozymes as efficient antibiotics for enhanced bacterial inhibition. Angew Chem Int Edit 58:16236–16242

52. Wei F, Cui X, Wang Z, Dong C, Li J, Han X (2020) Recoverable peroxidase-like Fe_3O_4@MoS_2-Ag nanozyme with enhanced antibacterial ability. Chem Eng J 408:127240. https://doi.org/10.1016/j.cej.2020.127240
53. Deng H, Li X, Peng Q, Wang X, Chen J, Li Y (2005) Monodisperse magnetic single-crystal ferrite microspheres. Angew Chem 117
54. Li BL, Setyawati MI, Chen L, Xie J, Ariga K, Lim C-T, Garaj S, Leong DT (2017) ACS Appl Mater & Interfaces 9(18):15286–15296. https://doi.org/10.1021/acsami.7b02529
55. Yin WY, Yan L, Yu J, Tian G, Zhou LJ, Zheng XP, Zhang X, Yong Y, Li J, Gu ZJ et al (2014) High-throughput synthesis of single-layer MoS_2 nanosheets as a near-infrared photothermal triggered drug delivery for effective cancer therapy. ACS Nano 8:6922–6933
56. Abareshi A, ArshadiPirlar M, Houshiar M (2019) Photothermal property in MoS_2 nanoflakes: theoretical and experimental comparison. Mater Res Express 6:105050
57. Shao J, Xie H, Huang H, Li Z, Sun Z, Xu Y, Xiao Q, Yu XF, Zhao Y, Zhang H et al (2016) Biodegradable black phosphorus-based nanospheres for in vivo photothermal cancer therapy. Nat Commun 7:12967
58. Liang W, Luo X (2020) Theoretical studies of MoS_2 and phosphorene drug delivery for antituberculosis drugs. J Phys Chem C 124(15):8279–8287
59. Geng S, Wu L, Cui H, Tan W, Chen T, Chu PK, Yu X-F (2018) Synthesis of lipid—black phosphorus quantum dot bilayer vesicles for near-infrared-controlled drug release. Chem Commun 54:6060–6063
60. Zhang X, Wu J, Williams GR, Niu S, Qian Q, Zhu LM (2019) Functionalized MoS_2-nanosheets for targeted drug delivery and chemo-photothermal therapy. Colloids Surf B 173:101–108
61. Rushworth JL, Montgomery KS, Cao B, Brown R, Dibb NJ, Nilsson SK, Chiefari J, Fuchter MJ (2020) Glycosylated nanoparticles derived from RAFT polymerization for effective drug delivery to macrophages. ACS Appl Bio Mater 3(9):5775–5786. https://doi.org/10.1021/acsabm.0c00529. Epub 2020 Aug 11 PMID: 35021808
62. Zhang A, Li A, Zhao W, Yan G, Liu B, Liu M, Li M, Huo B, Liu J (2018) An efficient and self-guided chemo-photothermal drug loading system based on copolymer and transferrin decorated MoS_2 nanodots for dually controlled drug release. Chem Eng J 342:120–132
63. Chacko L, Poyyakkara A, Kumar VS, Aneesh PM (2018) MoS_2–ZnO nanocomposites as highly functional agents for anti-angiogenic and anti-cancer theranostics. J Mater Chem B 6(19):3048–3057
64. Wang F, Lv FY, Liu KL, Li ZH, Niu S, Zhang YY, Cong HL, Shen YQ, Yu B (2021) Synthesis of MoS_2 nanosheets drug delivery system and its drug release behaviors. Ferroelectrics 578(1):31–39
65. Li BL, Setyawati MI, Chen L, Xie J, Ariga K, Lim CT, Garaj S, Leong DT (2017) Directing assembly and disassembly of 2D MoS_2 nanosheets with DNA for drug delivery. ACS Appl Mater Interfaces 9(18):15286–15296
66. Vadivelmurugan A, Anbazhagan R, Arunagiri V, Lai J-Y, Tsai H-C (2019) Pluronic F127 self-assembled MoS_2 nanocomposites as an effective glutathione responsive anticancer drug delivery system. RSC Adv 9:25592

Chapter 15
Flexible, Self-Healing, and Wearable Electronic and Opto-Electronic Devices Based on 2D-MoS_2

Pratik M. Pataniya, Pooja Sharma, C. K. Sumesh, and Dattatray J. Late

1 Introduction

One of the intensively explored two-dimensional materials, MoS_2 (molybdenum disulfide) has gained substantial consideration in recent years due to its exceptional electronic, optical, and mechanical characteristics [1–3]. 2D-MoS_2 is one of the promising 2D material for flexible and wearable electronic devices owing to the high mechanical flexibility, thermal stability, and electrical conductivity [1, 4, 5]. In order to use MoS_2 for the fabrication of flexible devices, it is sometimes formed into nanosheets form, which are basically thin layers of the substance. Atomically thin layers or nanosheets can be utilized as the building blocks for flexible and high-performance electronic devices such as sensors, transistors, memory devices, electrochemical electrodes, etc. The high surface area (large surface area to volume ratio) of 2D-MoS_2, which can result in greater sensitivity of sensors and high-memory store capacity in memory devices, is one benefit of employing them in flexible electronics. In addition to these, flexible devices based on 2D-MoS_2 can be bent, twisted, or even rolled up without any apparent damage, which leads to easy implementation of these devices in technical or industrial applications [6]. 2D-MoS_2 possesses extremely high environmental and chemical stability, which lead to the potential applications in harsh environmental conditions like high humidity level, high temperature, highly acidic, or alkaline conditions. These properties lead to the applications of 2D-MoS_2

P. M. Pataniya (✉) · P. Sharma · C. K. Sumesh
Department of Physical Sciences, P. D. Patel Institute of Applied Sciences, CHARUSAT, Changa, Anand 388421, India
e-mail: pm.pataniya9991@gmail.com

D. J. Late
Department of Physics, Federal University of Lavras, Campus Universitário, PO Box 3037, Lavras, Minas Gerais, Brazil

D. J. Late and C. S. Rout (eds.), *Science and Technology of 2D MoS_2*,
Materials Horizons: From Nature to Nanomaterials,
https://doi.org/10.1007/978-981-97-7367-1_15

for the fabrication of humidity sensors, fire alarm, electrocatalytic or photocatalytic hydrogen gas production, memory devices, and supercapacitors.

MoS_2 is basically a layered material in which 2D layers of MoS_2 are stacked one upon another by van der Waals attraction. In each 2D-MoS_2 layer, central Mo-planes are sandwiched between two S-atom planes by comparatively stronger co-valent interactions [5]. Most important characteristic of MoS_2 is that the electronic structure can be significantly modified by changing the thickness from mono-layer to few-layer-to bulk [7, 8]. For example, a bulk MoS_2 is the indirect band gap semiconductor with energy gap of 1.3 eV, while the monolayer MoS_2 has a direct band gap of 1.8 eV [1, 7]. As far as opto-electronic applications are concerned, monolayer MoS_2 is more efficient as compared to bulk or few layer MoS_2. In bulk MoS_2, the electronic transition involves the absorption of photon as well as phonon, while in the case of mono-layer MoS_2, only photon is required for the electronic transition from valance band to conduction band.

2 Photodetectors Based on 2D-MoS_2

2.1 Principle of the Self-Powered Photodetectors and Role of 2D-MoS_2

The self-powdered electronic and opto-electronics devices including photodetectors, gas sensors, etc. based on 2D-MoS_2 are highly efficient because they do not require any electrical power supply. For example, self-powered photodetectors refer to photodetectors that can produce their own power from the light they detect, without the need for an external electric power source [9, 10]. The self-powered photodetectors based on 2D-MoS_2 have shown the great potential in light detection over broad spectral range of visible light [5, 11] and self-built potential in MoS_2 channel. MoS_2, or molybdenum disulfide, is a two-dimensional material with excellent electrical, optical, and mechanical properties, making it an attractive material for use in self-powered photodetectors [9, 10]. The basic principle behind self-powered photodetectors based on MoS_2 is the photovoltaic effect. When light is absorbed by the MoS_2 material, electrons are excited from the valence band to the conduction band, creating a potential difference across the material. This potential difference can be used to generate a current, which can be measured and used as a signal. To improve the performance of self-powered photodetectors based on MoS_2, various strategies have been developed, such as improving the crystal quality of the material, optimizing the device structure, and integrating the material with other materials to create hybrid structures. Moreover, photoresponsivity, specific detectivity, and noise equivalent power (NEP) are critical parameters for assessing the device's ability to detect light.

2.2 Self-Powered Photodetectors Based on Semiconductor Heterostructures Involving 2D-MoS_2

Hao Wang et al. demonstrated experimentally the self-powered opto-electronic devices based on Sb_2Te_3/MoS_2 van der Waals p–n junction (Fig. 1a, b) with extremely low dark current and excellent photosensitivity. This optimized diode structure enables the zero bias photodetection with current ON/OFF ratio (defined as the ratio between photo-generated current and dark current) of 5.6×10^3. They achieved the ultra-low dark current of 2.4 pA and significantly higher photoresponsivity of 170 mA W^{-1} [12]. Nengjie Huo et al. demonstrated the two different configurations, vertical and planner device-based multilayer MoS_2-WS_2 heterostructures, with high photoresponsivity of 1.42 A/W, field-effect ON/OFF ratio of about 10^5 and high electron mobility of 65 cm^2/Vs [13]. Figure 1c, d shows the schematic of the device based on MoS_2/ Si_3N_4 heterostructure. In this research, monolayer-MoS_2 deposited by low pressure chemical vapor deposition directly on the Si_3N_4/SiO_2 layers. The thickness of the Si_3N_4 and SiO_2 layers is 250 nm and 1 μm, respectively. Marin et al. reported that the MoS_2-based photodetectors suffer from the slow photoswitching action with ride time and decay time greater than 10 s due to photogating mechanism which manifests itself as a change in the threshold voltage due to charge transfer to adjacent molecules [14]. In order to get the fast photoswitching action and to avoid photogating effect, the effective way is to introduce a multilayer h-BN flake with a thickness of ~11 nm between MoS_2 and Si_3N_4 layers. This modified device structure manifests the response of photodetector faster than 20 Hz. The photo response of the device can be recorded at the different chopper rotation frequency. Figure 1e shows the ratio of the DC photocurrent (I_{DC}) obtained for constant illumination, and oscillating photocurrent (I_{AC}) at the frequency of the chopper, measured by the lock-in amplifier [14]. The photodetector based on monolayer MoS_2/h-BN photodetector exhibits constant photocurrent up to 1 kHz frequency. It indicates that the photodetector is faster than 1 kHz and it has response time of less than 1 ms. MoS_2/h-BN photodetector exhibits photoresponse at different gate voltage (Fig. 1f).

2.3 Metal–Semiconductor-Metal-Type Semiconductors Based on 2D-MoS_2

The metal–semiconductor-metal-type photodetectors (MSMPs) can be fabricated by depositing the metal electrodes on the atomically thin film of 2D-MoS_2 [15]. Basically, there are two types of MSMPs, depending on the type of junction formed between active semiconductor and metal: (i) Ohmic contact and (ii) Schottky contact. Generally, the Ohmic contact are the low-resistance electrical contacts that allow current to flow easily between metal electrodes and semiconductor material and the current through the active semiconductor channel has a linear dependence on the

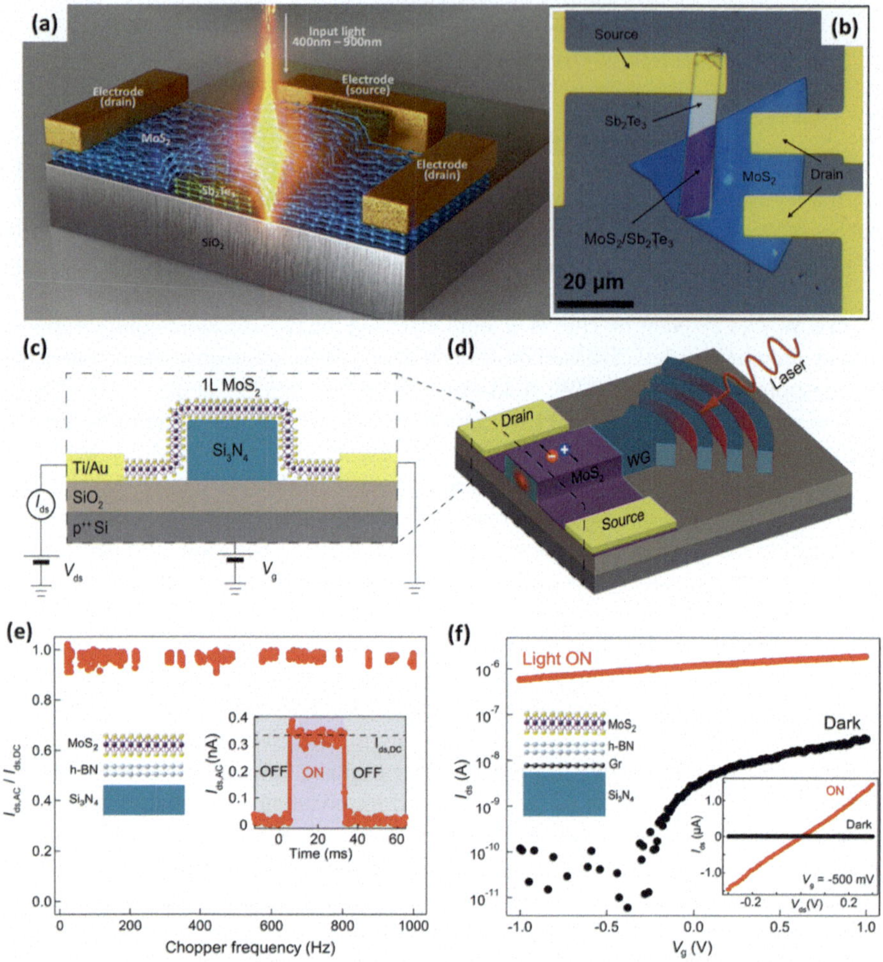

Fig. 1 **a** Schematic diagram of the Sb_2Te_3/MoS_2 van der Waals p–n junction photodetector. **b** The optical microscope image of PN junction device. Reprinted with permission under the Creative Commons CC-BY-NC-ND 2 from {Nanophotonics, 11 (2022) 5113–5119.} [12], Copyright©2022 the author(s), published by De Gruyter, Berlin/Boston. Schematic diagram of the MoS_2/Si_3N_4 heterostructure **c** side view, **d** top view, **e** analysis of response speed of MoS_2/h-BN/Si_3N_4 photodetector, and **f** Photoresponse of MoS_2/Si_3N_4 heterostructure at different gate voltage. Reprinted with permission under the Creative Commons CC-BY-NC-ND 2 from {Npj 2D Mater. Appl. 3 (2019) 14.} [14], Copyright©2019, by the Author(s), Springer Nature publication

applied bias. Unlike an Ohmic contact, a Schottky contact forms a rectifying junction that exhibits a nonlinear current–voltage characteristic. In a Schottky contact, a potential barrier at the junction forms between the metal and semiconductor due to their different work functions.

The liquid phase exfoliated MoS_2 nanosheets have excellent optical features, which can be utilized for fabrication of large area photosensitive devices like metal–semiconductor-metal photodetectors and photoelectrochemical type photodetectors. Bhakhar et al. have shown the high yield, size-controlled, and cost-effective synthesis 2D-MoS_2 nanosheets for Ag-MoS_2-Ag type photodetector with photoresponsivity of 6.93 mAW^{-1} under illumination of 670 nm (red) light of power intensity 3 mW cm^{-2} [16]. Even, the photoresponsivity of the MoS_2-based photodetectors can be enhanced by atomic doping of transition metals. Recently, the 5-fold enhanced photoresponsivity is reported by optimized doping concentration of indium in MoS_2 with fast photoswitching with response time of 110 ms [17]. Saenz et al. demonstrated the Mo-MoS_2-Mo type low Schottky barrier MSMPs for detection of light over broad spectral range from UV-to-IR. As shown in Fig. 2a, in this research, mechanically exfoliated 2D-MoS_2 sheets is directly transferred on the pre-patterned sputtered Mo-contacts on SiO_2 substrate. Figure 2b shows the photoconductive mechanism in 2D-MoS_2-based MSMPs, in which the charge carriers can be excited from valance band to conduction band when the light of energy greater than the band gap of mono-layer MoS_2. As shown in Fig. 2b, the top of the valance band is at K-point of the Brillouin zone while the minimum of the conduction band is at Γ-point, which indicates that the optical excitation of carriers is indirect. The photo carriers generated by light absorption are collected at the metal electrodes and generate the photocurrent in semiconductor channel. The collection of the photo carriers depends on the type of the metal–semiconductor-metal contacts. Figure 2c shows the increase in the drain-source current (I_{DS}) with increase in the temperature from 5.4 to 350 K due to thermal excitation of carriers in Schottky barrier. The thermal excitation leads to more Ohmic conversion of MoS_2-Mo contact at 350 K as compared to weak Schottky at 5.4 K. Figure 2d shows that the photocurrent (I_{ph}) increases with increase in the photocurrent because the intense light possesses more number of photons, which leads to higher photoexcitation. An idea photodetector follows the power law I_{ph} α P^{β}, with $\beta = 1$. However, photodetectors exhibit power β less than 1, depending on the concentration of defects on channel of photodetectors.

3 Flexible and Wearable Piezo-Electric Devices Based on 2D-MoS_2

Wearable electronic devices based on MoS_2 have emerged as a promising technology due to the material's excellent mechanical flexibility, high sensitivity, and tunable bandgap. The devices can be fabricated using various techniques including chemical vapor deposition (CVD), physical vapor deposition (PVD), pulse laser deposition, and solution-based techniques [1, 5]. A large area film of 2D-MoS_2 can be prepared on many flexible substrates like PET, PDMS, cellulose papers, etc. The adhesion of film

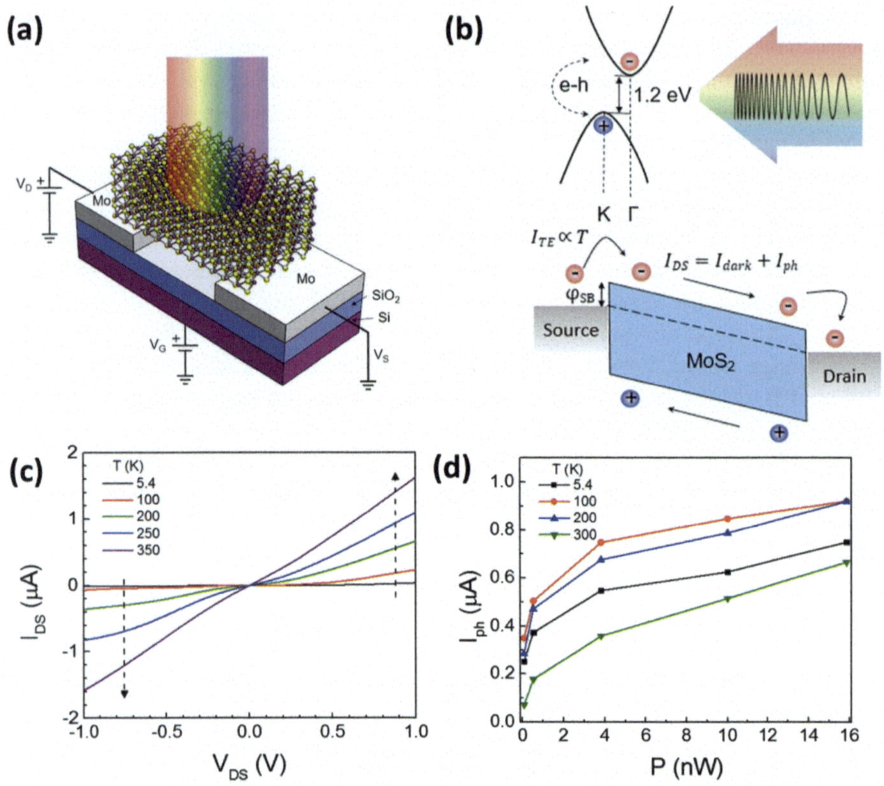

Fig. 2 **a** Schematic of two-terminal configuration of our device fabricated with 100 nm Mo bottom contacts formed on thermally oxidized (thickness of SiO_2 ~ 270 nm) Si substrates, **b** The photocurrent generation mechanism, **c** Temperature-dependent electronic response, I_{DS} -V_{DS} Characteristics, and **d** photoresponsivity with temperature and power intensity of light. Reprinted with permission under the Creative Commons CC-BY-NC-ND 2 from {Rep. 8 (2018) 1276.} [15], ©2022 by the authors

is the crucial factor for the fabrication of the flexible devices. In addition, wearable electronic devices require additional considerations for biocompatibility, comfort, and durability. Wearable photodetectors based on MoS_2 have potential applications in various areas such as healthcare, sports, and safety. For example, they can be used for monitoring vital signs, measuring exposure to UV radiation, and detecting hazards in industrial environments.

3.1 Reports on the Flexible and Wearable Piezo-Electric Devices Based on 2D-MoS_2

There are many reports on the development of pressure and strain sensors based on 2D-MoS_2. For example, Seong Jun Kim et al. have reported the fabrication of three-dimensional graphene porous networks combining 2D-MoS_2 for high sensitivity and durability of the devices for detection of motion signals such as neck bending and eye blinking. They achieved the highest pressure sensitivity of 6.06 kPa^{-1} [18]. Vaibhav Rana et al. reported the piezoresistive pressure sensor functionalized by polycrystalline molybdenum disulfide, which can be utilized for broad pressure range 0–20 kPa [19]. The piezoresistive effect refers to the change in electrical resistance of a material when it undergoes mechanical deformation, such as when pressure is applied [20]. Generally, the magnitude of the change in electrical resistance is proportional to the applied pressure, allowing the piezoresistive sensor to provide accurate pressure measurements. Additionally, MoS_2 is a biocompatible material, making it suitable for use in medical applications such as monitoring blood pressure. Alexey Tarasov demonstrated piezoresistive strain sensors based on flexible MoS_2 field-effect transistors made from a highly uniform large-area tri-layer film. They measured the variation of the band gap with applied pressure using the optical spectroscopy. They also demonstrated the optimization of the strain sensitivity over one order of magnitude via modulating the MoS_2 Fermi level with an applied gate voltage [21]. An atomically thin film of MoS_2 can be prepared on flexible polymer substrates like polydimethylsiloxane (PDMS) by magnetron sputtering, which can be utilized for fabrication of wearable piezoresistive sensors with a piezoresistive coefficient as high as 866.89 MPa^{-1} [22]. Recently, Dandan Xu et al. proposed the flexible novel iontronic pressure sensor based on monolayer molybdenum disulfide (MoS_2) [23]. As shown in Fig. 3, the sensors demonstrated the ultra-high-pressure sensitivity of 89.75 kPa^{-1} owing to excellent mechanical properties of mono layer MoS_2 and the large intercalation capacitance. Encouragingly, the pressure can function over broad pressure range up to 722.22 kPa. Because of the robust nature, the sensors is easily hand-held and flexible in nature [23].

4 Self-Healing Sensors for Wearable Electronics

Self-healing sensors are sensors that have the ability to repair themselves when cracks or damages are formed by any physical stresses. This can be accomplished by utilizing a variety of materials and technologies that can identify sensor degradation and start a repair procedure. Generally, self-healing sensor is based on polymers with the presence of a capsule of healing agent. The healing agent fills in the area of damage and restores the sensor's functionality when the capsules rupture when the sensor is damaged. Numerous potential uses for self-healing sensors exist, especially in environments where sensors are potentially subjected to harsh conditions like

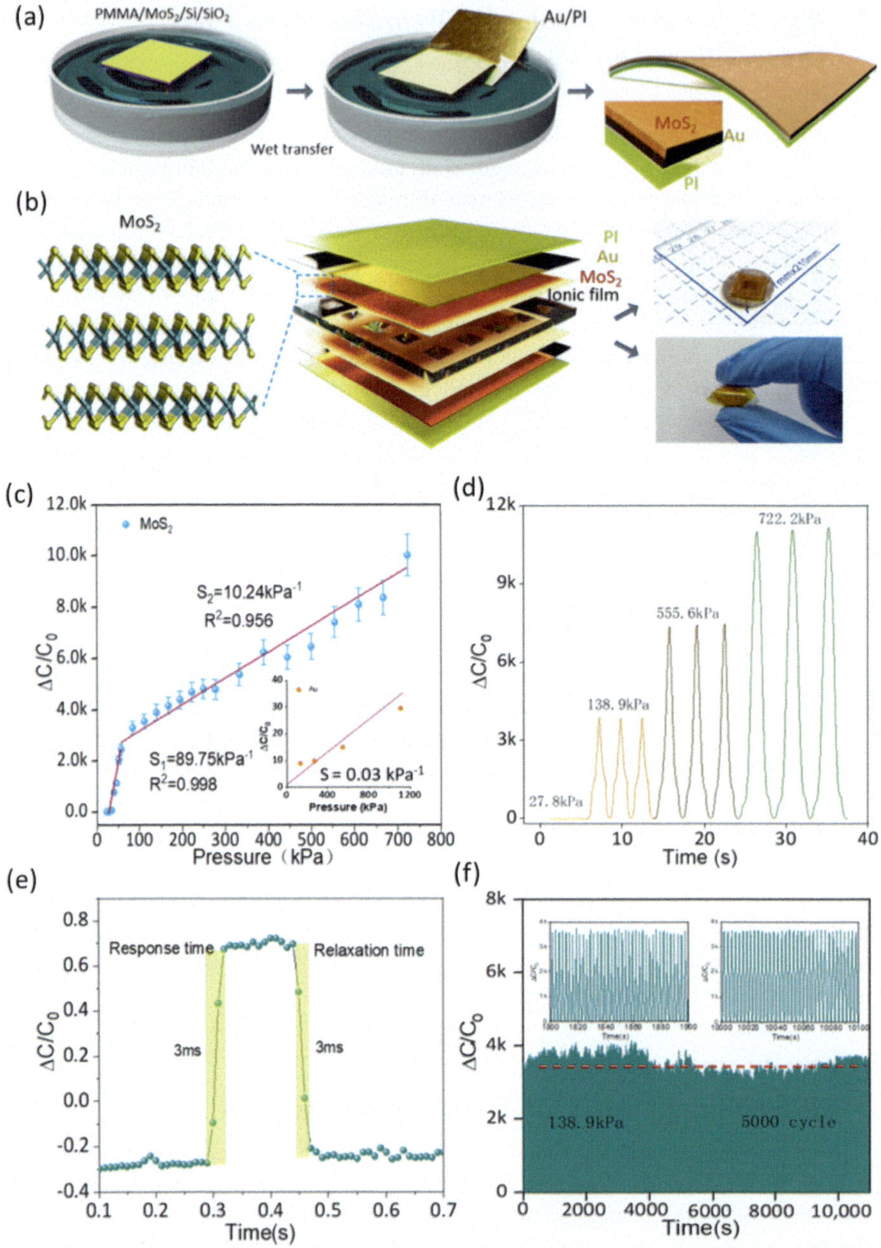

Fig. 3 Schematic diagram of pressure sensor and fabrication process. **a** Wet transfer process of MoS_2 monolayer. **b** Digital optical image and flexible feature of iontronic pressure sensor and the corresponding diagram layout of each layer in the sensor, **c** the performance of the flexible sensor in pressure range up to 800 kPa, **d** continuous pressure test of the sensor. **e** Response and relaxation time of the flexible sensor. **f** Durability test of the sensor under applied pressure of 138.9 kPa for 5000 cycles. Reprinted with permission under the Creative Commons CC-BY-NC-ND 2 from {Micromachines, 13 (2022) 660} [23], ©2022 by the authors

high temperatures or corrosive materials. They can be utilized in medical devices, aerospace and automotive applications, and industrial monitoring systems.

4.1 Report on Self-Healing Sensor Functionalized by on MoS_2-Based Hydrogel

Wenya Xu et al. reported the fabrication of devices based on Molybdenum disulfide (MoS_2) nanosheet-based hydrogels with light-triggered self-healing property (Fig. 4a). In this research, the MoS_2-based hydrogel is prepared by using the MoS_2 nanosheets, gelatin (Gel) and poly (ethylene glycol) (PEG) cross-linker [24]. Actually, the healing of the cracks in PEG-hydrogel under IR illumination is triggered by rise in its temperature. As shown in Fig. 4b, the increase in temperature of the GPM-hydrogen strongly depends on the content of MoS_2 nanosheets. As the MoS_2 loading increased from 0.04 wt% to 0.72 wt% leads to the final temperature increased from 32.5 °C to 44.9 °C. The healing efficiency greatly enhanced by the presence of MoS_2 in hydrogel, which may be due to effective absorption of IR by MoS_2 Fig. 4c, d [24]. Wenya Xu et al. reported the healing efficiency of about 90% at first cycle and it drops to about 53% after 10 cycles of damage.

The response of the GPM-hydrogen greatly depends on the content of the loading of MoS_2 (Fig. 5a). Owing to robust nature of the strain sensor based on MoS_2 can withstand a strain of up to 70% and bending angle up to 180° with high response (Fig. 5b, c). The high response of GPM sensor for compression leads to its application for wearable devices to monitor the movement of human body part and it can help to monitor the health of the human body. These types of flexible, non-toxic, and self-healable sensors can be easily mounted on body parts (**Figure (d–f)**). Overall, the development of self-healing sensors is promising for the area of research and innovation, with the potential to create more durable and consistent sensors that demand less frequent maintenance or replacement.

5 Flexible and Hand-Held Electrodes for Electrochemical Applications

Flexible electrodes have become increasingly important in electrochemical applications due to their unique properties and potential advantages over traditional rigid electrodes. Here are some requirements for flexible electrodes in electrochemical applications: (1) The electrode should be able to bend and conform to various shapes and surfaces, without cracking or breaking. This is particularly important in applications where the electrode needs to be applied to curved or irregular surfaces. (2) The electrode should be chemically stable under the conditions of the electrochemical reaction. This includes resistance to corrosion and degradation due to exposure to the

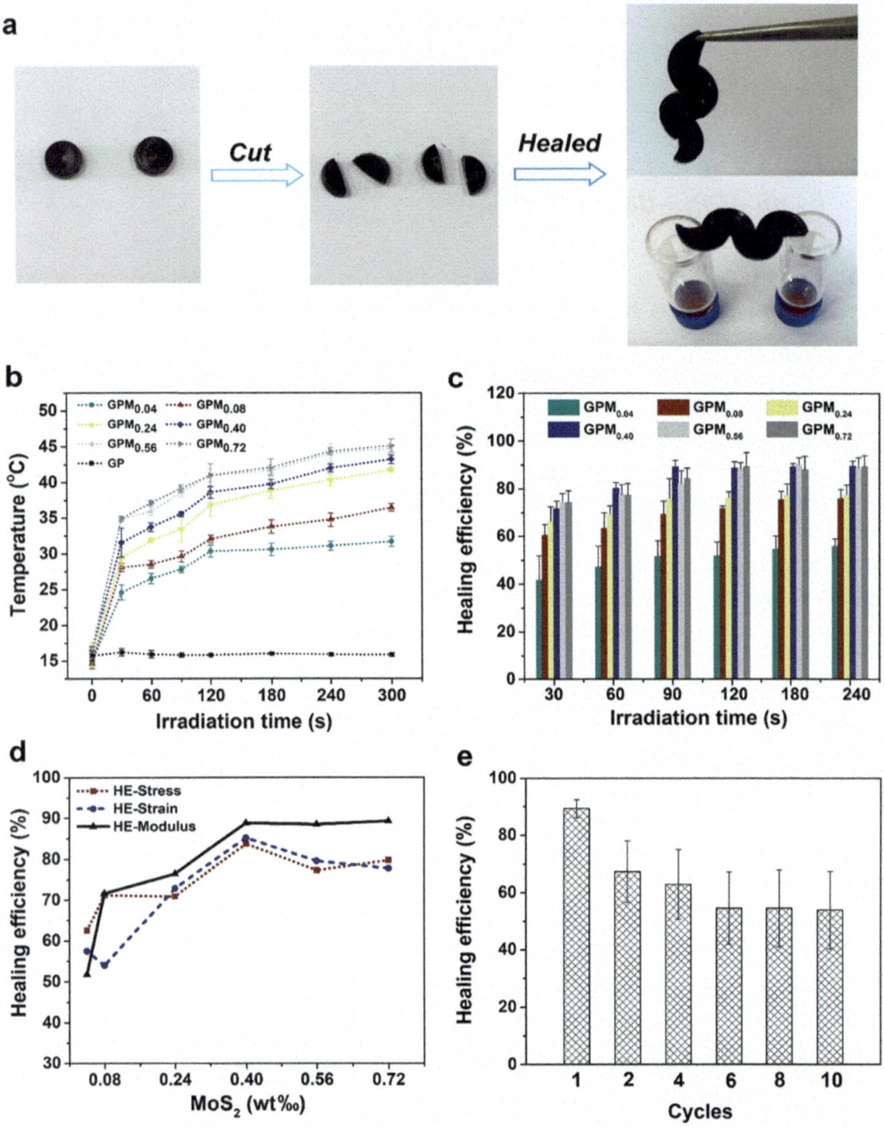

Fig. 4 **a** The digital image of self-healing process for the GPM hydrogel, **b** temperature change with time for GPM hydrogels with different content of MoS_2 under NIR irradiation. **c** Healing efficiencies for GPM hydrogels with different content of MoS_2 under NIR irradiation. **d** Comparison of healing efficiencies calculated by different methods, **e** the healing efficiencies of repeated healing cycles of GPM hydrogels. Reprinted from {J. Colloid Interface Sci. 586 (2021) 601–612} [24], Copyrights@2021, Elsevier Publication

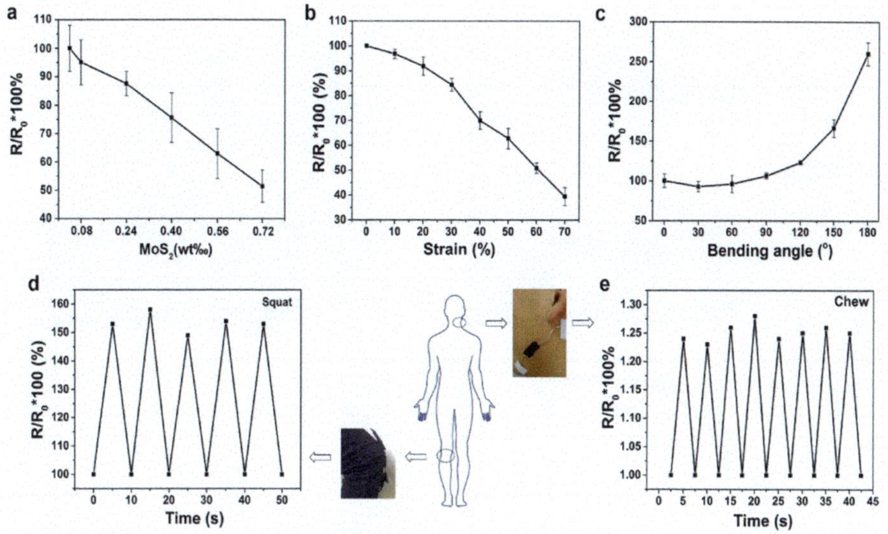

Fig. 5 **a** The response of sensors for the GPM hydrogels with different content of MoS_2, **b** response of the GPM hydrogel under compression, **c** response of the GPM hydrogel under bending, response of the wearable GPM hydrogel in response to **d** squatting, and **e** chewing. Reprinted from {J. Colloid Interface Sci. 586 (2021) 601–612} [24], Copyrights@2021, Elsevier Publication

electrolyte. (3) The electrode material should have sufficient mechanical strength to withstand bending and stretching without breaking or losing electrical contact with the electrolyte. (4) The electrode should be scalable and cost-effective to produce, particularly for large-scale industrial applications.

5.1 Reports on the MoS_2-Based Electrodes and Electrochemical Applications

Pataniya et al. reported the fabrication of large area, flexible, and hand-held electrodes from the liquid phase exfoliated MoS_2 nanosheets. The thickness of liquid phase exfoliated 2D-MoS_2 nanosheets can be tuned by various parameters such as exfoliation time, liquid (e.g. ethanol, water, N-methyl 2-pyrolidon, Di-methyl formamide, etc.), ultrasound power, temperature of the exfoliating medium, etc. As shown in Fig. 6a, b, the paper-based bio-degradable, porous, and flexible electrodes can be prepared using MoS_2 nanosheets and utilized for various electrochemical applications including water electrolysis, photoelectrochemical water splitting, supercapacitors, electrodes for batteries, etc. For example, these MoS_2-Paper electrodes have excellent electrochemical applications in green hydrogen generation, as shown in Fig. 6c, d. The electrodes are stable due to chemical stability of MoS_2 nanosheets in highly acidic

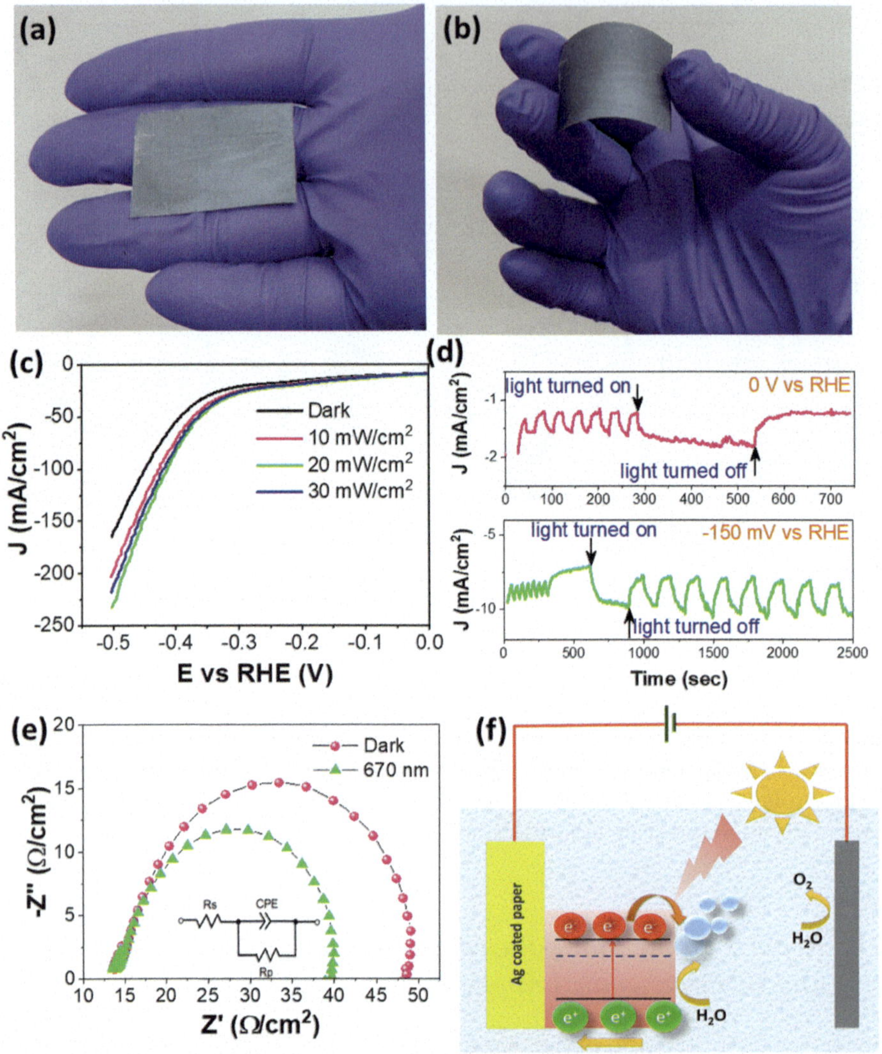

Fig. 6 **a**, **b** Digital images of MoS_2 functionalized paper-based electrodes, **c** linear sweep voltammetry curves for MoS_2 functionalized paper-based electrodes for green hydrogen generation, **d** variation of current with periodic illumination of 670 nm light at 0 and −150 mV vs RHE potential in electrocatalytic cell, **e** impedance spectra in dark and illuminated condition, **f** Schematic of charge transfer mechanism in electrocatalytic cell in illuminated condition [27]. Reprinted with permission from {ACS Appl. Energy Mater. 4 (2021) 4815–4822}. Copyright (2021) American Chemical Society

or alkaline environments of water splitting. Additionally, they possess unique photo-enhancement in water electrolysis owing to light sensitive behavior of 2D-MoS_2. Figure 6e shows the electrochemical impedance spectra, in which semi-circle indicates the strong interface formation between the MoS_2 nanosheets and electrolyte. The diameter of the semi-circle is the measure of charge-transfer resistance between electrode and electrolyte, which is found to fall under illumination of MoS_2-paper electrode, indicating strong light dependent electrochemical reaction on 2D-MoS_2. Owing to high chemical and environmental stability, S. ArunKumar et al. reported the Ni, Fe, and Co doped MoS_2 nanosheets for their potential application in corrosion protection of mild steel in an aggressive chloride environment [25]. Xin Guo et al. reported the preparation of a flexible, three-dimensional (3D) and durable electrode, which is composed of MoS_2 nanoparticles grown on bacteria-cellulose-derived carbon fiber foam (MoS_2/CNF) [26].

References

1. Li X, Zhu H (2015) Two-dimensional MoS_2: properties, preparation, and applications. J Mater 1:33–44. https://doi.org/10.1016/j.jmat.2015.03.003
2. Pataniya PM, Patel V, Sumesh CK (2021) MoS_2/WSe_2nanohybrids for flexible paper-based photodetectors, Nanotechnology 32. https://doi.org/10.1088/1361-6528/abf77a
3. Patel M, Pataniya P, Vala H, Sumesh CK (2019) One-dimensional/two-dimensional/three-dimensional dual heterostructure based on MoS_2-modified ZnO-heterojunction diode with silicon. J Phys Chem C. https://doi.org/10.1021/acs.jpcc.9b05134
4. Zhang W, Zhang P, Su Z, Wei G (2015) Synthesis and sensor applications of MoS_2-based nanocomposites. Nanoscale 7:18364–18378. https://doi.org/10.1039/c5nr06121k
5. Singh AK, Kumar P, Late DJ, Kumar A, Patel S, Singh J (2018) 2D layered transition metal dichalcogenides (MoS_2): synthesis, applications and theoretical aspects. Appl Mater Today 13. https://doi.org/10.1016/j.apmt.2018.09.003
6. Nalwa HS (2020) A review of molybdenum disulfide (MoS_2) based photodetectors: from ultra-broadband, self-powered to flexible devices. RSC Adv 10:30529–30602. https://doi.org/10.1039/d0ra03183f
7. Mak KF, Lee C, Hone J, Shan J, Heinz TF (2010) Atomically thin MoS_2: a new direct-gap semiconductor. Phys Rev Lett 105:2–5. https://doi.org/10.1103/PhysRevLett.105.136805
8. Zhang X, Qiao X-F, Shi W, Wu J-B, Jiang D-S, Tan P-H (2015) Phonon and Raman scattering of two-dimensional transition metal dichalcogenides from monolayer, multilayer to bulk material. Chem Soc Rev 44:2757–2785. https://doi.org/10.1039/C4CS00282B
9. Xin Y, Wang X, Chen Z, Weller D, Wang Y, Shi L, Ma X, Ding C, Li W, Guo S, Liu R (2020) Polarization-sensitive self-powered type-II GeSe/MoS_2 van der Waals heterojunction photodetector. ACS Appl Mater Interfaces 12:15406–15413. https://doi.org/10.1021/acsami.0c01405
10. Kang Z, Cheng Y, Zheng Z, Cheng F, Chen Z, Li L, Tan X, Xiong L, Zhai T, Gao Y (2019) MoS_2-based photodetectors powered by asymmetric contact structure with large work function difference. Nano-Micro Lett 11:34. https://doi.org/10.1007/s40820-019-0262-4
11. Velusamy DB, Haque MA, Parida MR, Zhang F, Wu T, Mohammed OF, Alshareef HN (2017) 2D organic-inorganic hybrid thin films for flexible UV–visible photodetectors. Adv Funct Mater 27:1–9. https://doi.org/10.1002/adfm.201605554
12. Wang H, Gui Y, Dong C, Altaleb S, Nouri BM, Thomaschewski M, Dalir H, Sorger VJ (2022) Self-powered broadband photodetector based on MoS_2/Sb_2Te_3 heterojunctions: a promising

approach for highly sensitive detection. Nanophotonics 11:5113–5119. https://doi.org/10.1515/nanoph-2022-0413
13. Huo N, Kang J, Wei Z, Li S-S, Li J, Wei S-H (2014) Novel and enhanced optoelectronic performances of multilayer MoS_2–WS_2 heterostructure transistors. Adv Funct Mater 24:7025–7031. https://doi.org/10.1002/adfm.201401504
14. Gonzalez Marin JF, Unuchek D, Watanabe K, Taniguchi T, Kis A (2019) MoS_2 photodetectors integrated with photonic circuits. NPJ 2D Mater Appl 3:14. https://doi.org/10.1038/s41699-019-0096-4
15. Saenz GA, Karapetrov G, Curtis J, Kaul AB (2018) Ultra-high photoresponsivity in suspended metal-semiconductor-metal mesoscopic multilayer MoS_2 broadband detector from UV-to-IR with low Schottky barrier contacts. Sci Rep 8:1276. https://doi.org/10.1038/s41598-018-19367-1
16. Bhakhar SA, Patel NF, Zankat CK, Tannarana M, Solanki GK, Patel KD, Pathak VM, Pataniya P (2019) Sonochemical exfoliation and photodetection properties of MoS_2 nanosheets. Mater Sci Semicond Process 98:13–18. https://doi.org/10.1016/j.mssp.2019.03.017
17. Bhakhar SA, Pataniya PM, Chauhan BL, Patel NF, Solanki GK (2023) Self-powered photodetectors based on In_x Mo_{1-x} S_2 crystals. Opt Mater (AMST) 136:113409. https://doi.org/10.1016/j.optmat.2022.113409
18. Kim SJ, Mondal S, Min BK, Choi C-G (2018) Highly sensitive and flexible strain-pressure sensors with cracked paddy-shaped MoS_2/graphene foam/Ecoflex hybrid nanostructures. ACS Appl Mater Interfaces 10:36377–36384. https://doi.org/10.1021/acsami.8b11233
19. Rana V, Gangwar P, Ramesh AK, Sharma T, Bhat KN, Nayak MM, Das S, Singh P (2022) High-performing polycrystalline MoS_2-based microelectromechanical piezoresistive pressure sensor. IEEE Sens J 22:18542–18549. https://doi.org/10.1109/JSEN.2022.3198761
20. Pataniya PM, Bhakhar SA, Tannarana M, Zankat C, Patel V, Solanki GK, Patel KD, Jha PK, Late DJ, Sumesh CK (2021) Highly sensitive and flexible pressure sensor based on two-dimensional $MoSe_2$ nanosheets for online wrist pulse monitoring. J Colloid Interface Sci 584:495–504. https://doi.org/10.1016/j.jcis.2020.10.006
21. Tarasov A, Tsai M-Y, Taghinejad H, Campbell PM, Adibi A, Vogel EM (2015) Piezoresistive strain sensing with flexible MoS_2 field-effect transistors. In: 2015 73rd annual device research conference, pp 159–160. https://doi.org/10.1109/DRC.2015.7175604
22. Pang X, Zhang Q, Shao Y, Liu M, Zhang D, Zhao Y (2021) A flexible pressure sensor based on magnetron sputtered MoS_2. Sensors 21. https://doi.org/10.3390/s21041130
23. Xu D, Duan L, Yan S, Wang Y, Cao K, Wang W, Xu H, Wang Y, Hu L, Gao L (2022) Monolayer MoS_2-based flexible and highly sensitive pressure sensor with wide sensing range. Micromachines 13:660. https://doi.org/10.3390/mi13050660
24. Xu W, Wang W, Chen S, Zhang R, Wang Y, Zhang Q, Yuwen L, Yang WJ, Wang L (2021) Molybdenum disulfide (MoS_2) nanosheets-based hydrogels with light-triggered self-healing property for flexible sensors. J Colloid Interface Sci 586:601–612. https://doi.org/10.1016/j.jcis.2020.10.128
25. ArunKumar S, Jegathish V, Soundharya R, JesyAlka M, Arul C, Sathyanarayanan S, Mayavan S (2017) Evaluating the performance of MoS_2 based materials for corrosion protection of mild steel in an aggressive chloride environment. RSC Adv 7:17332–17335. https://doi.org/10.1039/C7RA01372H
26. Guo X, Cao G, Ding F, Li X, Zhen S, Xue Y, Yan Y, Liu T, Sun K (2015) A bulky and flexible electrocatalyst for efficient hydrogen evolution based on the growth of MoS_2 nanoparticles on carbon nanofiber foam. J Mater Chem A 3:5041–5046. https://doi.org/10.1039/C5TA00087D
27. Pataniya PM, Sumesh CK (2021) Paper-based flexible and photosensitive electrodes for electrochemical hydrogen evolution. ACS Appl Energy Mater 4:4815–4822. https://doi.org/10.1021/acsaem.1c00377

Chapter 16
Recent Development in MoS_2 Through Density Functional Theory

Riddhi Kadrekar, Ratna Chauhan, and Dattatray J. Late

1 Introduction

The MoS_2 belongs to the category of transition metal dichalcogenides (TMDCs). After graphene, MoS_2 was the next 2D material that gained research attention considering its commercial potential. MoS_2 has vast and varied industrial applications like lubrication, catalysis, photovoltaics, optoelectronics, energy storage, etc. Drastic transition from being a semiconductor with an indirect bandgap (bulk MoS_2) to a direct bandgap (monolayer MoS_2) combined with a display of strong photoluminescence makes it a promising device for next-generation FET and optoelectronic devices. Appreciable carrier mobility (~350 cm^2V^{-1} s^{-1}) and massive ON/OFF ratios (~10^8) make MoS_2 monolayer useful for application as efficient switches and logic circuits. Weak interlayer interactions allow easy sliding of layers assisting in lubrication. High breaking strength and stiffness comparable to steel make it suitable for flexible electronics. In addition, the tunability of MoS_2 by strain engineering, band engineering, and application of electric field makes it commercially attractive [1, 2].

MoS_2 exists in different phases, 1T, 2H, 3R phases and most synthesis methods provide a mixture of more than one phase or its polytypes. It is known that 2H, 1T', 1T'', 1T''' structures co-exist in experimentally synthesized samples. However, identifying the electronic and mechanical properties and exploring their tunability

R. Kadrekar (✉)
Amity School of Applied Science, Amity University, Mumbai, Maharashtra 410206, India
e-mail: riddhi.physics@gmail.com

R. Chauhan
Department of Environmental Science, Savitribai Phule Pune University, Pune 411007, India

D. J. Late
Federal University of Lavras, Campus Universitário, PO Box 3037, Lavras, Minas Gerais, Brazil
e-mail: datta099@gmail.com

D. J. Late and C. S. Rout (eds.), *Science and Technology of 2D MoS_2*,
Materials Horizons: From Nature to Nanomaterials,
https://doi.org/10.1007/978-981-97-7367-1_16

experimentally is a challenge owing to the necessity of highly sophisticated experimental setups to obtain a single (pure) phase. Lack of high-quality single-phase samples hinders the experimental study of potential applications and underlying physics. It is here that density functional theory (DFT) calculations aid the development of MoS_2-based fine-tuned smart functional material with exotic properties. For example, DFT calculations have predicted that a spontaneous symmetry breaking of the metallic state of the 1T-MoS_2 monolayer, opens up a bandgap and leads to a display of ferroelectric behavior [3, 4]. Similarly, Lin et al., theoretically explored a new porous MoS_2 (P-MoS_2), which shows metallic behavior but can be converted back to semiconductor by application of strain [5]. The current chapter explores numerous such predictive and confirmatory theoretical studies which lead to an in-depth understanding of MoS_2.

The plan of this chapter is as follows.

- Section 1 includes an introduction to DFT as a tool to investigate MoS_2
- Section 2 includes details of the crystal structure of MoS_2.
- Section 3 explains the general properties of MoS_2.
- Section 4 includes a brief discussion of exotic properties of MoS_2.

2 Crystal Structure of MoS_2

MoS_2 is made up of a tri-atomic thick layer. Each layer consists of two hexagonal planes of S atoms separated by a plane of Mo atoms (S-Mo-S). The Mo and S atoms in individual layers are covalently bonded while weak van der Waal's (vdW's) forces form inter-layer bonding. It is this weak van der Waal's bonding that makes mechanical exfoliation of individual MoS_2 layers possible. The covalently bonded S-Mo-S layers make the structure mechanically and electrically anisotropic. MoS_2 has three polymorphs with Mo coordination being either trigonal prismatic (2H, 3R) or octahedral (1T). The digits in polymorphs (1T, 2H, 3R) signify the number of S-Mo-S units in the unit cell while the letters T, H, and R represent structure symmetry; trigonal, hexagonal, or rhombohedral, respectively. Figure 1 establishes all three polymorphs along with their corresponding lattice parameters. Of these polymorphs, 2H-MoS_2 is a naturally occurring and thermodynamically stable phase while 1T and 3R are the most popular metastable phases. 1T-MoS_2 being metastable undergoes distortion to form polytypes like 1T' 1T" and 1T"' [6]. Details of these superlattices of 1T-MoS_2 are also included in Fig. 1.

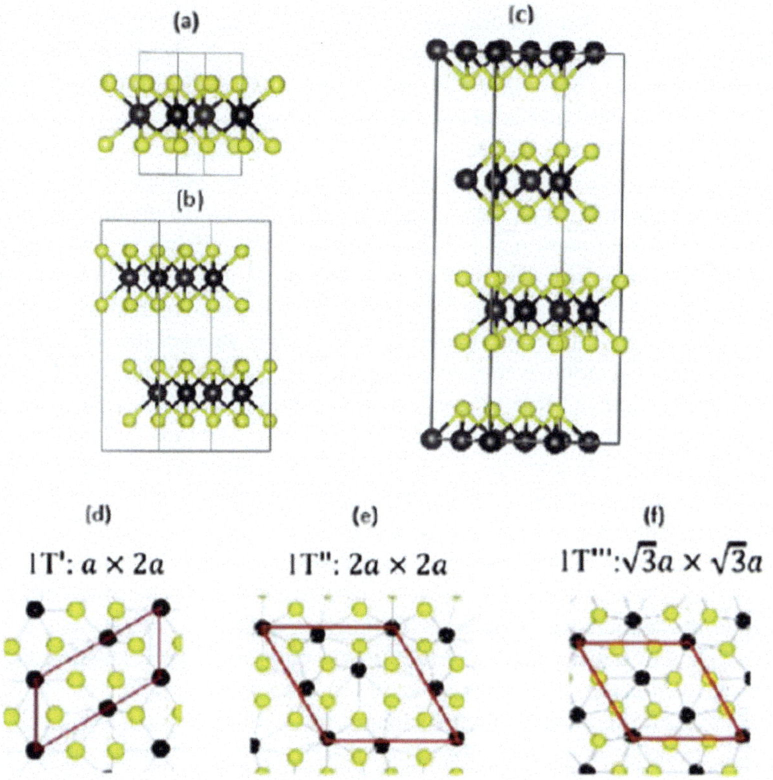

Fig. 1 The crystal structure of MoS_2 **a** 1T-MoS_2 ($P\bar{3}m1$, a = b = 3.185 Å, c = 12.389 Å, α = β = 90°, γ = 120°) **b** 2H-MoS_2 (P63/mmc, a = b = 3.185 Å, c = 12.389 Å, α = β = 90°, γ = 120°) **c** 3R-MoS_2, (a = b = 3.192 Å, c = 18.281 Å, α = β = 90°, γ = 120°). The 1T-MoS_2 polytypes **d** a × 2a superlattice (1T') **e** 2a × 2a superlattice (1T'') (f) √3a x √3a superlattice (1T''')

3 General Properties of MoS_2 Polymorphs

3.1 Electronic Structure Properties

Of all the above-mentioned, polymorphs and polytypes, 2H-MoS_2 is thermodynamically the most stable structure. 1T-MoS_2 is dynamically unstable and undergoes distortion during its relaxation likely forming 1T'-MoS_2, which is predicted to be highly functional. Hence, detailed theoretical studies focus on 1T'-MoS_2 rather than 1T-MoS_2. Zhao et al. studied the evolution of electronic structure in their bulk, monolayer and bilayer form [7]. Bulk 1T' and 3R-MoS_2 exhibit metallic characteristics while 2H-MoS_2 displays semiconducting properties. Reason for this being, in 2H-MoS_2, the S-atoms overlap along the c-axis resulting in relatively strong repulsive forces. While the S-atoms do not overlap along the c-axis in 3R and 1T'-MoS_2

resulting in relatively strong attractive forces. Due to the strong attractive forces, the overlapping of wavefunctions is stronger in 3R and 1T'- MoS_2causing metallic behavior.

Figure 2 shows the band structure for monolayer, bilayer, and bulk structures of 1T', 2H, and 3R-MoS_2. 3R-MoS_2 exhibits metallic character for all monolayer, bilayer, and bulk structures. For 2H-MoS_2, as the layers change from monolayer to bulk, there is a shift from direct bandgap of 2.219 eV to an indirect bandgap of 1.441 eV (HSE06 functional) [7]. The direct to indirect shift is attributed to the energy levels degenerating and shifting near the Fermi level. The direct to indirect shift along the Γ-M–K-Λ_{min}-Γ direction was studied in detail by Padilha et al., using HSE06 functional with GW approach [8]. The studies showed that the conduction band (CB) edge at K-point is made up of p_x and p_y S orbitals and $d_z{}^2$ Mo orbitals. The bands that form the CBM at K-point do not couple by symmetry and show n-fold degeneracy, n being the number of layers. The valence band (VB) at Γ-point is made of p_zS orbital and $d_z{}^2$ Mo orbital. Each layer in VB contributes to one band having strong interaction amongst them. When the number of layers is increased degeneracy is lifted splitting the band and pushing the VBM at Γ to higher energies resulting in a transition of direct to indirect bandgap. Bandgap studies using different approaches lead to slightly different band gap values, however, they all predict the reduction in band gap with the increase in the number of layers.

DFT studies implicate that 1T'-MoS_2 is the only polytype for which the band structure is metallic for a number of layers, $n > 2$; semi-metallic for bilayer and semiconducting for monolayer. The calculations of Zhao et al. predict that 1T'-MoS_2 monolayer shows a small energy bandgap of 0.076 eV (HSE06 functional) due to energy degeneracy present at the valence band in the Γ-Y region [7]. These calculations also show a metallic to semiconducting shift for 1T'-MoS_2 as the number of layers is reduced. Clearly, 1T polytypes make an interesting study that need detailed investigation. Zhuang et al., performed a comparative DFT study of 1T, 1T', 1T'', 1T'''-MoS_2 monolayers using PBE functional and projector-augmented wave method [9]. The comparison of formation energies of 1T-polytypes with 2H-MoS_2 monolayer showed that the energy difference with 2H-MoS_2 is maximum for 1T-$MoS_2$and minimum for 1T'-MoS_2. Figure 3 shows the band structure for 1T, 1T', 1T'', 1T'''-MoS_2inclusive of the spin–orbit coupling (SOC). The three bands crossing the Fermi level in 1T-MoS_2 establish the metallic nature of 1T-MoS_2. The distortion of 1T-MoS_2 to form 1T', 1T'', 1T''' structures results in opening of bandgaps in the metallic 1T-MoS_2 structure. 1T' structure shows a very small bandgap of 50 meV while 1T'' and 1T''' show a bandgap of ~ 0.14 eV and 0.57 eV, respectively [9]. Zhuang et al. also noted an interesting fact that 1T'-MoS_2 monolayer shows a Dirac cone without the inclusion of SOC but the inclusion of SOC splits the Dirac cone opening a bandgap of 50 meV. This fact is a major reason for the exotic properties displayed by 1T'-MoS_2 structure and will be discussed in detail in the later part of the chapter.

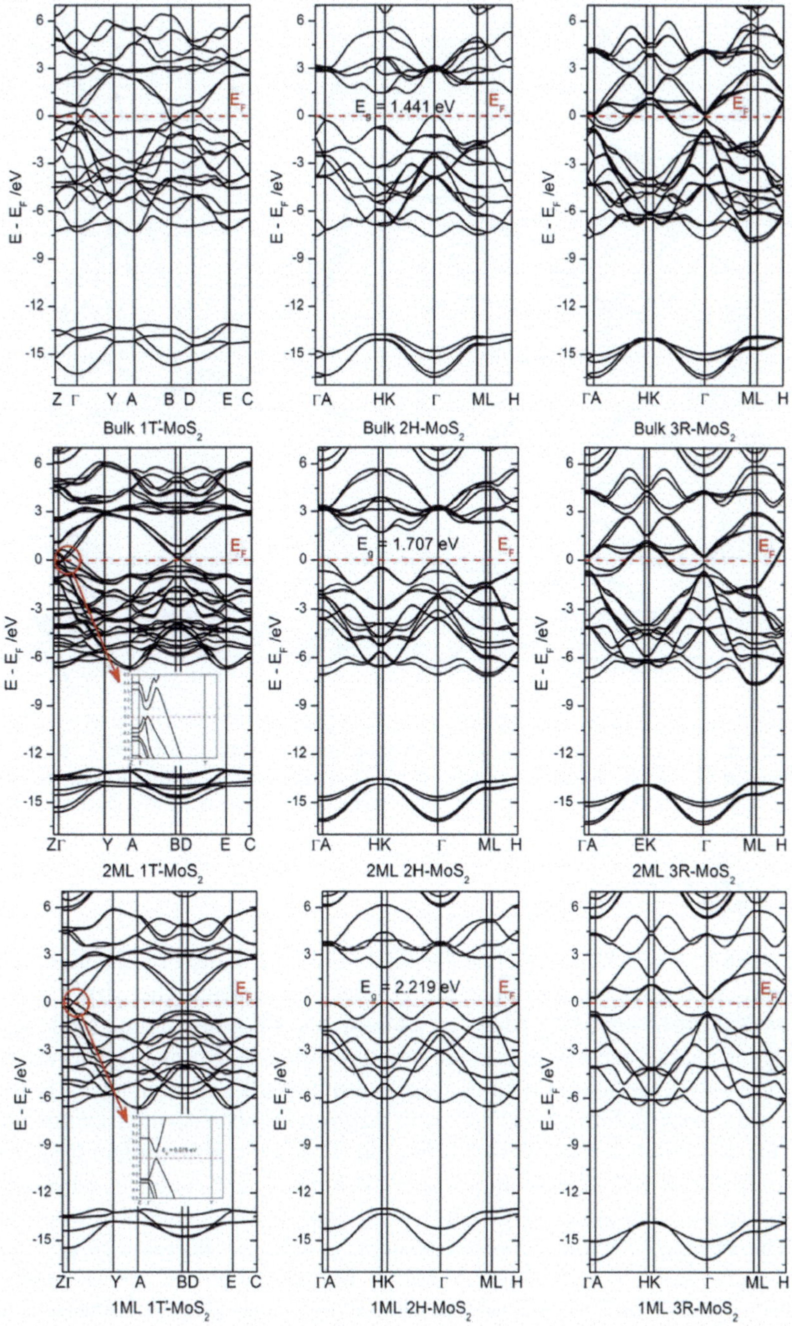

Fig. 2 Electronic band structure for monolayer (1ML), bilayer (2ML) and bulk with 1T'-MoS_2, 2H-MoS_2 and 3R-MoS_2 crystal phases. Reprinted with permission from Zhao et al., **Catal. Sci. Technol.10**, 1867–1879(2018). Copyright 2020 Author(s), licensed under a Creative Commons 4.0 Attribution License

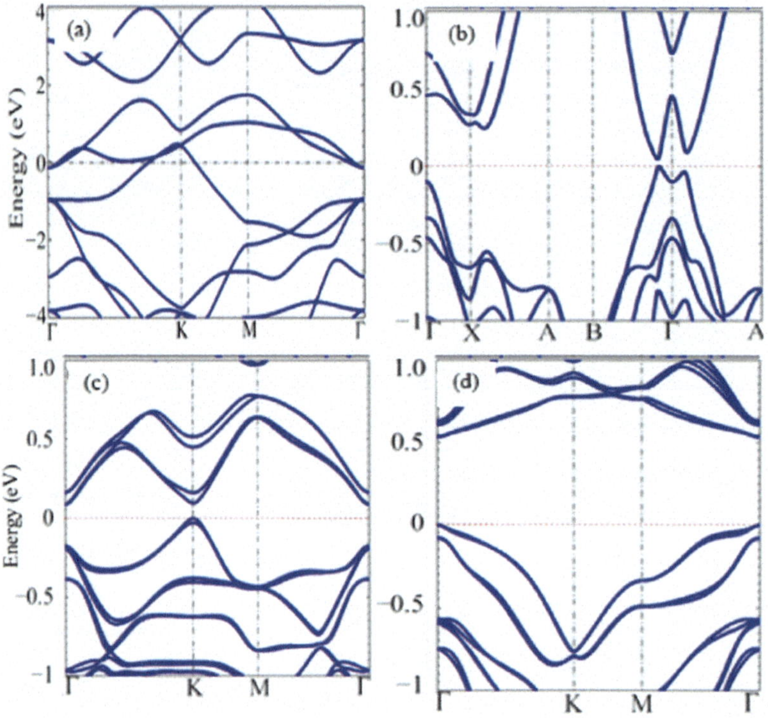

Fig. 3 Electronic band structure of various polytypes of monolayer MoS_2 **a** 1T-MoS_2, **b** 1T'-MoS_2 **c** 1T''-MoS_2 **d** 1T'''-MoS_2. Reprinted with permission from Zhuang et al., **Phys. Rev. B 96**, 165,305 (2017). Copyright 2020 Author(s), licensed under a Creative Commons 4.0 Attribution License

3.2 *Carrier Mobility*

Carrier mobility has a huge effect on the potential applicability of a material in catalysis, opto-electronics, photovoltaic, etc. Yun et al. calculated the carrier mobility for Γ-M-K-Γ direction of 2H-MoS_2 by adopting the GGA approximation [10]. The mobility of charge carriers is a band structure-dependent property and as established previously, the band structures are different for bulk and monolayer 2H-MoS_2. Hence, the carrier mobilities also vary with the layer number. As the structure moves from bulk to monolayer, the VBM shifts from Γ-point to K-point and CBM shifts from midpoint of K-Γ to K-point. The CBM for bulk and monolayer has similar effective masses and hence CBM shift with change in layer number is insignificant. The change in VBM, however, prompts a dramatic decrease in m_e* which should lead to an increase in mobility. This increased mobility is experimentally not observed, reason being, the charge carriers have a contribution to VBM of Γ-point through vdW-interlayer interaction while contribution to VBM of K-point is through strong covalent interaction. Simultaneous consideration of both these facts leads to a conclusion that carrier mobility reduces from bulk to monolayer [10].

3.3 *Optical Properties*

Optical properties gauge the photoelectric functionality of a material. Optical properties like static and high frequency dielectric constants, static and high frequency refractive indices, absorption coefficient and reflectivity can be predicted using DFT studies. Zhao et al. studied the variation of all these properties with layer number [7]. The static dielectric function and refractive index of 1T', 2H, and 3R-MoS_2 structures display a monotonic increase with increase in layer number and finally converge close to the bulk value. 1T'-MoS_2 does not display convergence with bulk values. The high frequency dielectric function and refractive index decrease monotonically with increase in layer number. However, there is a large difference between their convergence values and bulk values. The values of absorption coefficient and reflectivity for all the three crystal planes also increase monotonically with increase in number of layers. 2H-MoS_2 displays the largest while 3R-MoS_2 displays the smallest value of absorption coefficient and reflectivity. Also, the band edge for all the three shows a gradual blue shift with increase in layer number. The absorption and reflectivity spectra for 1T' and 2H-MoS_2 have band centered at ~180 nm while 3R-MoS_2 has band centered at ~220 nm. The DFT obtained values for all the above-mentioned parameters for bulk MoS_2 are listed in Table 1 [7].

4 Exotic Properties of MoS_2

DFT is well-known for accurately predicting the band structure, carrier mobility, optical properties, etc. However, realistic atomistic modeling is also capable of predictive studies. DFT provides accurate mechanistic understanding for unusual material behavior. The current section sheds light on few such unique properties predicted by DFT and displayed by MoS_2.

Table 1 Optical properties for 1T', 2H, and 3R-MoS_2

	1T'-MoS_2	2H-MoS_2	3R-MoS_2
Static dielectric function ε (0)	18.84	5.48	12.31
Static refractive index η (0)	4.34	2.34	3.51
High frequency dielectric function $\varepsilon(\infty)$	0.61	0.62	0.67
High frequency refractive index $\eta(\infty)$	0.78	0.79	0.82
Absorption coefficient (cm^{-1})	2.567×10^5	2.920×10^5	1.894×10^5
Reflectivity	0.480	0.510	0.417

4.1 Semiconductor–Metal Transition

Effect of application of in-plane (in the direction of S-Mo-S sheets) biaxial strain on 2H-MoS_2 monolayer and bilayer was studied by Scalise et al. [11] Interestingly, they predicted semiconductor to metallic transition in 2H-MoS_2 on application of strain. Tensile strain less than 2% results in a direct to indirect band gap shift in monolayer. For strain between 2 and 10%, there is progressive reduction in energy band gap. For ~10% strain, the system becomes semi-metallic beyond, which it shows a clear metallic transition. Similar reduction in bandgap is also observed in case of bi-layer and the semiconductor to metallic shift takes place around 8%. On the other hand, compressive strain up to 5% shows increase in bandgap and direct to indirect shift for monolayer. A compressive strain around 15% shows a semiconductor to metallic transition. Similar strain studies were also performed by Yun et al. They re-established the indirect to direct bandgap shift from bulk to monolayer and observed the strain that induces a change in lattice parameter 'a' beyond—–1.3 to 0.3% of the optimum value leads to an direct to indirect band gap shift for 2H-MoS_2 monolayer [10].

While the tensile stress that increases the lattice parameter by ~9.8% leads to a metallic band structure. Figure 4 illustrates the strain dependence of band gap energies for monolayer 2H-MoS_2 and also provides a few representative strains affected band structures.

4.2 Ferroelectrics

Ferroelectrics are insulators that exhibit a macroscopic electric polarization arising from the spontaneous ordering of electric dipoles which can be controlled by external electric or stress field. Realistic applications require a combination of ferroelectricity and semiconducting behavior with high carrier mobility and small band gap. Shirodkar et al. [3] predicted that 1T'''-MoS_2 was a robust ferroelectric phase with non-negligible electric polarization. Berry phase calculations revealed that 1T'''-MoS_2 shows spontaneous polarization along the z-axis while the in-plane polarization vanishes. They established that ferroelectricity in 1T'''-MoS_2 monolayer was a consequence of strong electron–phonon coupling and symmetry of K_3 modes having instability that originated from the degeneracy of Fermi surface. They also observed a barrier of 0.13 eV in the electrostatic potential at the domain wall, which could be utilized to alter the electron transport in the 1T'''-MoS_2 channel. They proposed that an XNOR gate can be developed by using FET with 1T'''-MoS_2 as a channel and two gate electrodes in series while NAND and OR gate can be realized in an FET device with three gate electrodes. To sum up, the current in semiconducting 1T'''-MoS_2 channel can be controlled by changing its dipolar structure with gate field, opening its use as novel dipolectronic devices. Such a DFT work is capable of stimulating experimental investigation in search of ferroelectricity in 1T'''-MoS_2 structure.

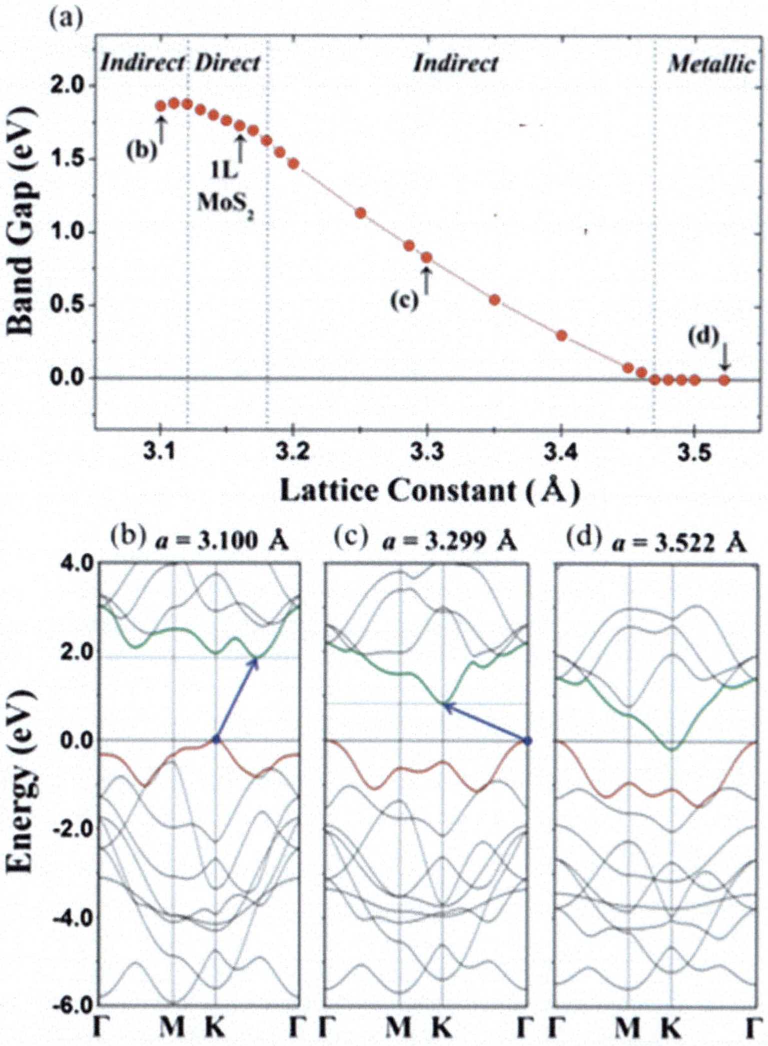

Fig. 4 **a** Strain dependence of band gap energies on monolayer MoS_2 (a = 3.160 Å). Representative band structures for **b** compressive stress and **c**, **d** tensile stress. Reprinted with permission from Yun et al., **Phys. Rev. B 85**, 033,305 (2012). Copyright 2020 Author(s), licensed under a Creative Commons 4.0 Attribution License

Ferroelectric semiconductor comprising of heavy ions, due to the presence of a polar axis, shows k-dependent spin splitting in band structure also known as SOC-induced Rashba effect. As monolayer 1T'''-MoS_2 displays ferroelectric phase simultaneously maintaining a small band gap of ~1 eV, it is an ideal candidate to study the interplay between ferroelectric polarization and Rashba-like spin splitting effect.

Bruyer et al. [4] thus studied the possible co-existence of ferroelectricity and spin-polarized states in distorted 1T-MoS_2. They concluded that spin splitting and Rashba-like spin polarization features are reversed by switching the ferroelectric polarization thus allowing full control of the spin degrees of freedom by an external electric field.

4.3 Quantum Spin Hall Effect

Intriguing physical phenomena can result from non-trivial topology. The basic mechanism behind non-trivial topology is band inversion between two orbitals that can be driven by chemical bonding, spin–orbit coupling and applying pressure/electric field. The study of band inversions in topological materials explains topological phase transition and determines the spin textures of the surface states. Based on DFT studies, Qian et al. proposed that monolayer 1T'-MoS_2 possesses properties of large gap quantum spin hall (QSH) insulator also known as large gap topological insulators (TI). QSH insulators are materials that have an insulating bulk but conducting edge states topologically protected from backscattering by time reversal symmetry. Monolayer 1T'-MoS_2 displays a camel-back shape near Γ-point with an inverted gap (2δ) of ~0.6 eV and a fundamental band gap of ~0.08 eV. The studied structure displays band inversion symmetry, which leads to appearance of two Dirac cones in the absence of spin–orbit coupling. The consideration of spin–orbit coupling opens up a band gap of 0.08 eV at Dirac point leading to QSH insulator. Qian et al. [12] not only established that a vertical electric field induces topological phase transitions but also suggested that such transitions have potential applications in low-power quantum electrodynamics, spintronics and may also enable topological quantum computing based on Majorana fermions.

Nie et al. [13] used a simple tight binding model to prove that similar to 1T'-MoS_2, defected 2H-MoS_2 could also host QSH effect. This defected 2H-MoS_2 is known as a TMD haeckelite with square-octagonal lattice, i.e. MoS_2-4–8. They established that nearest hopping parameter t2 was vital to trigger topological phase transitions and could be tuned through lattice strain effect. They observed that uniaxial strain either along the a-axis or the b-axis was beneficial for band inversion. Their work extended the searching range for topological materials.

5 Conclusions

It is evident that MoS_2 holds significant promise for next-gen technologies owing to their fascinating electronic, mechanical, and optical properties. The different possible atomic arrangements of the MoS_2 polymorphs make MoS_2 highly tunable and applicable in catalysis, spintronics, optoelectronics, photovoltaics, energy storage, field effect transistors, etc. Detailed DFT studies have established the applicability of

MoS_2 structures and composites in electrochemical energy conversion [14]; photocatalytic water splitting [15], CO_2 reduction [16], dye degradation [17]; hydrogen storage [18], sensing [19], etc. Considering the enormity of already completed MoS_2 studies based on DFT, a detailed discussion on MoS_2 applications is not included.

Screening and tuning materials are computationally less tedious and cost-effective. However, to experimentally achieve the actual application potential of MoS_2 polymorphs necessitates fine control over structure, morphology, and crystallinity during the synthesis process. High purity and high yield synthesis of desired MoS_2 phase and form is the most significant experimental challenge. Enormous DFT work on MoS_2 has already been done with a few missing links, which will definitely be completed in the near future stimulating novel experiments.

References

1. Kumar VP, Panda DK (2022) Next generation 2D material molybdenum disulfide (MoS_2): properties, applications and challenges. ECS J Solid State Sci Technol 11:033012
2. Krishnan U, Kaur M, Singh K, Kumar M, Kumar A (2019) A synoptic review of MoS_2: synthesis to applications 128:274–297
3. Shirodkar SN, Waghmare UV (2014) Emergence of ferroelectricity at a metal-semiconductor transition in a $1T$ monolayer of MoS_2. Phys Rev Lett 112:157601
4. Bruyer E, Di Sante D, Barone P, Stroppa A, Whangbo M-H, Picozzi S (2016) Possibility of combining ferroelectricity and Rashba-like spin splitting in monolayers of the $1T$-type transition-metal dichalcogenides MX_2 ($M = \text{Mo, W}$; $X = \text{S, Se, Te}$). Phys Rev B 94:195402
5. Lin X, Li W, Dong Y, Wang C, Chen Q, Zhang H (2016) Two-dimensional metallic MoS_2: a DFT study. Commun Mat Sci 124:49
6. Shang C, Fang YQ, Zhang Q, Wang NZ, Wang YF, Liu Z, Lei B, Meng FB, Ma LK, Wu T, Wang ZF, Zeng CG, Huang FQ, Sun Z, Chen XH (2018) Superconductivity in the metastable 1T' and 1T''' phases of MoS_2 crystals. Phys Rev B 98:184513
7. Zhao Z-Y, Liu Q-L (1867) Layer-dependent properties of MoS_2 nanosheets with different crystal structures study by DFT calculations. Catal Sci Technol 2018:8
8. Padilha JE, Peelaers H, Janotti A, Van de Walle CG (2014) Nature and evolution of the band-edge states in MoS_2: from monolayer to bulk. Phys Rev B 90:205420
9. Zhuang HL, Johannes MD, Singh AK, Hennig RG (2017) Doping-controlled phase transitions in single-layer MoS_2. Phys Rev B 96:165395
10. Yun WS, Han SW, Hong SC, Kim IG, Lee JD (2012) Thickness and strain effects on electronic structures of transition metal dichalcogenides: 2H-MX_2 semiconductors ($M = \text{Mo, W}$; $X = \text{S, Se, Te}$). Phys Rev B 85:033305
11. Scalise E, Houssa M, Pourtois G (2012) Valery Afanas'ev and André Stesmans, Strain-Induced Semiconductor to Metal Transition in the Two-Dimensional Honeycomb Structure of MoS_2. Nano Res 5(1):43
12. Qian X, Liu J, Fu L, Li J (2014) Quantum spin Hall effect in two-dimensional transition metal dichalcogenides. Science 346(6215):1344–1347
13. Nie SM (2015) Zhida Song, Hongming Weng and Zhong Fang, Quantum spin Hall effect in two-dimensional transition-metal dichalcogenide haeckelites. Phys Rev B 91:235434
14. Santra S, Streibel V, Sharp ID (2022) Emergingnoble metal-free Mo-based bifunctional catalysts for electrochemical energy conversion. Nano Res 15(12):10234
15. Thomas N, Mathew S, Nair KM, O'Dowd K, Forouzandeh P, Goswami A, McGranaghan G, Pillai SC (2021) 2D MoS_2: structure, mechanisms, and photocatalytic applications. Mater Today Sustain 13:100073

16. Ouyang T, Fan W, Guo J, Zheng Y, Yin X, Shen Y (2020) DFT study on Ag loaded 2H-MoS_2 for mechanism of improved photocatalytic reduction of CO_2. Phys Chem Chem Phys 22:10305
17. Lalithambika KC, Shanmugapriya K, Sriram S (2019) Photocatalytic activity of MoS_2 nanoparticles: an experimental and DFT analysis. Appl Phys A 125:817
18. Archana Sharma A, Husain M, Srivastava A, Shahid Khan Mohd (2018) DFT study of Ca-adsorbed MoS_2 monolayer for hydrogen storage application. Adv. Mater. Proc 3(1):25
19. Qian G, Peng Q, Zou D, Wang S, Yan B, Zhou Q (2020) First-principles insight into Au-doped MoS_2 for sensing C_2H_6 and C_2H_4. Front Mater 7:3389

The manufacturer's authorised representative in the EU is Springer Nature Customer Service Centre GmbH, Europaplatz 3, 69115 Heidelberg, Germany. If you have any concerns regarding our products, please contact ProductSafety@springernature.com

Printed and bound by CPI Group (UK) Ltd, Croydon, CR0 4YY
01/10/2026
02225962-0011